Towards a Convergence Between Science and Environmental Education

In the *World Library of Educationalists* series, international scholars themselves compile career-long collections of what they judge to be their finest pieces—extracts from books, key articles, salient research findings, major theoretical and/or practical contributions—so the world can read them in a single manageable volume. Readers thus are able to follow the themes and strands of their work and see their contribution to the development of a field, as well as the development of the field itself.

Internationally recognized for his research on environmental education, science engagement, learning outside the classroom and teacher identity and development, in this volume Justin Dillon brings together a thoughtfully crafted selection of his writing representing key aspects of his life and work leading to his current thinking on the need for a convergence of science and environmental education. The chapters are organized around 7 themes: On *Habitus;* On methodological issues; Developing theories of learning, identity and culture; Challenges and opportunities—science, the environment and the outdoors; Classroom issues—the emergence of Science|Environment|Health; Science engagement and communication; Science, environment and sustainability.

Justin Dillon is Professor of Science and Environmental Education and Head of School, University of Bristol Graduate School of Education, UK. After taking a degree in chemistry from Birmingham University, he trained as a teacher at Chelsea College and went on to teach in six secondary schools in London. His research originally focused on teaching and learning about chemistry in England and Spain. More recently he has focused on science learning outside the classroom, particularly in museums, science centres and botanic gardens in the UK, Europe and elsewhere. Together with two colleagues at King's College London, he co-ordinated the ESRC's Targeted Initiative on Science and Mathematics Education (TISME) and he was a member of the highly influential ASPIRES project. Dillon served as elected President of the European Science Education Research Association from 2007 to 2011. He co-edits the *International Journal of Science Education*, is a trustee of the Council for Learning Outside the Classroom, was Chair of the London Wildlife Trust for many years, and has co-edited a number of books including the *International Handbook of Research on Environmental Education*. He was given 'The Outstanding Contributions to Research in Environmental Education Award' by the North American Association for Environmental Education in 2013.

World Library of Educationalists Series

Thinking and Rethinking the University
The selected works of Ronald Barnett
Ronald Barnett

China through the Lens of Comparative Education
The selected works of Ruth Hayhoe
Ruth Hayhoe

Educational Experience as Lived: Knowledge, History, Alterity
The selected works of William F. Pinar
William F. Pinar

Dysconscious Racism, Afrocentric Praxis, and Education for Human Freedom: Through the Years I Keep on Toiling
The selected works of Joyce E. King
Joyce E. King

A Developing Discourse in Music Education
The selected works of Keith Swanwick
Keith Swanwick

Struggles for Equity in Education
The selected works of Mel Ainscow
Mel Ainscow

Faith, Mission and Challenge in Catholic Education
The selected works of Gerald Grace
Gerald Grace

Towards a Convergence Between Science and Environmental Education
The selected works of Justin Dillon
Justin Dillon

Towards a Convergence Between Science and Environmental Education

The selected works of Justin Dillon

Justin Dillon

Routledge
Taylor & Francis Group

LONDON AND NEW YORK

First published 2017 by Routledge

2 Park Square, Milton Park, Abingdon, Oxfordshire OX14 4RN
52 Vanderbilt Avenue, New York, NY 10017

Routledge is an imprint of the Taylor & Francis Group, an informa business

First issued in paperback 2019

Library of Congress Cataloging in Publication Data
Names: Dillon, Justin, author.
Title: Towards a convergence between science and environmental
    education : the selected works of Justin Dillon / by Justin Dillon.
Description: New York : Routledge, 2017. | Series: World library
    of educationionalists series | Includes bibliographical references
    and index.
Identifiers: LCCN 2016018986 | ISBN 9781138844292 (hardback) |
    ISBN 9781315730486 (ebook)
Subjects: LCSH: Environmental education. | Science—Study
    and teaching.
Classification: LCC GE70 .D55 2017 | DDC 333.7071—dc23
LC record available at https://lccn.loc.gov/2016018986

ISBN: 978-1-138-84429-2 (hbk)
ISBN: 978-1-138-34532-4 (pbk)

Typeset in Bembo
by Apex CoVantage, LLC

For all my colleagues past and present—this book would not be possible without you.

Contents

Preface

In putting this collection together I have tried to tell the story of my academic life highlighting the challenges and the collaborations as well as showing the steady evolution of my thinking. I have written about a number of diverse topics which, on reflection, hang together albeit loosely. Much of my writing has been about science education and environmental education and more recently I have been looking at how these two fields might converge. So, this book can be seen as snapshots from a long journey which is still unfinished.

I have benefited from working with a number of colleagues who have provided inspiration and provocation as well as friendship and love. Without them this book could not have been written but more importantly I would not be the person that I am now—my intellectual and personal life would be substantially poorer.

Choosing what to put in this volume was incredibly difficult and I've had to leave out material that I am proud of contributing to. I'm thinking here of the work of the Aspires project at King's College London as well as some of the outputs of the many European projects that I've taken part in. I've also had to leave out more recent collaborations with Steve Alsop from York University, Canada, on engaging with narwhals and with two dentists from King's—Brian Davies and Albert Leung—who were very open-minded about crossing disciplinary boundaries. So this collection is incomplete. I hope, though, that reading it might help aspiring academics to see how ideas develop over time and how collaboration in research and writing can provide opportunities for personal growth and intellectual challenge.

The book is divided into seven themes. Section 1, 'On *Habitus*', draws on five chapters and papers to illustrate key aspects of my life and work that help to contextualise the other sections. Section 2, 'On methodological issues', establishes the research paradigms within which my work has developed over time. Section 3, 'Developing theories of learning, identity and culture', contains three papers which illustrate some of the key theoretical ideas that have framed my work. They also illustrate the value of supervising and publishing with doctoral students. Section 4, 'Challenges and opportunities—science, the environment and the outdoors', contains two synthesis papers which show how new knowledge can be made from critiquing existing studies and why learning outdoors has so much to offer. I've also included an empirical study that shows what can be done in terms of teaching students outdoors and a critique of reasons why schools say they cannot take classes beyond the classroom. Section 5, 'Classroom issues—the emergence of Science | Environment | Health', illustrates how a new way of thinking about inter-related dimensions emerged from an initial interest in how students could be taught about risk. The section also contains an empirical study of teachers and how they taught controversial issues which led to the development of new ideas about the role of environmental education in developing scientific literacy in its broadest sense. Section 6, 'Science engagement and communication', illustrates how I moved into the realm of science engagement and communication. Finally, Section 7, 'Science, environment and sustainability', sets out my position on the relationship between science, the environment and sustainability. It contains a critique of simplistic thinking about the relationship between science and environmental education, a critique of education for sustainable development and my current thinking on the need for a convergence of science and environmental education.

Acknowledgments

I am glad to acknowledge all my co-authors and collaborators—this is as much our book as mine. Two friends and colleagues, John K. Gilbert and William Scott, helped me to structure the book in their role as reviewers for Routledge. There is one person, though, who deserves an award for her determination, patience and fortitude, Naomi Silverman, my long-suffering publisher. Thank you Naomi, you are old school.

Introduction

The invitation to put together this collection arrived while I was working at King's College London where I expected to see out my academic career. However, some months later, during the extended period when I was gathering the permissions to reproduce the various chapters and papers in this volume, I moved to take up a new post at the University of Bristol. While for some people, moving institutions is normal if not the norm, for me it was very unusual. I started working at King's in 1989 and left in 2014—almost 26 years in one institution, and my standard response to questions about 'moving on' was 'But where would I go?' I have since wondered whether the process of going through my academic life's work, deciding what stories I wanted to highlight and what I thought I had to say that's worth reading now, made me more susceptible when Bristol approached me.

I don't remember anything about the most significant event that happened to me. I was adopted at an early age by Jean and Stuart Dillon who took in Robert Boyd and renamed him Justin Dillon. Actually they renamed me Justin Simon Dillon, but after publishing a few papers in the 1990s I stopped using my middle name. Both my parents taught though neither had been to university and neither was qualified to teach at the school that they encouraged me to attend in the 1960s and 1970s. Their experience of and interest in education made them sceptical of the comprehensive (that is non-selective) school that most of my primary school peers had decided to go to at the age of 11 so they encouraged me to go to a school which was slightly further away.

Newcastle High had been a minor independent (private) school some years before I joined and although it was government funded when I started in 1967 it was still very traditional. It was assumed that most of the boys would go onto study science or mathematics at university. And so I did. The fact that my father taught biology and rural science at school might suggest that, in my case, nurture beats nature hands down.

After an undistinguished time studying chemistry at Birmingham University I applied to do a one-year postgraduate course in science education at the Institute of Education, part of the University of London. They turned me down—'better candidates available' it says on my application form that

I unearthed from King's College London's archives some years ago. Chelsea College, my second choice, gave me a chance. In 1984, Chelsea merged with King's and Queen Elizabeth College, which is how my files came to be accessible to me later in my career.

Discovering the World of Science Education Research

I first met a science education researcher in 1979 when I was interviewed by Dr Jan Harding for a place on Chelsea College's one-year postgraduate certificate in education (PGCE) course. The Centre for Science Education was a converted engineering factory with massive floors able to withstand the weight of heavy machinery. The Centre contained the most vibrant set of science educationists that has probably ever existed in one place. As well as Jan, there was Paul Black, John Head, Joan Bliss, Guy Claxton, Bob Fairbrother, John Barker, Martin Monk, John Harris, Michael Shayer and, many, many more. Philip Adey had just left to work overseas but was set to return. The Centre, and the people working there, changed my life beyond recognition. Without that experience I would not be writing this text nor, I suspect, would I have worked at King's.

During the 1970s, many of the staff had been heavily involved in curriculum development through the Nuffield Science projects. New initiatives were starting up, not the least of which was the Assessment of Performance Unit (APU) which had had its first public outing at the ASE conference in 1978. Jan Harding and John Head's work on gender were beginning to have an impact on science education researchers beyond the UK. Science education research was, if not in its infancy, not far into adolescence. The first edition of the *European Journal of Science Education* (which later became the *International Journal of Science Education*) was published in 1979, the year that I began my teacher training course.

Experimenting with Teaching and Research

By the end of 1980 I was established in London and had begun to learn how to teach. I had secured a job as an assistant teacher at the John Roan School in Greenwich. The school was going through a difficult period. A girls' grammar (selective) school, a boys' grammar school and an old secondary-modern school were merging to form a split-site comprehensive. Staff travelled from site to site by taxi and in one day you could be teaching a selective all-girls class, a selective all-boys class and a mixed comprehensive class. The head of science at the boys' school was Brian Matthews who went on to work at Goldsmiths College and later King's and at the girls' school it was Wendy Riddle who had had a long association with Chelsea College.

Brian had a vision of science education that was radical and encouraged debate and dialogue in the science department. For example, he advocated tape-recording ourselves teaching to see if we used sexist language. During my four years at the John Roan School I did a small amount of simple educational research. I carried out a survey of pupil attitudes towards science—something that I must have picked up from my pre-service Chelsea days. I also attended my first international conference, the International Conference on Chemical Education in Maryland, USA, in 1981. I gave a table-top presentation of our work trialling the Independent Learning Project for Advanced Chemistry (ILPAC) which had been set up by the Inner London Education Authority (ILEA). The early starts and late finishes of a US conference surprised me, UK conferences seemed much more laid-back in comparison.

The ILEA ran a series of day and evening courses at their science centres (for teachers and technicians) which provided opportunities to keep up-to-date with a range of research and developments. As well as supporting a network of people committed to innovative science education, the centres were the base for the Science Support Team which I joined in 1984. My role in the team was to teach examination classes in schools with short-term shortages of chemistry teachers. During the time that I was in the team I taught in five different London schools and was almost immediately offered a post in one of them, Eltham Hill School in southeast London.

After a year in post as Head of Chemistry I started a part-time MA at King's. At the time, the ILEA not only paid all the fees, it also provided two weeks' study leave for two years. Times are, sadly, different now.

Mastering Science Education

In the intervening years since I had done my initial teacher training, Chelsea College had merged with King's and Queen Elizabeth College so I was reunited with the tutors who had been so inspirational back in 1979–80. Enrolling for the MA was one of the best decisions of my life. As well as studying research methods, I was able to develop a much greater awareness of the sociology of science and of technology education. My dissertation, which I completed in 1987, was on technology, culture and gender. I convinced colleagues at Eltham Hill that we should introduce technology into the curriculum. I don't know whether we were the first all-girls school to introduce technology in London but it would probably never have happened if I had not done the MA.

Shortly after finishing the course I was encouraged by one of the ILEA science inspectors to apply for the post of Head of Science at Kingsdale School in south London. During my career, I had mentored several student teachers from Chelsea, and then King's, and had done the odd session on their PGCE courses. My interest in remotely-accessed databases (such as Dialog) and earth science and, specifically, in mineral process chemistry (a Nuffield A-level Special Study) also led to me being invited to go on a residential field-course with a group of King's PGCE students led by Rod Watson who had taken Jan Harding's post when she retired. Rod and I got to know each other through the South London Chemistry Teachers' Centre of which I was Secretary for many years and which was hosted at King's. At the beginning of the 1988/9 school year, Rod rang me at Kingsdale to say that he had been successful in getting funding from the Nuclear Electricity Information Group to scope the possibility of an environmental database project and asking whether I'd like to work with him on the project. I instantly said that I was interested.

Joining King's

Rod only had money to pay for someone for a term in the first instance. My head teacher agreed that I could be seconded for the spring term and then, if the project did not carry on, I could return to Kingsdale. However, another chemistry colleague announced that she had got a new post and that she was leaving at the end of December. The head asked me what I'd say if she reneged on her decision to let me go on secondment. I told her that I would resign and that's what happened. So, having given up a permanent job as head of the science department, I found myself, at the age of 31, on a one-term contract.

Rod and I were operating a job-share, in effect, with both of us working 50% on the National Environmental Database and 50% on the science education PGCE. Although I had done some research in schools, it hadn't been very substantial and it had only been published in the ILEA *Science News*. We managed to get several low-level papers out of NED but it was more curriculum development than research. However, Rod had other irons in the fire and one of them was a successful bid to forge a link with a colleague at the University of Malaga, Spain, Teresa Prieto.

Burning Questions

The funding for the link was very small and came from a Spanish organisation and from the British Council. However, it allowed us to visit each other's universities and provided a first and significant step into the world of educational research. Rod had worked in Spain as a school-teacher earlier in his career and realised that some useful comparisons could be made between the two education systems. The topic that was chosen for study was children's ideas about combustion. The abstract of my first academic paper (Prieto, Watson & Dillon, 1993) explains what we did:

> A questionnaire survey of 300 14- and 15-year-old pupils in England and Spain was carried out to investigate pupils' general ideas about the process of burning and their ideas about specific types of combustion, using open-ended and structured response questions. Pupils' responses were analysed and categories were defined from a classification scheme previously reported by Andersson (1990). A possible model for progression of pupils' ideas about combustion is discussed.

You can see from this abstract where science education research was at the time. A number of researchers were focusing on children's ideas and attempting to map out how they developed with time. The international comparison meant that we could compare a country where school science education prioritised practical work (England) with a country that didn't (Spain).

Rod Watson was a very thorough researcher and I learned a lot from him. The size of the King's team meant that we were able to focus our time on more than teaching on the PGCE course and, almost as importantly, we had several staff with experience in each of the sciences. This latter fact meant that Rod and I could talk to each other about chemistry education as ex-chemistry teachers.

Rod knew the value of publishing in international peer-reviewed journals. The Prieto *et al.* (1993) paper was published in *Research in Science Education* and two subsequent papers (Watson, Prieto & Dillon) were published in the *Journal of Research in Science Teaching* (1995) and *Science Education* (1997). Our relatively small study spawned several other publications based on conference proceedings (Watson & Dillon, 1996; Watson, Dillon & Miguens, 1991; Watson, Prieto & Dillon, 1995). The Watson, Dillon and Miguens paper is in Portuguese and was written in collaboration with one of Rod's PhD students, Manuel Miguens. The first that I knew of it was when Rod gave me a copy and said 'you've got a new publication'.

Papers count more than edited books and chapters when it comes to promotion and appraisal in the UK's research assessment exercise. I make this point at the start of an autobiographical chapter in Ken Tobin and Michael Roth's book, *The Culture of Science Education: Its History in Person* (which forms Chapter 1 of this collection).

> I should not be doing this. My priorities should be obtaining grants from research councils and writing papers for peer-reviewed journals. In the UK Research Assessment Exercise (RAE), which determines the 'visions' and practices of all universities and most of their staff, book chapters in edited volumes usually do not count for much. During my eighteen years at King's, the RAE has steered academic life, reduced collegiality, and seriously affected research practices.
>
> (Dillon, 2007, p. 311)

On Controversy

Another small-scale research study that yielded some publications in 'good' journals owed its origins to the 'Earth Summit' which took place in Rio de Janeiro in 1992. I'd been invited to attend a workshop on environmental education and had met a number of environmental educators many of whom have remained friends for life including Chris Oulton who was then at the University of Bath. Some years later, after his move to University College Worcester, Chris approached Marcus Grace from Southampton University and me with an invitation to work together on a project called 'Unlocking Controversial Issues' (UCI).

The funding for the UCI project came from the Countryside Foundation for Education (CFE), a small charity that had been set up to promote an understanding of the countryside as a living, working environment. Teaching about the countryside can include a range of controversial issues including factory farming, GM food, fox-hunting and organic agriculture. CFE set out to support teachers faced with teaching these controversial topics. With the advent of citizenship in the curriculum, CFE felt that teachers would need more support for the teaching of controversial issues. We were asked to provide insights into teachers' views about the teaching of controversial issues which could then be used to provide guidance on what strategy CFE might adopt in order to better support teachers, student teachers and teacher educators.

This small-scale study led to several papers and a small degree of notoriety. As well as a paper describing the study which was published in the *Oxford Review of Education* (Oulton *et al.*, 2004), we wrote another for the *International Journal of Science Education* which suggested a reconceptualisation of the teaching of controversial issues (Oulton, Dillon & Grace, 2004). I summarised the key points of our work for an invited paper in the *Development Education Journal* (Dillon, Grace & Oulton, 2004). Towards the end of 2004 I gave a talk on 'Teaching Controversial Issues' at the British Association Festival of Science in Exeter. King's external relations team sent out a press release, that I had written suggesting that:

> Teachers should be open with their own biases rather than pretend to be neutral and students should be asked to take any bias into account when making up their minds on a topic. Taking a neutral stance is not a good strategy for teaching children how society works.

And, I added:

> The traditional approach to leading a discussion on a controversial science subject is for the teacher to take a neutral role . . . We believe that this strategy is wrong and that it is unethical to pretend to pupils that teachers have no opinion.

The *Guardian*'s Mark Crow (2004) picked up on the story and reported it as 'New research has lambasted science teaching in the UK as "unethical" for its failure to acknowledge the importance of personal bias'. I think 'lambasted' is somewhat inaccurate.

Other Writing

When I started my PGCE at Chelsea College in 1979, educational research in the UK was relatively new. It was more often than not carried out by teacher educators who found time to do relatively small-scale studies. By the time

I joined the staff in January 1989, it was increasingly clear that the job involved teacher education and research. While some colleagues had been employed because of their experience in curriculum development, the next generation of science educators would need to be research active. The pressure to engage in scholarly activity beyond training teachers meant that colleagues were increasingly open to writing for both an academic and a professional (that is, a teacher) audience.

An inspection of our PGCE course by Her Majesty's Inspectorate pointed out that our students did not seem to do much reading. There did not seem to be a suitable textbook for trainee teachers available so Meg Maguire and I decided to produce our own. The first version was set of papers written by colleagues which was photocopied, bound and sold to students. The Open University Press (OUP) took an interest and encouraged us to expand the collection and produce a more professional version. And that was how *Becoming a Teacher* was born. The first edition, published in 1997, sold well enough for OUP to come back and ask us to do a second edition which was published in 2001. A third edition came out in 2007 and the fourth, and dare I say, final edition, was published in 2011. As an exercise in collegiality, *Becoming a Teacher* provided an opportunity for an assortment of world-leading researchers—Stephen Ball, Paul Black and Louise Archer, for example, to publish in the same volume as our PGCE tutors. If I had to choose one scholarly work to symbolise my academic career it would be *Becoming a Teacher*.

The experience of editing *Becoming a Teacher* led me to get involved in a number of other edited works including *The Re-emergence of Values in Science Education* (2007) and *The Professional Knowledge Base of Science Teaching* (2011) (both with Debbie Corrigan & Dick Gunstone). Both these books, and the 2015 volume, *The Future in Learning Science: What's in it for the learner?* (Corrigan *et al.*) were written using a strategy that I have found immensely productive. Authors meet at a writing workshop to critique each other's chapters before going away to lick their wounds and rewrite or, ideally, just polish their contributions. I've been to three of these workshops at Monash University's campus in Prato, near Florence, and they have all resulted in much better end products than might otherwise have been the case. Sadly my move to Bristol meant that I had too much work to do to take part in the 2015 workshop which preceded the ESERA conference in Helsinki.

We used a similar approach when putting together the second edition *of Good Practice in Science Teaching: What Research Has to Say* (Osborne & Dillon, 2010). We were not able to use that approach in producing *Science Education Research and Practice in Europe* (Jorde & Dillon, 2012), a book described by Edgar Jenkins in his review for *Studies in Science Education* as 'a disappointing, even puzzling, publication'. Oh well. More satisfying, although equally challenging to put together, was *Understanding Student Participation and Choice in Science and Technology Education* (Henriksen, Dillon & Ryder, 2015) which pulls together the findings of the European Union funded Interests and Recruitment in Science (IRIS) project.

In the field of environmental education, I worked with Marianne Krasny on *Trading Zones in Environmental Education: Creating Transdisciplinary Dialogue* (2013) which became the first volume published in the '[Re]thinking Environmental Education' series which I co-edit with my old friend, Connie Russell for Peter Lang. Prior to that I had edited *Engaging Environmental Education: Learning, Culture and Agency* with Bob Stevenson. Bob, Arjen Wals, Michael Brody and I collaborated on the exceptionally ambitious but ultimately satisfactory *International Handbook of Research in Environmental Education* (2013) which has 51 chapters written by authors from six continents. Putting the Handbook together strained a number of friendships and, I suspect, lost the editors some friends. However, as the Handbook, the first ever in the field, was commissioned by the American Educational Research Association, it puts environmental education firmly on the high table of scholarly activity.

Philip Adey and I enjoyed working together on *Bad Education: Debunking Myths in Education* (2012). We were able to draw together a stellar cast of contributors who produced what I consider to be an excellent and much needed volume. Sadly, during the production of the book, Philip was diagnosed with multiple system atrophy, a particularly debilitating condition. One evening, after we'd dined at the Athenaeum Club in London, Philip struggled to climb the stairs and for a moment his brave face slipped to reveal the depths of his frustration at his rapid physical decline. He confided that he wanted to see his new grandchild who was soon to be born into the world, to see all his family at Christmas and to get *Bad Education* published. He achieved all three goals before taking his own life on 31 January 2013.

At this point I should also acknowledge the contribution of Rosalind Driver to our work at King's. Ros joined us in 1995, succeeding Paul Black as chair of science education. She had already had a major influence on science education in the UK and internationally and she was a natural successor to Paul who had established King's as one of the strongest research groups in Europe in science education and in assessment. Soon after she joined us Ros was diagnosed with cancer and died in October 1997. Her obituary, in the journal *Studies in Science Education* indicates why her death was so tragic.

> In a very real sense Ros Driver saw research as a team enterprise and was always keen to encourage and support those whom she worked with. The delight of collaborating with Ros was the sheer passion and enthusiasm that she brought to whatever project she was working on. Endowed with the gift of being not only an eloquent speaker, but also a good listener, she took an avid interest in colleagues' ideas and work, always willing to argue the point, but always offering the reassuring support so vital to sustain research work through the many dark hours and difficult periods. Moreover, she led by example—a maelstrom of energy and hard work that carried those fortunate enough to be her colleagues with her, and gained their highest respect and commitment.
>
> (Osborne, Leach & Scott, 1997, p. 3)

One of Ros's major contributions was to support the development of the European Science Education Research Association which was founded in Leeds in the UK in 1995. She managed to attend the first conference organised by ESERA in Rome in 1997 but she succumbed to her illness soon afterwards.

Some years later, with the support of the Nuffield Foundation, Jonathan Osborne, Robin Millar and I organised two seminars that brought together a number of ESERA members to take a critical look at science education in Europe. The report which emerged (Osborne & Dillon, 2008) has been cited over 800 times and, I hope, has been influential in driving policy and practice in Europe. I was honored to be elected as president of ESERA for a four-year term in 2007.

Some Lessons

The combustion work and the controversial issues study both support a conclusion that you don't always need a lot of money to carry out research. What you do need are colleagues in your own institution, somewhere else in the UK or overseas who share a commitment to investigating a topic of mutual interest. The vast majority of my publications are co-authored which indicates that I enjoy working with others more than working alone. I'm quite happy to be working with people who know more than I do, who write and edit better than me, and who have sharper analytical minds—the personal benefits usually outweigh the embarrassment!

In terms of publishing, the message from the two studies and from everything else that I've published is that persistence pays. However, it's not as simple as that. I spend many hours reviewing papers for journals and it's usually time well spent. Not only do you keep abreast of recent research but you learn what good writing looks like. The acceptance rate of the major science education research journals is pretty low but one has to bear in mind that a large number of papers are rejected almost immediately because they have not followed some of the basic criteria that journal editors use. If you follow the guidelines carefully and have something interesting to say, you'll at least get a couple of reviews from established academics that will help you to improve your work.

The Collection

I've organised the contributions into seven sections. Section 1 illustrates key aspects of my life and work that help to contextualise the other sections. One of the key themes emerging from this section is how much my writing has benefitted from working with other colleagues. Section 2 establishes some of the research paradigms within which my work has developed. I do think it important that new researchers develop their understanding of methodologies and methods and one way of doing that is to write specifically about them. Section 3 looks at the theoretical frameworks within which I have worked.

Much of my work has focused on the relationship between learning, identity and culture. In Section 4 the focus is on the development of my ideas about the value of learning outdoors. The section draws on critical reviews of the literature as well as on empirical studies. Section 5 takes more of a curriculum focus and illustrates how the idea of Science | Environment | Health emerged from a long-standing interest in the wider purposes of science education. In more recent years I have written more about science engagement and Section 6 provides three examples of theoretical and empirical work that illustrate some aspects of my current thinking. Section 7 sets out my position on the relationship between science, the environment and sustainability. Two of the three contributions are from journals aimed at teachers with the third being published in *Science*, a journal primarily read by scientists. Throughout my career I have tried to write for audiences beyond the science and environmental education academic communities.

Looking Forward

The move from King's to Bristol has impacted on my research and writing time. Leading a university department is immensely challenging and the learning curve is steep. At this stage in my career I am more likely to be asked to edit individual books or book series, handbooks and encyclopedias. I also find myself being asked to write forewords and editorials. The temptation to say 'yes' to opportunities is almost irresistible and I find myself working on a new project—the Routledge Science Education Series, with two good friends, Steve Alsop from York University in Canada and Marianne Achiam from Copenhagen University. While this series is embryonic, another series, with Peter Lang, [Re]thinking Environmental Education is going from strength to strength with contributions from new scholars as well as experienced ones.

I feel very privileged to have worked with some of the brightest and best science and environmental educators over a number of years. I could not have written most of the contributions in this volume without their influence and ideas. And for that I am truly thankful.

References and Further Reading

Adey, P., & Dillon, J. (eds) (2012). *Bad Education: Debunking Myths in Education*. Milton Keynes: Open University Press.

Corrigan, D., Dillon, J., & Gunstone, R. (eds) (2007). *The Re-emergence of Values in Science Education*. Rotterdam: Sense.

Corrigan, D., Dillon, J., & Gunstone, R. (eds) (2011). *The Professional Knowledge Base of Science Teaching*. Dordrecht: Springer.

Corrigan, D., Dillon, J., Gunstone, R., & Jones, A. (2015). *The Future in Learning Science: What's in it for the Learner?* Dordrecht: Springer.

Crow, M. (2004). Call for a new approach to science teaching. *Guardian*. Retrieved from http://guardian.co.uk/education

Dillon, J. (2007). An organic intellectual? On science, education, and the environment, in K. Tobin and W. M. Roth (eds), *The Culture of Science Education. Its History in Person.* Rotterdam: Sense, pp. 311–322.

Dillon, J., & Maguire, M. (eds) (1997). *Becoming a Teacher* (1st ed.). Milton Keynes: Open University Press.

Dillon, J., & Maguire, M. (eds) (2001). *Becoming a Teacher* (2nd ed.). Milton Keynes: Open University Press.

Dillon, J., & Maguire, M. (eds) (2007). *Becoming a Teacher* (3rd ed.). Milton Keynes: Open University Press.

Dillon, J., & Maguire, M. (eds) (2011). *Becoming a Teacher* (4th ed.). Milton Keynes: Open University Press.

Dillon, J., Grace, M., & Oulton, C. (2004). Some critical reflections on the teaching of controversial issues in science education, *Development Education Journal*, 10(3), 3–6.

Dillon, J., Morris, M., O'Donnell, L., Reid, A., Rickinson, M., & Scott, W. (2005). *Engaging and Learning with the Outdoor Classroom: The Final Report of the Outdoor Classroom in a Rural Context Action Research Project.* Slough: National Foundation for Educational Research.

Dillon, J., Osborne, J., Fairbrother, B., & Kurina, L. (2000). *A Study into the Professional Views and Needs of Science Teachers in Primary and Secondary Schools in England.* London: King's College London.

Henriksen, E. K., Dillon, J., & Ryder, J. (eds) (2015). *Understanding Student Participation and Choice in Science and Technology Education.* Dordrecht: Springer.

Jorde, D., & Dillon, J. (eds) (2012). *Science Education Research and Practice in Europe.* Rotterdam: Sense.

Kendall, S., Murfield, J., Dillon, J., & Wilkin, A. (2006). *Education Outside the Classroom: Research to Identify What Training Is Offered by Initial Teacher Training Institutions.* Research Report 802. London: DfES.

Krasny, M., & Dillon, J. (eds) (2013). *Trading Zones in Environmental Education: Creating Transdisciplinary Dialogue.* New York: Peter Lang.

Osborne, J., & Dillon, J. (2008). *Science Education in Europe: Critical Reflections.* London: Nuffield Foundation.

Osborne, J., & Dillon, J. (eds) (2010). *Good Practice in Science Teaching: What Research Has to Say* (2nd ed.). Milton Keynes: Open University Press.

Osborne, J., Leach, J., & Scott, P. (1997). Obituary: Professor Rosalind H. Driver (1941–1997). *Studies in Science Education*, 30, 1.

Oulton, C., Day, V., Dillon, J., & Grace, M. (2004). Controversial issues—teachers' attitudes and practices in the context of citizenship education, *Oxford Review of Education*, 30(4), 489–507.

Oulton, C., Dillon, J., & Grace, M. (2004). Reconceptualizing the teaching of controversial issues, *International Journal of Science Education*, 26(4) 411–23.

Prieto, T., Watson, J. R., & Dillon, J. S. (1993). Pupils' understanding of combustion, *Research in Science Education*, 22, 331–40.

Rickinson, M., Dillon, J., Teamey, K., Morris, M., Choi, M. Y., Sanders, D., & Benefield, P. (2004). *A Review of Research on Outdoor Learning.* Preston Montford, Shropshire: Field Studies Council.

Stevenson, R., & Dillon, J. (eds) (2010). *Engaging Environmental Education: Learning, Culture and Agency.* Rotterdam: Sense.

Stevenson, R. B., Brody, M., Dillon, J., & Wals, A. E. J. (eds) (2013). *International Handbook of Research in Environmental Education.* New York: Routledge.

Watson, J. R., Dillon, J., & Miguens, M. (1991). Uma Experienci Pedagogica Educacao Ambiental Sem Fronteiras, *Aprender*, 14, 49–53.

Watson, J. R., Prieto, T., & Dillon, J (1995). The effect of practical work on pupils' understanding of combustion, *Journal of Research in Science Teaching*, 32(5), 487–502.

Watson, J. R., Prieto, T., & Dillon, J (1997). Consistency of students' explanations about combustion, *Science Education*, 81, 425–444.

Watson, R., & Dillon, J. (1996). Progression in Pupils' Understanding of Combustion, in G. Welford, J. Osborne and P. Scott (eds), *Research in Science Education in Europe*. London: Falmer, pp. 243–253.

Section 1

On *Habitus*

Instead of a chronological organisation I've gone for a thematic grouping of the papers and chapters. I hope that, as a result, readers will be able to get a sense of how my own personal development is reflected in my writing. This section, then, draws on five chapters and papers which I believe illustrate key aspects of my life and work that help to contextualise the other sections.

Shortly before I turned 50, Ken Tobin and Michael Roth asked me to contribute to a compilation of autobiographical studies written by academics at different stages of their careers. Some science educators find Ken and Michael's work challenging and at odds with their own. It is, though, important that any field looks at itself and its practices using different lenses. My contribution, 'An organic intellectual? On science, education, and the environment' attempts to show how I moved from being a school-teacher to a researcher. It was a pleasure to write and helped me to understand my own career better. The title reflects a comment that my appraiser, Stephen Ball (another contributor to the World Library of Educationalists series) once made about me. Not knowing its genealogy, I thought it meant a scholar interested in the environment—but I later discovered that it describes someone who links people and ideas together. I think that's very fair. Certainly this volume is the product of many collaborations with a number of academics from around the world.

When I joined King's it was clear that I was expected to enrol for a PhD. However, it took me a long time to complete it and if I could change one thing in my career it would be to have finished it much quicker. Before I'd finished it—Paula Abder-Fraser whom I had met at various international conferences, asked my to contribute to an edited book, *Professional Development in Science Teacher Education: Local Insights with Lessons for the Global Community*. I think I was the token Brit. The chapter, 'Managing teacher development: the changing role of the Head of Department in England' links my experience as a head of department in two schools to an empirical study into the needs and wants of teachers of science which I carried out with a number of colleagues in the late 1990s on behalf of the UK Council for Science and Technology.

I have always been interested in the links between science education and environmental education. When I was at university in the 1970s, environmental issues such as acid rain and biodiversity loss were beginning to be identified. I began to read the environmental education (EE) research literature and was lucky enough to be mentored by Arthur Lucas, one of the key figures in the early days of the field. Arthur, when he was my head of department at King's, supported my attendance at a workshop on the subject in 1992 which took place at the same time as the Rio Earth Summit. I see that time as being life-changing in many ways and I doubt that my career would have ended up as it has done if I had not gone to Brazil.

The environmental education research community is smaller than the science education field although there are a number of people with their feet in both camps. The EE community was for many years dominated by US academics who took a relatively narrow view of what counted as research. A number of non-US researchers and some more liberal US researchers have worked hard to broaden the field. 'Towards creating an inclusive community of researchers: the first three years of the North American Association for Environmental Education research symposium' published in the journal *Environmental Education Research* traces the history of some of those developments and is co-authored by a number of long-term friends and colleagues: Ron Meyers, Michael Brody, Marianne Krasny and Martha Monroe (all from the US), Paul Hart and Connie Russell from Canada and Arjen Wals from the Netherlands.

Part of the development of the community involved a focus on promoting discussion of work in practice rather than complete studies. Another dimension, reflected in the paper, 'The messy process of research: dilemmas, process, and critique' was an opening up of the field in terms of its practices so that new academics could get a sense of the reality of carrying out research. The paper, published in the *Canadian Journal of Environmental Education*, emerged from an innovative session which took place at an EE conference.

Perhaps the most career influencing project that I was involved with was the National Science Foundation funded Centre for Informal Learning and Schools, a collaboration between King's, The San Francisco Exploratorium— one of the world's first hands-on science centres—and the University of California Santa Cruz. This long-term initiative had various aspects, one of which was support for a number of excellent doctoral students to study at King's. We realised after a while that the idea of 'informal learning' was nonsensical—learning is learning is learning. However, CILS provided opportunities to study science learning in museums, science centres and botanic gardens—and that experience opened up a number of new avenues for my own work and that of the research group that I inherited in 2010 when Jonathan Osborne left us for Stanford. The rather inelegantly titled chapter, 'Broadening views of learning: developing educators for the 21st century through an international research partnership at the Exploratorium and King's College London' was published in *The New Educator* and is co-authored by the dynamic and inspirational Bronwyn Bevan.

1   An Organic Intellectual?

On Science, Education, and the Environment

Justin Dillon

Dillon, J. (2007). An organic intellectual? On science, education, and the environment. In, K. Tobin and W.-M. Roth (eds), *The Culture of Science Education: Its History in Person*. Rotterdam: Sense Publishers, pp. 311–322.

The contested and diverse relationships between science, education, and the environment have provided the context for my own personal and professional development throughout my career as science teacher and academic. With that in mind, I examine the nature of scholarship and quality in educational research through the lens of the day-to-day work of a "brain worker."

On the Increasing Industrialisation of Academic Life

> Education is the silver bullet. Education is everything. We don't need little changes. We need gigantic revolutionary changes. Schools should be palaces. Competition for the best teachers should be fierce. They should be getting six-figure salaries. Schools should be incredibly expensive for government and absolutely free of charge for its citizens, just like national defense. That is my position. I just haven't figured out how to do it yet.
>
> ("Sam" in "Six Meetings Before Lunch," *The West Wing*)

I should not be doing this. My priorities should be obtaining grants from research councils and writing papers for peer-reviewed journals. In the UK Research Assessment Exercise (RAE), which determines the "visions" and practices of all universities and most of their staff, book chapters in edited volumes usually do not count for much. During my eighteen years at King's, the RAE has steered academic life, reduced collegiality, and seriously affected research practices. The British scholar A. C. Grayling, writing in 1997, lamented the "increasing industrialization of academic life," noting that "Most scholars are academics now, and not all academics are intellectuals":

> In the new climate of research ratings, the cultivation of intellectual virtues, and the organic rather than forced pace of enquiry, is discounted. So the intellectual scholar, a person occupying a place apart, is a rarer creature now, even though there are many more universities.
>
> (1997, n.p.)

I don't feel in "a place apart" so I guess that I haven't reached the category of "intellectual scholar" yet but it seems like a reasonable goal to aim for. Some years ago, Stephen Ball described me in my appraisal as "an organic intellectual." Not being familiar with Antonio Gramsci's writings, I took it to be a kind reference to my scholarship in the field of environmental education, whereas it was actually meant to signal my ways of working as being collegial rather than isolationist.

Notions of what counts as an intellectual differ depending on who is doing the defining, Gramsci thought everyone was an intellectual, or at least had the potential to be so. In line with the thinking of Edward Said, Henry Giroux argues that educators are "public intellectuals" challenging the *status quo*. Giroux argues that such public intellectuals engage in intellectual practices that refuse:

> both the instrumentality and privileged isolation of the academy, while affirming a broader vision of learning that links knowledge to the power of self-definition and the capacities of administrators, educators, and students to expand the scope of democratic freedoms, particularly as they address the crisis of the social as part and parcel of the crisis of both youth and democracy itself.
>
> (Giroux, 2003, n.p.)

Such public intellectuals, Giroux argues, "interpret and question power rather than merely consolidate it." One reason that I became a teacher and, subsequently, an academic is to promote social and environmental change. So, while I *should* be writing another paper for a peer-reviewed journal or filling in a grant application form, I would argue that I *ought* to be writing for a wider audience, questioning and interpreting power.

Typically, the invitation to write this chapter came while I was attending a conference. I was in a wet and windy San Francisco, attending the annual meeting of the National Association for Research in Science Teaching (NARST). I say "typically" because conferences seem to be the place where one can engage with people and ideas for hours on end without the distractions that usually get in the way of being an academic. It's at conferences where you meet like minds and decide that you really should edit a book, write a paper or put in a research proposal. It's also at conferences where you're on public display and one can offend and discourage both deliberately and accidentally. Joy Carp, who used to work for Kluwer, once told me how she could recognize academics' nationalities by the way that they asked questions at conferences: U.S. scholars, she said, tended to ask gentle, congratulatory questions; Germans asked technical questions; and the British would begin with, "Well, that was a very interesting presentation" before asking a question that would completely expose the lack of theory, rigor, or whatever needed to be exposed.

In typical U.S. style, NARST hands out awards each year for a range of "bests." This year the award for "Outstanding Contribution to Science

Education" went to David Treagust from Curtin in Australia, someone I have known for many years and one of my referees in my recent, successful bid for promotion to Senior Lecturer (what every country in the world apart from the UK would call Associate Professor). David, a former president of NARST, made a gracious acceptance speech acknowledging the value of collegiality throughout his career. I bumped into him a few days later at the annual meeting of the American Educational Research Association (AERA) and congratulated him on his award. The first thing he said, in that unique Australian/Yorkshire accent, was "I meant it, you know?" reiterating the value of colleagues around the world. Both AERA and NARST continue to provide an opportunity to stimulate and catalyze my professional interests in science and environmental education. More importantly they provide an opportunity to be collegial—to meet, eat, and drink with friends and colleagues from around the world. They provide an opportunity to listen, talk and think—three aspects of the job that never get their fair share.

On Names

I was actually born Robert Boyd. I was adopted a few months after being born in June 1957. More adventurous parents might have chosen "Sputnik" as my middle name not Simon. But Justin Simon Dillon I became, sometime around about October 1957. My parents were both educators: my father trained as a teacher of arts and crafts during the World War II and my mother taught what was termed at the time "remedial reading." They both had had other jobs earlier in their lives, my mother had worked in the pottery industry (I was born in Stoke-on-Trent, in "the Potteries") and, at one time they ran a chemist shop (a pharmacy). As science teachers were in short supply in the 1960s, my father retrained as a biology teacher and spent most of his days in "secondary modern" schools, teaching students who had failed the 11-plus examination and who were, generally, given a second-class education. Without a science degree, my father's chances of teaching in a grammar school, working with those students who had passed the 11-plus, were slim if not non-existent. I passed the 11-plus—actually, being born in June, I took it when I was ten. Unlike most of my primary school friends, I went to Newcastle-under-Lyme High School, about thirty minutes away from home by public transport. It was my parents who decided that the other, closer grammar school was definitely not where I was going to spend my formative years. School choice has been an issue in England for much longer than people would have you believe.

On Learning about Science and the Environment

Until now I had never given much thought about when my interest in the environment began. Although there is a literature on "significant life experiences," I have never found the idea that particular events in one's early years result in life-changing decisions particularly convincing—a case of *post hoc ergo propter hoc* if ever there was one. I do, however, remember having a nature table at home—full of pinecones, leaves, and other souvenirs and specimens. I had stick insects, the odd rabbit, hamsters, guinea pigs, and the like, but I was not a dutiful owner.

Teachers in the 1960s and 1970s were not well paid. For a few years my father bred Mongolian gerbils as a sideline. They make ideal pets—awake during the day, clean, bright and they do not bark, bray, hoot, or otherwise disturb the peace. Local pet shops bought them in substantial numbers, and at one time, I think we had more than forty cages in the back bedroom. Unfortunately, my father's attempt to boost the family income came to a fairly sudden demise. Mongolian gerbils breed with alacrity: demand, and thus price, diminished to the point at which it was no longer a financially viable concern.

The post-Sputnik angst and the ambitions of the Labour Party to modernize Britain (the so-called "white heat of the technological revolution") led to physical science and technology becoming more prominent in the public consciousness. The rise of science has not been free from critique and criticism and this partly stemmed from concerns, often justified, about environmental impacts of new products and processes. Rachel Carson's *Silent Spring* was published in 1962, when I was five, and during the 1960s and 1970s, growing numbers of the public were becoming anxious about where science and society were headed. Environmental issues were increasingly reported in the media and the UK created the first Ministry of the Environment in the world in 1970 (which had been designated European Conservation Year).

Environmental education, which was to become my intellectual home some years later, was in its infancy in the 1970s with international gatherings in Belgrade (1975) and Tblisi (1977) setting the scene for future developments in the field. I did not therefore have any environmental education myself in any formal sense although the informal education provided by the media was probably highly influential in shaping my views.

My formal science education, post-11, involved a couple of years of "science" followed by a year of separate biology, chemistry, and physics. At the end of the third year in secondary education (at the age of 14) I had to select what options I was going to choose to go along with compulsory English, mathematics and religious knowledge. At this point, biology and I parted company for several years: I opted to take physics and chemistry instead. In a boys' school in the 1960s and 1970s, biology was perceived as a "girls'" subject and one which few adolescent boys would risk choosing (even if their father was a biology teacher). Chemistry and physics seemed to leave more doors open—and medical schools did not require school-level biology, not that I ever really considered being a doctor.

Two years later I emerged with a good crop of "Ordinary level" exam passes. As with the vast majority of my peers, I stayed at school and went onto take four "Advanced levels," chemistry, physics, mathematics, and general studies for two years. Again, being a boys' school, science subjects were the norm. I was actually better at English than I was at science in those days but the job prospects for English graduates at the time were not as good as those of science graduates.

Mathematics teaching at the school varied substantially and any love of the subject that I had faded fast. Chemistry was taught well and, in the post-Sputnik era, was supported by new teaching materials from the Nuffield Foundation. Nuffield Advanced Chemistry involved a substantial amount of practical work—the *sine qua non* of English science education. The cost of teaching Nuffield Chemistry must have been quite high—lots of ground-glass-jointed boro-silicate glassware and other state of the art apparatus.

Physics was taught by several staff at least one of whom was sufficiently acerbic and insular to make the decision of which subject to study at university relatively straightforward. I went to the University of Birmingham, sufficiently far away from my parents to allow independence but close enough in case of emergencies. In those days, the vast majority of students lived away from home, paid no fees and received a grant from their local authority to cover rent and food. Things have changed substantially since then.

From what I can gather, Birmingham University's chemistry department was typical of UK chemistry departments. The teaching varied from reasonably effective to appalling. One academic was famed for never making eye contact with the students during fifty-five-minute lectures. In general, the lectures and the laboratory work bore little relationship to each other. The contrast between chemistry at school and chemistry at university was marked: At school, your work was marked regularly and teachers had effective teaching styles and knew who you were. At university the opposite was true. I am afraid that, in my experience, the quality of university pedagogy is substantially lower than that provided in schools.

Growing concerns about the environment in the 1970s led to university courses focusing on the impacts of chemistry. One of the two essays that I wrote in four years at university focused on the environmental impacts of lead in petrol, an issue that was making the headlines at the time. The opportunity to research a topic, drawing on scientific papers and relating the chemistry to public health and government policy, was a highlight of my undergraduate studies. The other essay that I wrote, in 1976, was on gender and science, part of the Inter-faculty Studies course that undergraduates had to take. 1970 had seen the publication of *The Female Eunuch,* by Germaine Greer turning her into a "public intellectual," an academic who was able to shape and influence society through her writing and speaking.

On Becoming a Teacher

During my final year at Birmingham I decided to apply for teacher training. I could not envisage working for a chemical company and it was increasingly unlikely that I would do a master's or a PhD in chemistry. I was turned down by my first choice of institution, the London Institute of Education ("better candidates available") but was accepted at Chelsea College, also part of the University of London.

The preservice year (1979–80) was probably the turning point in my life. The Centre for Science and Mathematics Education was pre-eminent in the UK and many of the staff had been heavily involved in development of Nuffield Science. Two eminent figures in science education, Paul Black and Jon Ogborn, were joined by Arthur Lucas, one of the "founding fathers" of environmental education, at the end of my one-year course in June 1980. Jan Harding and John Head proved instrumental in developing my thinking about gender in science education and the personal response to science. John Barker, an inspirational biology educator and oenophile turned into a life-long friend and Bob Fairbrother initiated me into the complex world of assessment. Such a concentration of talent (in such an unprepossessing building—a converted factory) would have been hard to find anywhere in the world. However, my father, a trained and committed science teacher, had probably never heard of them or their contributions science education, working as he did in the second tier of science teaching. The distance between science teachers and science education researchers was greater then than it is now.

The one-year preservice course involved teaching at two schools, in class-room groups (three to four student teachers with a small number of students) and solo. About half the course was spent in school. My first teaching post, which I took up in September 1980, was as an assistant teacher of chemistry at the John Roan School in the southeast of London. The head of science at the school was Brian Matthews, who later went on to become the main science educator at Goldsmiths College in London. Brian led an innovative department and was committed to a socially critical form of science edu-cation. The first-year science course, for pupils aged eleven, included such activities as trying to identify the contents of tins without ever being able to see inside or open them—an attempt to focus students on thinking about the nature of science.

During my early years teaching science in London I attended some meetings of the British Society for Social Responsibility in Science, led by Joan Solomon. The radical science strand within British science has always been on the fringes and was never in a position to have much influence. Nevertheless, the experience shaped my view of science, the environment and scientists. I realized that my understanding of chemistry improved radically as a result of teaching the subject. Both Nuffield O-level and A-level courses offered a range of options that students could choose. I was able to teach historical topics at O-level—making sodium by electrolysis—and mineral process chemistry at A-level, extracting metals from mineral ores. Together with friends from Chelsea College days, I began to organize field trips to some of the UK's most beautiful and geologically interesting places, ironically, not far from where I was born.

During my time at the John Roan School, I carried out my first piece of rudimentary research in science education, looking at the attitudes of pupils towards science and scientists, a topic of concern even then. After four years at the school, I was recruited to the Inner London Education Authority's Science Support Team. The team contained several innovative science educators (including Jonathan Osborne) and engaged in curriculum development as well as providing cover for examination classes in schools. I gained experience of teaching in six schools across London and was involved in making a video about the iron and steel industry in southern Wales as part of a major review of the secondary science curriculum in the days before the implementation of the national curriculum.

One of the schools that I was supporting, Eltham Hill, offered me the head of chemistry position, which I was happy to accept. I led a group of committed chemistry teachers, all of whom were older than me, and, together, we made chemistry a popular subject, not an easy task in an all-girls school in the 1980s. I also argued successfully for the implementation of technology as a separate subject in the school—one of the first all-girl schools in London to teach the subject.

During my time at Eltham Hill I took a two-year, part-time master's degree in science education at King's College. Doing the MA was one of the wisest choices that I have ever made. It opened up my eyes and brain to a broad range of literature and enabled me to meet teachers from around London and beyond. As well as teaching courses in research methods, recent developments in science education and curriculum studies, I wrote a 15,000-word dissertation on "Technology, Culture and Gender," thus pulling together several interests into one piece of work.

After four years at Eltham Hill I was encouraged to apply for the vacant head of science position at Kingsdale School, labeled by the *London Evening Standard* as the Inner London Education Authority's "School of shame" as a result of its poor record for behavior and attainment. Situated in a leafy suburb, many of Kingsdale's pupils came from poorer areas several miles away. This was, by far, the hardest job of my life and despite the skill and collegiality of the staff—it was one of only two London schools to have a bar—the school struggled to maintain any semblance of a decent education for all its students. The Authority's response to the challenging situation that we were in was to employ "Inspectors Based In Schools" (the IBIS team), an interesting innovation that did not last beyond the break up of the ILEA in 1988.

On Becoming an Academic

After a year at Kingsdale, Rod Watson, who had succeeded Jan Harding as the main chemistry educator at King's, contacted me. Rod had already encouraged me to get involved in the PGCE chemistry course as a mentor of student teachers and through assisting with geological field trips. When he received a grant to set up a National Environmental Database (NED) project, he asked me to join the team at King's, sharing the teaching of the preservice course and working on the NED project. However, it was quite a risk—I gave up a permanent post for a contract that lasted three months with only the possibility of an extension.

The difference between working as a schoolteacher and working at a university was quite remarkable. Sunday afternoon was free of the stress revolving around the difficulties of teaching in inner-city schools and the regimented life that they force on staff and students alike. Going to work became a pleasure. As Dylan Wiliam, a colleague at the time, put it: "When I was a teacher, all the reading and writing I did for an MA was a hobby—now I get paid to do it all the time." Another colleague complained to our then-head-of-department Arthur Lucas that she was so busy "doing" things that she did not have the time to think. Arthur replied that that was the one thing he paid her to do; and that is primarily how I see myself—as someone who is paid to think.

A recent review on the death of the English intellectual notes that the term "intellectual" has several meanings including the sociological one—"brain worker." "Lecturer," my current job title, is almost completely inappropriate as I rarely lecture—virtually all the teaching I do is leading workshops, seminars, and discussions or conducting supervision.

Arthur Lucas was highly influential in encouraging me to devise and teach a master's course on environmental education. That in turn led me to read more literature and to develop a line of work that has dominated my career so far. Arthur was a natural leader and would spend time wandering the corridors late at night to see who was working and how life was going. His influence on my academic life and my confidence was profound.

On Beginning Research

The first funded research that I was involved in came out of a very small project initiated by Rod Watson. Rod had taught in Spain earlier in his career and had developed a range of contacts in Spanish universities. With the aid of a small grant from the British Council and a Spanish funding agency, we were able to set up a link with the University of Malaga. Together with Teresa Prieto, Rod and I carried out questionnaire surveys of students looking at their understanding of everyday processes such as burning. The project lasted for three years and resulted in publications in *Research in Science Education,* the *Journal of Research in Science Teaching* and *Science Education.* The studies also provided data for use in pre-service teaching and in research methods workshops. I have never engaged in such a cost-effective study since!

A disadvantage of being on temporary contracts is that you feel obliged to say "yes" to any teaching or other work that comes along. As a result, Martin Monk—with whom I shared an office—and I taught far more than we should have and did far more consultancy work than was the norm. Martin played a key role in my early development as an academic. He had been an outstanding teacher when I did my MA and continued to provide support and guidance whenever we worked together. His knowledge of the history and philosophy of science seemed limitless and he had a better understanding of psychology and sociology than the rest of us in the science education group.

A significant amount of my time was spent on preservice teacher education—teaching and visiting students in school. By then, it was a government requirement that preservice teachers spent twenty-four weeks of their thirty-six-week course in school. Although in some ways laudable, the net result was a major inequity between the quality of mentorship received by our students. As a science group, we decided to produce a book of theory-based practical activities that was published by Falmer Press as *Learning to Teach Science*. Written entirely by King's science staff, the book was edited by Martin and me. Both Martin and I were clear about what we wanted and edited other peoples' and each other's contributions—quite heavily in retrospect. This initial experience of editing gave me the confidence to develop critical lines of thinking about science education and science teacher training.

In response to comments from inspectors that our preservice teachers did not appear to do much reading during their one-year course, Meg Maguire and I set about putting together an in-house collection of papers on educational issues. The collection of papers was enough to convince the Open University Press of the potential for a book, written entirely by King's staff, aimed at the preservice teacher education market. The first edition of *Becoming a Teacher,* published in 1995, contained twenty-three chapters. The tone of the book was set out in the introduction:

> In putting together this book we have tried to emphasize the three Rs: reading, reflection and research. Good teachers are able to learn from their experiences, reflecting on both positive and negative feedback. The best teachers are often those who not only learn from their experience but also learn from the experiences of others. Reading offers access to the wisdom of others as well as providing tools to interpret your own experiences. We have encouraged the authors contributing to this book to provide evidence from research to justify the points that they make. We encourage you to reflect on that evidence and on the related issues during the process of becoming a teacher.

The book sold well and OUP commissioned a second edition, published in 2001 and a third edition is due out in 2007. I have sworn never to do another edition of the book after each edition has been published—editing the writing of one's colleagues can be difficult, embarrassing, and time-consuming.

Increasing Informality

Rosalind Driver's appointment, in 1995, to the Chair of Science Education at King's strengthened the science unit that already contained well-known figures such as Philip Adey, Paul Black, Arthur Lucas, Jonathan Osborne, and Michael Shayer. Ros was a "person" person, interested in people as well as ideas, and able to engage with anyone she met. Though her time at King's was relatively short, she brought a sense of clarity and determination and played a key role in the department as well as in the science unit. She joined John Head as one of my PhD supervisors and offered firm and wise guidance.

Ros' illness and untimely death in 1997 diminished the world of science education immeasurably. Earlier in the year she had been the recipient of NARST's Outstanding Contribution to Science Education award. She received a standing ovation from the audience in Chicago. I was privileged to be asked to contribute to an edited collection highlighting the impact of her work on the science education community *Improving Science Teaching Through Research,* which was written in an innovative and collegial manner: the authors produced first drafts of their chapters in advance of a writing workshop held in Leeds. The rest of the authors shred each chapter, sometimes quite finely: the end product is certainly much better than the first drafts that we bought to the table. I have used this technique of producing a book subsequently together with Dick Gunstone and Debbie Corrigan from Monash (King's and Monash are twinned). This time the writing workshop preceded the 2005 *European Science Education Research Association* (ESERA) conference in Barcelona.

Ros had played a key role in the creation of ESERA, which was founded at the European Conference on Research in Science Education held at Leeds in April 1995. The final conference session involved vigorous discussions about "Europe," "science education," and "research," which have different meanings in different European countries. John Gilbert ably chaired the session, managing to cajole and coax agreement among a very diverse group of science educators. Rick Duschl—the only U.S. science education academic at the meeting—and I counted the votes for the first ESERA board.

Jonathan Osborne and I spent a good deal of time at the 1998 NARST conference trying to identify potential candidates for the vacant chair of science education. One of several names that kept coming up was Rick's, then at Vanderbilt University. He was appointed to the Chair of Science Education in 1999.

A year after Rick's appointment, he was approached by colleagues in the US who wanted King's to join in a consortium with the University of California Santa Cruz and the San Francisco Exploratorium to bid for funding from the U.S. National Science Foundation. A year later, the Centre for Informal Learning and Schools (CILS) was created, with a grant of almost $11 million. The creation of CILS radically affected the nature of the science unit at King's. Eight PhD students and three postdoctoral researchers all looking at aspects of learning in schools, museums, and science centers have broadened the focus of the science group's interests.

As a result of CILS, our relationships with institutions such as the Science Museum and the London Natural History Museum (NHM) have become much stronger and mutually beneficial. CILS has also facilitated collaborations with other departments within the university. Together with a colleague from our mechanical engineering department, Mark Miodownik, I have just been awarded £82,000 by the UK Engineering and Physical Sciences Research Council to support an innovative project called "What can the matter be?" The aim of this project is to investigate ways in which engagement with contemporary culture enhances the public's appreciation of science and engineering. We intend to develop tours of the Tate Modern, which will cover three themes: chemistry/images, materials/form, and engineering/installations. An MP3 tour—for iPods and the like—will be downloadable from the Internet.

Another outcome of CILS was the opportunity to evaluate the *Permanent EuropeaN resource Centre for Informal Learning* (PENCIL). PENCIL consists of fourteen institutions from twelve countries working together developing and sharing good practice. Our task is to evaluate the progress of six of the pilot projects and to look at the development of the network as a whole.

Researching Education and the Environment

I was lucky enough to be able to attend NARST and AERA virtually every year. I attended the early meetings of AERA's Ecological and Environmental Education Special Interest Group in the 1990s, which drew together some of the most well known environmental educators from Australia, Canada, the UK and the US. Over time I developed a strong collaboration with colleagues at Bath University and the *National Foundation for Educational Research* (NFER) in the UK that also have benefited some of our doctoral students.

In recent years I have worked with colleagues at both institutions on a range of research projects funded by government agencies and non-governmental organizations. Two of the projects involved literature reviews which provided me with the need and the time to read widely in the field of outdoor education. If I had to choose one publication that I think represents my best piece of collaborative work it would be the *Review of Research into Outdoor Learning* published by the Field Studies Council (Rickinson et al., 2004). Literature reviews, which again don't seem favored by the RAE, actually provide opportunities for scholarship of the highest order: reading, interpreting, and synthesizing hundreds of studies is an inordinately challenging task.

Subsequently, Bath, King's, and the NFER were commissioned to carry out a study into *Learning in the Outdoor Classroom*. This action research project coincided with me developing cataracts in both eyes resulting in a severe deterioration in my eyesight and a slowing in my work. The wonders of modern science and medicine coupled with a free public health service have transformed my life.

Editing, Scholarship and Quality

In 2005, a year after I became Secretary of ESERA, itself a great honor, I was invited to become one of the editors of the *International Journal of Science*

Education. The journal has five editors working with the Editor-in-Chief, John Gilbert. With fifteen editions of 225 pages each year, the workload on the editors and the reviewers is substantial. The growth in the number of science education journals and the pressure on academics to publish or perish has its critics. A. C. Grayling describes the situation:

> Coteries of dons write in impenetrably specialist codes for internal consumption only, in hundreds of journals and monographs. *(Peccavimus omnes)*

> (1997, n.p.)

Editors and reviewers are seen as the gatekeepers of scholarship in the community of practice that is science education, trying to improve or, at least, maintain, the quality of science education research. This might not be the ideal system for encouraging change and diversity among the field. Editors are not elected by their peers and reviewers may hide behind the cloak of anonymity. It is by no means a level playing field. Jim Shymansky, NARST's outgoing President at the 2006 meeting where I was invited to write this chapter, raised the issue of journal article length during his awards-luncheon speech. His point, tellingly made with data from the *Journal of Research in Science Teaching,* was that papers have got longer and that the total number of papers being published in *JRST* has not increased concomitantly. Shymansky's beef was that papers are too long; but the real issue is whether the days of paper-based journals are numbered. The answer is almost certainly "yes," the issue of article size thus becomes almost redundant.

More recently, I have been involved as a coach in the ESERA Summer Schools, in which sixty or so doctoral students from around Europe get together for a few days of workshops, lectures and discussions about science education. It is activities such as these that help to develop the community of practice that is science education. The challenge of understanding and critiquing the work of colleagues from varied cultural backgrounds provides intellectually stimulating opportunities from which everyone seems to benefit. ESERA has grown now to be a strong organization able to organize biennial conferences and summer schools and with the financial reserves to support colleagues from the central and eastern parts of Europe, where science education is fledgling and poorly funded.

I have been lucky in working with excellent doctoral students of my own while, ironically, struggling to find time to complete my own doctorate which looks at the role of middle managers in science teachers' professional development. Not having a PhD is a significant barrier to promotion in the UK system. As one of my colleagues once said, having it is not the issue; it's not having it that is the problem.

So, after 26 years as a science and environmental educator, I look back at what has been a challenging, enjoyable, and eventful career. There is, I hope, much more to come. The day before I concluded this chapter, I was attending a meeting of a group of science educators from across Europe who had been brought together by the Nuffield Foundation. We were looking at the challenges facing science teachers and science education generally as we move in to the twenty-first century. So, what have I learned? Well, "Sam" was right, "Education is the silver bullet. Education is everything. We don't need little changes. We need gigantic revolutionary changes," still. And they will only come about when we as a community have enough evidence, commitment and influence to make them happen.

2   Managing Teacher Development

The Changing Role of the Head of Department in England

Justin Dillon

Dillon, J. (2002). Managing teacher development: the changing role of the Head of Department in England. In, P. Fraser-Abder (ed.), *Professional Development in Science Teacher Education: Local Insights with Lessons for the Global Community*. Abingdon: Taylor and Francis, pp. 172–186.

Teacher development is not something to be done solely in private behind closed doors, at least not if it is to be effective in changing teachers' practice (Joyce & Showers, 1988). The skills and knowledge required for effective teaching develop and grow in a range of settings and are constrained by a wide variety of factors ranging from personal circumstances to systemwide features such as the curriculum or assessment policies. In England, which has undergone substantial education reform, particularly since the mid-1980s, there have been many attempts to provide professional development opportunities for individuals, for groups of teachers, and for the whole teaching cohort. One feature that characterizes the climate in which teachers work is the focus on standards and outcomes rather than on processes. A government-funded body, the Teacher Training Agency, has produced a normative set of standards that are meant to guide teachers and professional development "deliverers." These standards are already the subject of interest of a range of countries around the world.

Since the mid-1980s, researchers have investigated the success of specific projects or courses and have suggested lessons for a wider audience, while others have looked at the general circumstances within which teachers operate. What seems clear is that teacher development *is* possible but is likely to be hampered by a range of barriers to change, many of which show no sign of disappearing. In this chapter I focus on two pieces of research with which I and colleagues have been involved that examine the current state of professional development in England at a systemwide and at a more school-focused level. One study involves a nationwide survey of science teachers' needs and wants, whereas the other is an in-depth study of the management of teacher development. I argue, using evidence from these and other studies, that normative models of teacher development are totally unrealistic and inappropriate for addressing the needs of teachers and schools.

Background

In England, education is compulsory for children ages 5–16. Most children attend primary schools (of which there are about 26,000 in the country) until they are 11 and then transfer to secondary schools (of which there are about 4,000). In the former, they are likely to be taught by the same teacher for all subjects for 1 year before moving onto another teacher in September. In secondary schools, children have different teachers for different subjects and could well have the same teacher at several times in their schooling. This chapter refers only to those teachers in state schools, those funded by the government, which look after around 93 percent of the country's children.

For various reasons, a national curriculum, made up of a range of subjects studied each year, was introduced in 1988. National assessment, which takes place at the ages of 7, 11, 14, and 16, is used to monitor attainment as well as to compare schools. Science is one of three core subjects that students must take throughout their years of compulsory education. Partly as a result of this "core subject" label, secondary science teachers, who are usually graduate scientists with an education qualification, have a relatively high status compared with their art, history, geography, and modern language colleagues.

Newly qualified secondary school teachers would expect to teach about 36 out of 40 periods (lessons) each week. A science teacher would normally be part of either a science department or a single-subject (biology, chemistry, or physics) department led by a Head of Department (HoD) (Departmental Chair in the United States). The HoD, as well as being paid for his or her responsibility, would expect to be teaching less than an ordinary classroom teacher—say, 32 periods each week. It has been argued that the pyramidal hierarchy of the English education system owes much to the influence of military service on a generation of young adults during and after World War II.

The surface features of the current role and responsibility of the HoD do not appear to have changed greatly over the years. However, a growing climate of accountability and managerialism in English schools has led to fundamental changes in the day-to-day existence and experience of many HoDs. In particular, the role of the HoD in teacher development, through monitoring, appraising, and setting targets, is now radically different. As an example of the climate of expectation that surrounds HoDs, one need look no further than the standards set by the Teacher Training Agency (TTA), a government funded body for subject leaders (the name given to heads of department in secondary schools and subject coordinators in primary schools). The standards define "key outcomes" of the job, which include "teachers who: work well together as a team; support the aims of the subject . . . [and] are dedicated to improving standards of teaching and learning" and "the ability to lead and manage people to work as individuals and as a team towards a common goal" (Teacher Training Agency, 1997, pp. 4, 6). The utility of normative models of management, which fail to problematize the roles and responsibilities of teacher leaders, is highly questionable. The imprecision and naïveté of such models are neither enabling nor informative. The reality of HoDs—the changing role and the consequent impact on the management of teacher development—are the foci of this chapter.

The Issues Affecting Science Teachers' Professional Development

Heads of Department can only work within the constraints of the system. The current status in England with respect to science teachers' experience and expectations toward professional development can be gauged from a study commissioned by the Office of Science and Technology in 1999 (Dillon, Osborne, Fairbrother, & Kurina, 2000). The study's findings are based on research with 20 focus groups involving more than 150 teachers from 50 schools in five regions of England and a questionnaire survey of a randomly selected sample of 1,973 primary and 735 secondary state schools in England. The study looked at a range of key issues: teachers' qualifications in science, their opinions of their preservice training, their current teaching profile, their experience of inservice training during the year, their opinion of that training and advice, and their desires for improvements in the quality of what professional development was offered to them.

In England, most secondary school teachers take a 1-year Postgraduate Certificate in Education (PGCE) at a university department of education. English science teachers, in line with those in some U.S. studies (see, e.g., Luft & Cox, 1998), feel well prepared for teaching some or all of the science curriculum at the end of their preservice courses. However, they are very critical of the support that they receive during their careers.

A particularly worrying feature for Heads of Department was that the teachers in the study were not engaged in a subject-related, classroom-based, systematic process of Continuous Professional Development (CPD) matched to their individual needs. Interestingly and critically, "many teachers in the focus groups did not fully appreciate the term CPD." Although in the 1980s teachers had 5 days of holiday removed and replaced with 5 days devoted to training, there was significant widespread dissatisfaction with the use made by schools of those days. Teachers felt that there was too much focus on administration or on whole school issues, such as literacy and numeracy, and too little focus on personal, professional development.

Another disturbing feature for Heads of Department was that teachers were critical of the existing "appraisal arrangements for identifying their individual strengths, and of just how little say they had in their individual CPD, or the courses that they did attend." Although most teachers had received some form of inservice education and training (INSET) during the preceding school year, it was mainly from colleagues in their own school rather than from other sources of inservice education and training. In their view, INSET at present was insufficiently focused on their individual needs, involved too little in the way of practical activity rather than pure theory, and too rarely permitted interaction with other teachers.

What the teachers *did* want were more opportunities, "not only to share experience and good practice with colleagues in their own school and in other schools, but also to compare their practice with others in order to identify their individual CPD needs." Specifically, they wanted higher quality INSET and more emphasis on classroom-focused support. In terms of topics, the teachers prioritized subject knowledge, pedagogy, pupil learning, and classroom management (Dillon *et al.*, 2000, p. 21).

Around half of the teachers in the study had made use of local government advisors (although 31% of those that had received advice rated it as "poor") and only 25 percent had been to their local teachers' center (of those, 42% described the help as "poor"). Of all our data, these negative findings were perhaps the most surprising.

The main professional association, the Association for Science Education (ASE), had provided advice for 56 percent of the teachers and was very positively reviewed. Teachers subscribe to the ASE and so might be more likely to review it positively than they would if it were a free service. The role of professional organizations is somewhat underresearched in England despite their importance in influencing both government policy and teacher practice.

One of the major issues in science education in England has been the effect of the move from separate science teaching (biology, chemistry, and physics) to "balanced science." Many teachers do not have an adequate background in all three science subjects. Among those teachers teaching science topics to students ages 14–16, 26 percent of those teaching biology topics did not have an A-level (taken at age 18) in the subject, nor did 13 percent of those teaching chemistry topics and 29 percent of those teaching physics topics. Furthermore, 39 percent of those teaching biology topics did not have a degree in biology, 51 percent of those teaching chemistry topics did not have degree in chemistry, and 66 percent of those teaching physics topics did not have a degree in physics. Not surprisingly, teacher confidence, one of the major factors affecting teaching quality, varied from topic to topic.

Another area in which successive governments have tried to innovate in science education has been in the use of computers. The Department for Education and Employment has spent significant sums of money equipping schools with computers and on setting up a National Grid for Learning (NGfL). Despite this investment, the level of use was reportedly rather low. From the questionnaires returned, the NGfL was used only "rarely" by 72 percent of teachers. Whereas 42 percent of primary teachers reported using computers "often" in their science teaching, only 9 percent of secondary science teachers could say the same.

The amount of money available for INSET varied widely from school to school. The mean total amount available in the secondary schools was £14,730. The amount of money available per member of staff ranged from £50 per teacher to £1500. The average amount was £304. The head (principal) or a senior colleague generally makes decisions about how INSET funds are allocated. Most heads often determine INSET needs in line with School Development Plans with individual requests next in line as a determining factor.

Head teachers differentiated between the needs of less and more experienced teachers. For inexperienced teachers of science, the head teachers rated Teaching Skills, How Students Learn, Raising Achievement, and Class Management as the most important. For more experienced teachers head teachers viewed Raising Achievement as by far the most important topic, with How Students Learn and Middle Management some way behind. The major constraints on the provision of adequate professional development, according to head teachers, were, unsurprisingly, a lack of time and a lack of finances.

We concluded that there was an urgent need to examine carefully the effectiveness of the current provision for *personal professional development* of teachers of science. In particular, more time and funding need to be allocated to personal, professional training rather than institutional imperatives. In England, there needs to be a major cultural shift within the profession to make subject-specific, classroom-based continuous professional development the norm. At present, for a significant number of newly qualified teachers of science, if not the great majority, their individual needs for subject-based CPD take second place to whole school issues, and any support they do receive diminishes rapidly after the first year of their careers.

In concluding, we argued that school managers have a responsibility for the creation and maintenance of a pro-CPD culture in a school. Management should have the capability to sustain and nurture the subject-related expertise of its teaching staff effectively within the existing very real constraints on science teachers' CPD that the study has highlighted, namely, the provision of time and funding, an appropriate reduction in workload and fatigue, and the necessary supply cover for teachers to attend external courses.

Government Initiatives

One strategy adopted by previous governments and reinforced by the current government is mandatory appraisal for teachers. The philosophy behind the scheme is that experienced teachers are able to identify the effectiveness and the competence of their colleagues and can suggest the most appropriate professional development for them. In practice, heads of department observe a teacher and then talk through the teacher's strengths and weaknesses, setting targets for improvement and suggesting strategies and professional development needs. In practice, appraisal has been a major failure (OfSTED, 1996). Specifically:

- The impact on teaching standards and learning has not been substantial.
- In a majority of schools appraisal has been too isolated from planning for inservice training (INSET) and for school development.

- The classroom observation of teachers has been variable in quality, and this has led to appraisal interviews being insufficiently focused on the improvement of teaching.
- Schools have chosen not to link appraisal to pay and promotion, even indirectly.

The emphasis in appraisal has too often been the institution and not the individual. A second issue has been the inappropriate and inadequate nature of staff development available for teachers. The current government, picking up on the last point listed by OfSTED, has deliberately linked appraisal with performance. Performance-related pay is likely to be one of the major controversies in education in England in the first decade of this century.

The Impact of Educational Reform on Teachers and on Their Professional Development

For several years I have been studying the impact of government reform on heads of department. The impact of the National Curriculum has been profound (see, e.g., Donnelly & Jenkins, 1999). For example, the pressure of an overcrowded curriculum has caused teachers to adopt a transmissive mode of teaching (Hacker & Rowe, 1997; Osborne & Collins, 1999).

Helsby and Knight (1997) identified three postreform changes that are relevant to the area of professional development: "an obvious and pressing need for new learning, to build or extend knowledge of subject matter; to develop pedagogical processes appropriate to the new requirements; and to foster mastery of teaching to, and assessing against, the AT [attainment target] statements" (p. 146). However, they go on to point out, "The changes in the formal structures of in-service education and support for teachers (INSET) which have accompanied the educational "reforms" of recent years, have seriously restricted the opportunities for personal, professional development" (p. 149). According to Helsby and Knight, INSET is now "heavily managed from the center within tight budgetary constraints" (p. 149). The onus for staff development now lies more with schools than at any time in the last 30 years.

The implementation of the National Curriculum coincided with responsibility for managing financial resources being devolved to schools. This resulted in a lack of any systemwide priorities at local or national level. For instance, Her Majesty's Inspectorate (HMI) (1992) commented, "Teachers attended a range of courses but with many schools' receiving devolved INSET funding, much of the INSET has been school-based. . . . Overall, however, the systematic identification and prioritization of INSET needs, both individual and departmental, was not sufficiently common" (p. 27).

There is also further evidence that schools and Local Education Authorities have not invested enough in training for middle management, particularly those aspects of the job that focus on staff development, appraisal, and teamwork. HMI (1992) concluded, "Schools and LEAs would be wise to invest a greater proportion of INSET provision in the training and support of Heads of Science" (p. 31).

The Reality of Teacher Development

I have previously outlined a range of factors that might encourage successful teacher development, including:

- Initial disturbance or dissatisfaction
- Time to reflect on existing strategies
- An evidence base of successful teaching strategies based on models of learning
- Feedback on classroom performance
- Encouragement from managers
- A feeling of personal growth
- A sense of ownership of innovation (Dillon, 2000)

Simply listing supposedly desirable factors is nothing more than an academic exercise unless there are mechanisms in schools for the factors to be facilitated. Without informed, committed, and skilled managers, the prospects that teachers will be able to engage in systematic, continuous, personal professional development are almost nil.

In a study conducted with Heads of Department in nine schools in a large city in Southeast England, I have found evidence that throws doubt on the likelihood that the culture of CPD is likely to change radically in the near future (Dillon, in preparation). In the light of the findings from that study, I will address some of the issues that affect the factors just listed.

Initial Disturbance or Dissatisfaction

Bell and Gilbert (1996), in their study of New Zealand science teachers, found that one of the critical stages in teacher development was the recognition that an aspect of personal practice was problematic. This echoes Nancy Davis' (1996) finding that a sense of dissatisfaction can lead to a desire to change an aspect of a teacher's practice.

However, in the current climate of accountability and managerialism, which pervades schools in England, it is unlikely that teachers will be eager to admit too readily to problems with their practice. Performance-related pay, introduced in September 2000, means that teachers are set targets on an annual basis. Admitting to failure could well mean that a teacher does not receive a pay rise. This is hardly the climate to foster open admission of problems.

As I have pointed out elsewhere (Dillon) "in departments which have experienced staff, successful students and an established social structure, it is very difficult to create a sense of dissatisfaction with existing practice" (pp. 94–109).

Time to Reflect on Existing Strategies

The overwhelming pressure teachers and their managers are faced with is that caused by lack of time. The busyness that characterizes schools in England is the result of almost constant change—curriculum, assessment, funding, to name but a few. Managers spend more time on administration than they do on person management. Few involved in education have the opportunities for reflection, which, as Dewey (1933) says, involves "a state of doubt, hesitation, perplexity, mental difficulty, in which thinking originates, and . . . an act of searching, hunting, inquiring, to find material that will resolve the doubt, settle and dispose of the perplexity" (p. 12).

One of the participants in my study, an experienced Head of Department, summed up what had happened to her before I interviewed her:

> Oh God, today has been a really bad day. I've been in tears today. That hasn't happened for a long time. It's been horrible. I've just felt that all day today I have been like . . . I get strings of children sent to me consistently for behaviour [. . .] I'd been on the go since eight this morning, . . . , consistently following up cases of bad behaviour, right, I taught two lessons this morning and then I observed Gerard after break. And that wasn't too bad and then it started really badly about 12:00 p.m. and I spent the whole of my own lunchtime dealing with this awful class and all they were saying was what rights they had. Ah, you know, the street talk and none of them owned up to this and it made me feel sick.

Lack of time to find out about new ideas limits the discourse of middle managers to repeating the rhetoric of their own managers: targets, performance, appraisal, and improvement. Managers' abilities to describe what is required in concrete terms at anything more than a general level ("students on task," "safe environment") can lead to frustration on the part of teachers who are vague about what is required and on the part of managers who feel helpless to provide real help.

An Evidence Base of Successful Strategies Based on Models of Learning

Evidence that teachers access educational research is hard to find, to say the least. This is not because researchers have ignored teachers as consumers of their findings (see, e.g., Adey & Shayer, 1994; Driver, Squires, Rushworth, & Wood-Robinson, 1994; Monk & Osborne, 2000; Novak & Gowin, 1984). In general, "teachers are forced to rely on local networks of informal contacts, either in-school or between schools" (Osborne & Dillon, 1999, p. 3).

It should be added that although a growing body of research has taken place into school effectiveness and school improvement, the validity of the findings has been questioned. There is a perception that there is an educa-tion management market that has so far failed to deliver a quality product (Gunter, 1997).

Feedback on Classroom Performance

Joyce and Showers' (1988) meta-analysis of the effectiveness of inservice edu-cation showed that success depends on *long-term,* classroom-focused coach-ing involving feedback. However, I found little evidence that teachers watch each other frequently or systematically, and when they do it is often part of a formal, short-term, appraisal system. As one HoD, Anne, put it:

> The plan . . . was that each member of staff was observed twice by two separate people. . . . I'd be looking at doing it termly. And really, if every-body was involved then it's a case of doing one and maybe being looked at once. I started off by doing "buddy pairs" . . . so that you saw and watched somebody else so that the two of you can work together. Then you can change the "buddy pairs" . . . or you might say, I can't work out how to do Science 1,[1] the investigative work, you might match up with somebody then who'd be a bit stronger.

Appraisal of teachers by managers, of which the above example is one model, became statutory in England in the 1990s, although it has not had the impact that was originally hoped. Formal appraisal usually involves focused obser-vation and other data gathering followed by discussion. In reality, because HoDs teach for the majority of the week, the opportunities for appraisal are limited. The idea of mutual observation is also becoming more common, although the time constraints for teachers without significant responsibility are even greater than for HoDs. Mike, one of the HoDs in my study, put it:

> I try to observe everybody at least once a year formally. There's the book monitoring: You can see what the sort of standard of the students' work is. There's the homework setting and the defaults from that. And I think you can pick up more when people put in no defaulters than when they put in a whole list. And so it's a question of saying, well didn't you get any defaulters last week? And then, if they say no, you say, well what did you set, you know, have you collected it in, got it marked. Why are you getting no defaulters and I'm getting about half-a-dozen?

For various reasons, many teachers have deployed a range of strategies to "block, deny or distort feedback from the rare observations that they receive from inspectors or other colleagues" (Dillon, 94–109). As I have pointed out earlier, "[T]he changes in the curriculum and the assessment policies in England, which have failed to shake the reliance on memorization as a learning strategy, have failed to create much dissatisfaction in teachers' own practice. Teachers are in danger of becoming resistant to criticism as a strategy for maintaining professional pride" (Dillon, 97).

During one of my interviews, an experienced HoD, Alan, was questioned as to how his school was preparing for a forthcoming inspection: "Well Senior Management say they're going to get all the "top bods"[2] in to tell us what to do but I think some of us, and I think myself included, are saying 'well you know, we'll do what we can but we're going to carry on doing our proper job and not let OfSTED get in the way too much.'" To me, the idea that experienced teachers and managers are simply told what to do is an example of the "discursive working of the new managerialism" (Reay, 1998, p. 188). The mechanistic, technocratic approach to schools and to education reform has set an agenda for change that is disempowering and disenchanting.

Encouragement from Managers

The importance of support for teacher development from senior management has been recognized for some time (see, for example, Adey, Dillon, & Simon, 1995). Getting the balance right between support and pressure is a key aspect of teacher development. In England, however, it does seem that the role of the HoD has shifted more toward pressuring rather than supporting. HoDs are now "proxy-managers"—carrying out school policy (and government policy) at a distance—using appraisal as a means of quality assurance (Dillon, 1997). As one of the HoDs indicated, the period of adjustment for this process has been minimal, the training variable, and the penalties for failure high, particularly for the less powerful groups in schools: "[We] are more accountable than [we] ever were before. Changes are taking place . . . overnight and we're having to . . . implement them more-or-less as we speak and there are so many changes that people are threatened by that . . . especially the young teachers.

A Feeling of Personal Growth

The aspects of teaching that appealed to me throughout my classroom career was what Huberman (1989) recognised as the opportunity for "tinkering" with resources and teaching styles. In Huberman's study, teachers that engaged in tinkering were likely to be "more satisfied" later in their career than those who did not adapt. Huberman also identified other factors that were predictors of "satisfaction"—the first being the ability to "change one's role when one begins to feel stale" and the other comes from a realisation that one's students are actually learning something. In the light of the issues raised earlier, opportunities for personal growth appear to be limited for many teachers. The evidence from my research is that teachers, particularly in small schools, are faced with increasing isolation and pressure, which are more likely to result in feelings of inadequacy rather than growth.

A Sense of Ownership of Innovation

One of the taken-for-granteds about teacher development has been the feeling that teachers needed to feel committed to change (to "own" it) for it to be implemented, although some research has questioned that necessity (Adey, Dillon, & Simon, 1995). Personal control over the change process should be a desirable situation if only on democratic and equity grounds rather than simply psychological grounds. My understanding is that teachers use a range of micropolitical processes to exert ownership on change, whether it is through interpretation of the curriculum or through deliberate refusal to participate in activities.

Gold and Evans (1998) express the view that "excessive micropolitical activity within a school may be indicative of blocked or ineffective decision-making routes" (p. 22). They go on to say that "whatever the cause of the excessive micropolitical activity, those with management responsibilities within a school need to be aware when they are overactive and to make some basic decisions about whether to use or ignore the unofficial structures." However, it must be borne in mind that: "Decision-making is not an abstract rational process which can be plotted on an organizational chart; it is a political process, it is the stuff of micro-political activity" (Ball, 1987, p. 26). This position does not seem to be recognized by those giving advice to HoDs (see, e.g., Gold, 1998). Normative models of teacher education are more likely to be subverted if teachers perceive themselves as being molded into particular forms.

Gender

The issue of gender in management, which was not one of the factors listed earlier, has been studied but is still little understood in terms of teacher development and its management in England. As more women become managers and more men become managed, there is a continuing need to appreciate the gender issues involved. My feeling is that research into gender issues needs to focus more than it has upon providing rich descriptions and critical analysis of the situations in which women find themselves as managers. Stereotypical views of women as managers are still too common. One of my participants articulated a view that is rarely found in the literature:

> I'm very careful how I deal with the men 'cos I think [. . .] and I'm not saying this in front of you in any way to insult male populations, but I feel that with men you've got to be careful. They're not used to women managing them a lot of the time and um . . . and I don't want to ruin their egos. I don't want to denigrate them and make them feel small. But sometimes they see me with the kids and how strong and tough I am with them. Not necessarily succeeding, but I am tough with them and I suppose maybe feel a little bit threatened by that. . . . So I do, I do deal

with the men differently, the woman I tend to have a great relationship with. They just accept everything. I have no problems [. . .]

Another female HoD, Kelly, spoke of the differences between men and women.

> I think women tend to be more flexible and more adaptable in terms of what they're prepared to accept and do sometimes see alternative viewpoints though there are obviously still some women who see something and decide against it also. But I mean there are big disadvantages having a woman because quite a few of them have got young children and so they have the commitment to childcare that men don't always have.

Final Comments

The focus of this chapter has been on the challenges faced by today's Heads of Department. However, I think that a range of issues found in England have a more widespread relevance. From my experience of education systems around the world, there are many similarities in the way professional development is conducted in global communities:

- The tendency to ignore contextual constraints on teacher change
- The indifference to teachers' needs and wants and the simplistic assessment of the effectiveness of teacher development
- The disempowering imposition of standards and procedures

What I have tried to do here is to use the English context as an exemplar rather than to say that England is unique in all aspects of teacher development. However, it is unique in some aspects. The suddenness of change from extreme hands-off to extreme hands-on in terms of the government's attitude toward the education system has rarely been witnessed elsewhere. The consequent grinding of gears has fragmented the teaching profession in a way that has not happened in many, if any, other countries. These contextual factors have led to a range of micropolitical realities that contrast with the mythology of professional development legislature and guidance.

The culture of accountability and managerialism has major implications for equity issues. The use of power and control to effect changes in the education system relies on traditional attitudes and structures coupled with the reviewing of individual agency as a means of (self)-improvement. The model of professional development that emerges from a study of English teachers is one of spasmodic, unplanned, and poorly funded support set in a virtual framework of standards and outcomes.

The success of the professional development program associated with CASE (Cognitive Acceleration through Science Education) (see Philip Adey's chapter in this volume) provides some lessons. CASE is based on a sound theoretical foundation, a clearly articulated and replicable teaching approach, trialled and tested support material, classroom-focused coaching, and feedback over a 2-year period. It is not cheap but it appears to work.

The extent to which models of professional development are transferable is clearly an issue in a book such as this. My own feeling is that although there are many differences between teachers and education systems from state to state and from country to country, some of the challenges are generic.

What can other countries learn from this work? The issue that should trouble researchers and professional developers alike is the seemingly constant reinvention of the wheel that seems to take place all too frequently. However, another message is clear—models of teacher development that are based on simplistic, normative, technicist ideas are unlikely to be effective. Teacher development needs support from managers, but if teaching is complex then managing teachers is even more so. Teachers and their managers need to see development as multifaceted, progressive, and dependent on individual personalities, psychology, and politics—both macro and micro. These lessons apply in education systems where there are few middle managers as well as in those, like England, where there are many.

Management is partly about confidence as well as competence. The inadequate training that Heads of Department receive, the lack of time that they have to manage, the ethical dilemmas facing them, the pressure on from parents, heads, inspectors, and colleagues all conspire to make the job difficult, unsatisfactory, and impossible to finish. The job of the Head of Science is possibly the most demanding of all middle managers, and there is little sign that that will change in the future.

Notes

1 This refers to a particular section of the National Curriculum followed in Anne's school.
2 That is, staff from the local education authority who have a role in monitoring standards in schools.

References

Adey, P. S., Dillon, J. S., & Simon, S. A. (1995). *School management and the Effect of INSET.* Paper presented at the European Conference on Educational Research, Bath.

Adey, P., & Shayer, M. (1994). *Really raising standards.* London: Routledge.

Ball, S. J. (1987). *The micro-politics of the school.* London: Routledge.

Bell, B., & Gilbert, J. (1996). *Teacher development: A model from science education.* London: Falmer Press.

Davis, N. T. (1996). Looking in the mirror: Teachers' use of autobiography and action research to improve practice. *Research in Science Education, 26*(1), 23–32.

Dewey, J. (1933). *How we think: A restatement of the relation of reflective thinking in the educative process.* Chicago: Henry Regnery.

Dillon, J. (1997). *Managing teacher development: The role of the Head of Department in England.* Paper presented at the European Association for Research in Science Education Summerschool, Barcelona.

Dillon, J. (in press).—Managing science teachers development. In R. Millar, J. Leach, & J. Osborne (Eds.), *Improving science education*: Buckingham: Open University Press, pp 94–109

Dillon, J. (in preparation). The Role of Middle Management in Science Teacher Development in Schools. Unpublished PhD thesis, King's College London.

Dillon, J., Osborne, J., Fairbrother, B., & Kurina, L. (2000). *A study into the professional views and needs of science teachers in primary and secondary schools in England.* London: King's College London.

Donnelly, J. F., & Jenkins, E. W. (1999). *Science teaching in secondary school under the National Curriculum.* Leeds: Centre for Studies in Science and Mathematics Education, University of Leeds.

Driver, R., Squires, A., Rushworth, P., & Wood-Robinson, V. (1994). *Making sense of secondary science.* London: Routledge.

Gold, A. (1998). *Head of Department: Principles in practice.* London: Cassell.

Gold, A., & Evans, J. (1998). *Reflecting on school management.* London: Falmer Press.

Gunter, H. (1997). *Rethinking education: The consequences of jurassic management.* London: Cassell.

Hacker, R. J., & Rowe, M. J. (1997). The impact of National Curriculum development on teaching and learning behaviors. *International Journal of Science Education, 19*(9), 997–1004.

Helsby and Knight (1997) Continuing Professional Development and the National Curriculum in Helsby, G. and McCulloch, G [eds], Teachers and the National Curriculum, London: Cassell, pp. 145–162.

Her Majesty's Inspectorate (HMI). (1992). *Science: Key stages 1, 2 and 3. A report by H M Inspectorate on the Second Year, 1990–91.* London: HMSO.

Huberman, M. (1989). *Teacher development and instructional mastery.* Unpublished paper presented at the International Conference on Teacher Development: Policies, Practices and Research, Toronto: Ontario Institute for Studies in Education.

Joyce, B., & Showers, B. (1988). *Student achievement through staff development.* New York: Longman.

Luft, J. A., & Cox, W. (1998). *Final report: A report on preservice and mentoring programs in Arizona for mathematics and science teachers.* Arizona Board of Regents: Eisenhower Mathematics and Science Program.

Monk, M., & Osborne, J. F. (Eds.). (2000). *Good practice in science teaching: What research has to say.* Buckingham: Open University Press.

Novak, J. D., & Gowin, D. B. (1984). *Learning how to learn.* Cambridge: Cambridge University Press.

Office for Standards in Education (OfSTED). (1996). *The appraisal of teachers 1991–1996.* OfSTED reference HMR/18/96/NS. London: Ofsted.

Osborne, J., & Collins, S. (1999). *Pupils' and parents' views of the role and value of the science curriculum.* Paper presented at the British Educational Research Association Conference, Brighton, September.

Reay, D., (1998). Micro-politics in the 1990's: staff relationships in secondary schooling, Journal of Education Policy, 13 (2) pp 179–196.

Teacher Training Agency (TTA). (1997). *National standards for subject leaders—annex.* London: TTA.

3   Towards Creating an Inclusive Community of Researchers

The First Three Years of the North American Association for Environmental Education Research Symposium

Ronald B. Meyers, Michael Brody, Justin Dillon, Paul Hart, Marianne Krasny, Martha Monroe, Constance Russell and Arjen Wals

Meyers, R. B., Brody, M., Dillon, J., Hart, P., Krasny, M., Monroe, M., Russell, C., & Wals, A. (2007). Towards creating an inclusive community of researchers: the first three years of the North American Association for Environmental Education research symposium. *Environmental Education Research*, 13(5), 639–661.

This article uses a series of interlinked, personal vignettes to discuss the first three years of the North American Association for Environmental Education research symposium, from the perspectives of the key organizers. Seven challenges in the field of environmental education research are identified in a recent historical context, and we illustrate how the symposium sought to address them. The challenges were, that: (i) environmental education research has been marginalized in some areas and not recognized in others; (ii) environmental education research and environmental education practice need to be brought closer together; (iii) environmental education research is still in early development of a professional perspective; (iv) environmental education research has to give a voice to early career scholars and graduate students; (v) environmental education research needs to enable discourse about both process and outcomes; (vi) environmental education researchers need social learning contexts to help develop professional identities and create more meaningful dialogue to address these challenges; and (vii) methodologies, theoretical frameworks and differences in beliefs in environmental education research need to be accommodated. The last challenge is seen as the most significant with which to continue to engage, in developing open, inclusive forums for researchers of environmental education.

Background

[Meyers] Creating, developing and sustaining rigorous and inclusive forums for researchers is a perennial issue in any research community. In 2002, several of the authors represented here sought to address this challenge in the context of the environmental education research field in the North American environmental education community, and more internationally. Michael Brody[1] suggested that such a forum at the North American Association for Environmental Education (NAAEE) should address seven challenges:

1   Environmental education research has been marginalized in some areas, and not recognized in others.
2   Environmental education research and environmental education practice need to be brought closer together.
3   Environmental education research is still in early development of a professional perspective.
4   Environmental education research has to give a voice to early career scholars and graduate students.
5   Environmental education research needs to enable discourse about both process and outcomes.
6   Environmental education researchers need social learning contexts to help develop professional identities and create more meaningful dialogue to address these challenges.
7   Methodologies, theoretical frameworks, and differences in beliefs in environmental education research need to be accommodated.

Each of these challenges could be readily met by a well designed and executed forum at the NAAEE annual meeting, yet the last of these would prove particularly difficult, complicating efforts to achieve the others.

In their introduction to a special issue of *Environmental Education Research* reviewing the first 10 years of the journal, Reid and Scott (2006a) note the broadening of research methods and methodologies used by those in the international field of environmental education research since the mid 1990s. In some quarters, this broadening had evolved both from and into growing philosophical and methodological schisms in the field. At times the schisms became too personalized, as can tend to happen with theoretical disputes in the sciences (Kuhn, 1970). My 10 years observation of the North American field suggests that it has began to expand beyond the more conservative, quasi-Popperian approaches to science used in its founding (Meyers, 2006), to include—and at times, contest—what Hart (2005) terms the 'posts' approaches to inquiry. The work to broaden what counts as environmental education research has been strongly challenged and responded to in kind, and this has lead, amongst other things, to hurt feelings as well as a lack of meaningful communication about the theoretical issues at hand.

In my view, these divides challenge our ability to address each of the issues identified by Brody through a research forum. Earlier in *Environmental Education Research*, Connell (1997) had noted the negative discussion of methodological differences and misrepresentation of them in the field during the preceding decade, calling for a more constructive approach. More recently, Russell (2006) has clearly described cases of a 'stinging' adversary method of methodological discussions in environmental education research, calling for 'collaboration for generous scholarship' (p. 407). While as Reid and Scott (2006a, p. 243) go on to note:

> The last ten to fifteen years of articles [in *Environmental Education Research*] show a marked deepening of interest in the development of research skills and competence, a greater exploration of the nature of research as a form of practice, and some welcome lessening of tensions between representatives of entrenched methodological camps that might have given the impression that they are always ready to re-engage in paradigm wars.

Given this situation, the intentions for a research symposium at NAAEE might be viewed as having two more general and deliberate goals: first, to quickly address the first six challenges identified by Brody; whilst second, laying the groundwork for continuing efforts to meet the seventh. Such goals entail providing a forum where researchers from diverse methodological perspectives would be encouraged to share their work in a critical, yet positive atmosphere, and build a community engaged in generous scholarship across methodological differences.

As Monroe notes (below), the first goal was supported by the NAAEE Research Commission and (relatively) easily met through the hard work of Richard Jurin, Joe Heimlich, Michael Brody, Martha Monroe, and others. The second goal has seen significant progress, but is much more challenging. The contributors to this article share their perspectives on these goals, and their experiences of the research symposium in addressing them.

In light of this, the article serves three main purposes. The first is to review an important evolution in the field of environmental education research—the formation of the NAAEE annual research symposium. Since 2003, the event has grown from a nascent idea to a vibrant meeting that attracts over 120 researchers (in 2006) from many countries who engage in generally respectful and sometimes vigorous critical dialogue within and across extremely diverse epistemologies, methodologies and methods. A second purpose is to illustrate the broadening of what counts as research in the field through this documentation and commentary. The third purpose is perhaps the most important and demanding: the collaboration of eight authors representing diverse epistemological and methodological positions to jointly develop this account—thus advancing our sense of being in a community of scholars with a shared mission of increasing our ability to understand and critically evaluate each others' work.

In what follows, readers will find a rich account of the experiences of participants involved in the development of the research symposium, through a series of interlinked, personal vignettes. We open a window with an unusual view of the personal nature of research activity, and the development of our growing sense of research community. My experiences, which reflect my particular interests in encouraging greater attention and productive dialogue about epistemology, methodology and methods in environmental education research, provide this introduction and a conclusion organized from my perspective—though the importance of these issues is shared by all the authors. Michael Brody was instrumental in helping to form the symposium and craft this paper, so I asked Michael and each contributor to share their views on their expectations, experiences, and hopes for the future. Inevitably and productively they shared more. We continue with Michael Brody.

Section 1: Getting Started

[Brody] In the 25 years that I have been involved in higher education in the US working in the field of education with science, ecological and environmental foci, I have gone from feeling marginalized and isolated to regarding myself as an active member of a forward-thinking, intelligent and friendly community of international environmental education scholars. This has not been an easy journey. The trail was neither well marked, nor was there experienced guides willing to lead the way on well-worn paths. Sustained involvement in the creation of the NAAEE annual research symposium represents my most recent attempt to help create and foster a closer coming together of colleagues to share our interests.

Personal/Professional Needs and Initial Solution

As a new faculty member in higher education, I began my work in environmental education in science education. My work in environmental education, especially young people's understanding of ecological concepts, was not part of the mainstream science education agenda. The 1980s saw the more widespread adoption of the constructivist paradigm in education. Many in the science education community would eventually embrace the emphasis on cognition and personal and social construction of knowledge in the early 1990s. However, at a critical time in my career, as I worked towards tenure, it was still difficult to present my work and find a professional venue with like-minded people. The need to share and interact led to the creation of the American Education Research Association (AERA) special interest group in Ecological and Environmental Education (EEE SIG). As the group grew through the 1990s, it served very well in gathering colleagues from around the world on an annual basis. Despite the fact that we shared the stage with the largest (impersonal) meeting of education researchers in the world (over 14,000 people at Chicago, 2007), we succeeded in creating a close community of supportive professionals complete with fieldtrips and shared housing.

The establishment and legitimization of environmental education research at AERA was an important process leading to my interest in working with NAAEE. It showed there was a community of people who wanted to share their work and discuss it critically and openly. As we headed into the new millennium it became apparent that the EEE SIG was not big enough to accommodate the growing environmental education research community. The SIG had a limited number of spaces for presenters and as the environmental education research community grew it began to encompass a wide range of abilities, interests and accomplishments. It was clear to me and others that it was not enough to just have the experienced researchers presenting each year. We needed to grow and be more inclusive, especially as it related to new and emerging scholars. At this point I asked myself, 'How can we create an annual professional venue for environmental education researchers in which everyone can contribute and learn?' and, 'If other research groups can have their own meetings, why can't we?'

Personal/Professional Desires and Return to Roots

I have been an on-again/off-again member of NAAEE from the 1970s. Initially, when I was a high school teacher, NAAEE annual conferences (when they were in my region of the USA) were a great opportunity to meet other practitioners to share teaching, learning and curriculum ideas and experiences. Later as a graduate student at both the University of New Hampshire and Cornell University, I combined my marine interests and association with the Shoals Marine Laboratory with my interests at NAAEE. I will never forget the NAAEE annual meeting at which I led a pre-conference field trip to the Isles of Shoals, 10 miles off the coast of Maine and New Hampshire. It was autumn, sea birds were migrating and after we turned off the boat's engines on the outward shores of Smuttynose Island to sit among thousands of birds, several participants on the field trip bought out their instruments to play and sing along with the world around us. I remember this and similar events as the inspiration for my work. However, although it was my inspiration, as I moved into the cognitive academic world of higher education, NAAEE did not meet my professional needs in the research community. As noted above, I drifted in the 1980s and 1990s to more research-oriented organizations such as AERA, leaving behind some important parts of my life and career. During this time, a number of things were happening with NAAEE.

[**Monroe**] NAAEE was launched in 1971 as the National Association of Environmental Education by a group of community college and university instructors who taught *about* the environment. They probably did not realize the term 'environmental education' was about to take on a life and legacy of its own. The organization quickly attracted a number of non-formal educators and formal administrators. Early leaders wisely used the association as an umbrella for member interest groups (labeled 'sections') in environmental studies, elementary/secondary and non-formal education. The early influence of environmental studies faculty promoted research in environmental studies as well as environmental education. For example, the eight papers selected from the 1983 conference for publication in the first NAAEE Monograph covered riparian lands conservation, environmental consciousness, perceived environmental control, results of using a water management simulation, perceptions of the urban environment, acid rain on crops, acid rain information for teachers, and environmental ethics.

The link between sections and the Association governance was integral to the success of NAAEE and helped serve the needs of a diverse set of members. Section Presidents served on the NAAEE Board for a three-year term, creating a Board primarily of section leaders.

The North American Commission for Environmental Education Research (NACEER) was formed in 1980 as the National Commission for Environmental Education Research with an appointed group of experienced researchers. They met during NAAEE conferences and were quite active in the 1980s. Members embarked on a number of projects that served the entire field of environmental education, such as compiling research (Iozzi, 1984) and evaluating Project Learning Tree materials. Because it was not officially part of NAAEE, the Commission was able to respond to need and opportunities without the encumbrance of the Association. But the closed membership and governance were a limitation. Also, these researchers were not afforded the same opportunities as other sections (e.g., to sponsor workshops, or serve as representatives of a research section on the Board). NAAEE members questioned how NAAEE could serve as the host to NACEER and whether a different structure would better serve NAAEE members.

Questions raised at the 1989 NAAEE conference (see Wals, below) about the types of research promoted by existing journals led Ian Robottom, Paul Hart and Rick Mrazek to organize a symposium for the 1990 conference, 'Contesting paradigms in environmental education research'. The symposium enabled a rich discussion of different research perspectives and an interest in expanding the discussion to include more participants. To that end, the NAAEE newsletter published a series of six articles on environmental education research and Mrazek edited a monograph, 'Alternative paradigms in environmental education research' (1993) with papers from 20 NAAEE members from five nations.

During this time, NAAEE was gaining strength from the non-formal sector as membership increased from 1000 to 2000 between 1990 and 1995. Toward the late 1990s environmental studies faculty found the conference less appropriate for their interests and were attracted to other associations (see Brody, above, and Dillon, below). Environmental education researchers continued to support NAAEE, but often found a more helpful and critical environment for research discussion in other conferences. By the early 1990s the association had moved to a Washington DC office, hired several staff, and used external projects to maintain the organization. When NAAEE received the multi-million dollar EETAP grant in 1995, the association focused on producing project deliverables that would directly enhance K-12 environmental education. Curriculum guidelines, teacher training, professional preparation, and materials evaluation were addressed in a series called *Guidelines for excellence* (NAAEE, 2005). Researchers involved in evaluation were in high demand, but other environmental education researchers found less in NAAEE to service their needs.

At the close of the EETAP project in 2000, NAAEE reduced staff and began to regroup. The economic downturn and the general fear of flying after September 2001 made conference attendance less possible for many, in turn reducing a membership base driven by conference attendance. The Board had grown to the unwieldy size of 28 people as more sections were added (such as Conservation Education and International) and was dramatically reshaped to an elected body of 13, severing the link between the sections and Board membership. The sections and the old Research Commission (NACEER) were reorganized into member interest groups and renamed Commissions, establishing the NAAEE Research Commission. They began operating under Charters that they created and the Board approved. The purposes of the Research Commission (RC), referred to as the RC, are to:

1   Provide a forum for members of the NAAEE who are interested in conducting, reporting, using, or supporting environmental education (EE) findings and/or results.
2   Advance efforts to link EE practice to the findings/results of EE research.
3   Provide expert assessment of various current and emerging issues concerning EE theory and practice for which research is needed.
4   Provide mentoring/networking opportunities for novice/experienced EE researchers.

The Board began a strategic planning process in 2002, developing it with input from nearly 100 people, and approved it in 2005. The NAAEE Strategic Plan re-emphasized the role of research in NAAEE, prominently featuring it in the vision, values, and goal statements. Paul Hart rejoined the Board in 2001 with an expressed interest in revitalizing research in NAAEE, including that of expanding the range of research methods and methodologies considered legitimate and useful for environmental education research and evaluation.

[Brody] As the demands on the AERA EEE SIG membership grew and my desire to gather more colleagues together in a bigger and better venue became more focused, I responded in the spring of 2003 to a call by Richard Jurin, then chair of the NAAEE Research Commission, to volunteer to be the next chair-elect of the Commission. I saw this as an opportunity to assume leadership responsibility and to promote my personal agenda of establishing an annual professional venue exclusively for environmental education researchers to get together and learn from each other. My vision was for an intimate venue of environmental education researchers meeting in ecologically wonderful places inspired to create new ways of thinking about our work. I was elected as the next chair-elect. As in many things, coincidence can play a significant role in any endeavour. The year I became the Research Commission chair was the year an established select group of international environmental education scholars chose to hold their international environmental education research meeting directly prior to the NAAEE annual conference. Paul Hart led the 7th Invitational Seminar on Environmental and Health Education held in Anchorage, Alaska, 5–7 October 2003. It was not coincidence, of course, that the seminar was held in Anchorage: how this came to be is described below by Paul Hart and others. This event became my model for the annual NAAEE research symposium. I especially appreciated the openness of the session in Anchorage, the inclusion of 'unfinished business' in environmental education research and the participation of many graduate students. Following the annual conference, with the support of my commission advisors and past Research Commission chairs (Tom Marcinkowski, Richard Jurin, Nick Smith-Sebasto and Joe Heimlich), I began planning the first NAAEE research symposium for 2005 in Biloxi, Mississippi. With the encouragement of the Research Commission advisors and the organizational management support of the NAAEE headquarters, I took on the task of planning the symposium . . .

Expected Outcomes for the First Year

[**Brody, cont'd.**] My first and foremost goal in the first year of the research symposium (2004) was to get enough attendees to make it fun and productive. At that time, I guessed that maybe 35 attendees would be sufficient. Secondly, I wanted to draw environmental education researchers back to NAAEE. I thought this would help reintegrate research into the organization as researchers who showed up for the symposium would stay for the NAAEE annual conference. Third, I wanted this to be a positive contribution to the NAAEE organization by helping to promote membership and its image.

[**Monroe**] NAAEE was very supportive of this experiment for several reasons. First, it could attract more people to the NAAEE annual conference, which broadens the membership base. Second, it would stimulate the Research Commission. These member interest groups are able to develop projects that meet member needs, and having one successful, well-organized commission could help encourage others by example. Third, because researchers tend to be at universities, and universities tend to have students, there was a possibility that the symposium could attract young professionals in environmental education, creating a foundation for a strong future. Fourth, strengthening the participation of researchers in NAAEE could help the organization in a number of ways, from our own data collection projects, to ways of thinking about improving environmental education evaluation and practice. Fifth, hosting the symposium directly prior to the annual conference meant that several types of exchanges between researchers and practitioners would be made easier. Presenters at the symposium were encouraged to stay for the NAAEE annual conference and present their work for practitioners. Practitioners interested in research would attend the symposium, utilizing what they learned, and sharing it with others as they also attended the NAAEE annual conference. Altogether, this strategy would help practitioners find and use research results and learn about innovations in research methods and methodologies. It may be a burden for one symposium to affect change at these three scales (commission, association, and field), but the possibilities and needs were present. The Research Commission's research symposium was an excellent example of members using the umbrella of NAAEE to create a professional development opportunity that served their needs. We held our collective breath as we went into the fall of 2004 and the first NAAEE pre-conference research symposium, 'Sharing our research interests, accomplishments and plans'.

Section 2: Year 1 (2004)—'Sharing Our Research Interests, Accomplishments and Plans', Biloxi

[Hart] As Michael Brody mentions in his introductory comments, NAAEE research seminars are rooted in ideas derived from a series of international seminars on research development in health and environmental education, the seventh of which was coordinated to immediately precede the annual NAAEE conference in Anchorage, Alaska in October 2003. Some of the ideas that continue to influence these international seminars, as well as initial thinking about NAAEE research symposia, seem worth reiterating here. The first thing to note is the simple point that having a research-focused sympo-sium before the NAAEE conference proper enables more people to attend than if it were a free-standing event.

Each of us, as researchers working more or less independently within rap-idly growing and somewhat disparate fields of inquiry such as environmental and health education, appreciates the value of engaging in somewhat regular face-to-face discussion about issues of concern. In the minds of those scholars who, with Jensen and Schnack, organized the first international seminar in Denmark, there were good reasons to open up the dialogue on the development of research issues, perspectives, trends, and debates. These have ranged from theoretical/conceptual preferences and priorities to practical, geograph-ical and cultural assumptions and preoccupations. Although we are enmeshed in the politics of research within each of our own jurisdictions, throughout their history, these seminars have been characterized by attempts at openness to ideas, and a generosity of spirit towards those putting ideas forward.

Over the past 14 years, these meetings across four continents have allowed us to reach beyond traditional boundaries of various kinds to find new ways of thinking about and conducting research in health and environmental education. We think those of us fortunate enough to have had several of these experiences look forward to the next.

The idea of NAAEE research symposia in the same spirit of openness and boundary crossing impelled both of us to respond to Michael's request to assist with the 2004 seminar in Biloxi. The notion not to present the results of completed work but to focus on issues of works-in-progress is difficult to maintain within traditional conference *modus operandi*. The notion of critical friendship within an atmosphere of mutual trust and respect for difference is demanding, and that of engaging both seasoned and fledgling students of inquiry across territories can be challenging, but worth the struggle, as Janet Dyment notes in her reflections on the 2003 seminar:

> As a graduate student, I . . . felt very welcomed to participate in discussions; I felt that my voice/stories/concerns/ideas/research were important. This inclusive atmosphere was a welcome change from some conferences . . . that have felt very hierarchical and exclusive. This non-hierarchical tone was maintained throughout formal and informal aspects of the seminar.
>
> (Hart *et al.*, 2004, p. 569)

We came to the 2004 Biloxi meeting concerned that, at the same time as issues of quality were overtaking the educational field more generally, we were asking health and environmental education researchers to broaden their awareness across methodological approaches such as action research and narrative inquiry that only respond well to quality criteria well beyond traditional conceptions of *rigor* and *mortis*.

We hoped that NAAEE participants would benefit from the type of event that permitted both seasoned and novice researchers to engage in formative discussion of emergent research ideas and methodologies as a means of improving the overall quality of research in environmental education. The idea was to not impose an agenda—only an organizational frame that would work to create conditions for active discussion of methods and issues. The seminar provided good discussions, well-engaged and received, about relatively new methodologies.

Educational research is extremely difficult to do well. Even to be able to read critically and intelligently across a variety of perspectives using different approaches demands extensive knowledge, creative applications and insight that are best served by forums of the sort that engage debates and discussions in an atmosphere of learning as a social and (cross) cultural process. How else can we address concerns about the fragmentation and balkanization of research activities in complex fields of education globally, as well as locally? As M. J. Barrett and I noted after the 2003 international symposium:

> Difference, in such situations, must not merely be tolerated, it must be cultivated. As researchers discuss methods and methodologies from diverse perspectives, comparison of methods may begin to shape itself within conceptual conversations about epistemological and ontological assumptions. Perhaps, by creating the conditions for such discussions, it will be recognized how diversity frames research orientations as particular socially constructed perspectives, each subject to scrutiny, and each directed at deepening the conversations about the enquiry process.
>
> (Hart *et al.*, 2004, p. 565)

The intentions for the NAAEE research symposium, like the international seminars that spawned them, provide many benefits to relatively isolated researchers who seek to test their ideas formatively, to engage ideas of theory and practice that keep the fields of environmental and health education on the edge of, and critical of, the discourses that drive us. The alternative seems rather bleak and uninspiring—and far less likely to allow the testing of ideas and images, multiple layers of meaning, interpretations and representations that best work together to improve the quality of our work. Perhaps almost the last word should be Charlotte Clarke's:

> . . . as I struggled to design my doctoral research study, I found myself surrounded by experienced researchers struggling to design their own work, and assisting their colleagues in the design process. I took home with me three important lessons. First, even experienced researchers . . . face thorny research dilemmas that they feel unable to address, and may never ultimately solve. Second, this particular community of academic colleagues shares a spirit of supportiveness that can supersede personal ego and permit public acknowledgement of research that is different, incomplete, ineffective, or completely stalled. Third, this group invites and welcomes young and aspiring researchers to join.
>
> (Hart *et al.*, 2004, p. 572)

. . . as her thoughts capture both the essence of the international seminars, and the spirit which we trust will continue to inform NAAEE's research symposium.

[Brody] I thought to myself, Who better to ask as introductory and concluding speakers for the inaugural research symposium than Paul Hart and William Scott, influential leaders in environmental education research? My vision of their contribution was multilayered. Surely they would bring legitimacy to a fledgling endeavour. I considered their combined years of experience, intellectual accomplishments and leadership in environmental education research as critical to the success of the symposium. My vision was of symposium 'bookends' holding upright and steady the contributions of the presenters. One at the beginning (Hart) to open doors and suggest pathways to rethink our efforts; the other (Scott) to bring closure and insight into our event, with both picking apart the most obvious, accepted and familiar into their components and rearranging them in strange and interesting ways. Their twin contributions would be essential to holding the symposium together.

My dream had been to get about 35 environmental education researchers together that first year. The attendance was closer to 50 with about 20 presentations. Everything went smoothly and the first symposium felt like a success. We began plans for 2005 and the second meeting in Albuquerque, New Mexico. Little did we know that our 2005 conference and symposium site would be leveled during [Hurricane] Katrina before we met again.

Section 3: Year 2 (2005)—Continuing the Dialogue.
The Truck Stops Here: Remembering Albuquerque

[Wals] I would like to put the symposium in Albuquerque in a personal and historical perspective. Having been at a number of NAAEE conferences in the late 1980s and early 1990s, I vividly remember earlier attempts to get 'research' on the agenda. At places such as Estes Park (1989), where the nature of research covered by the *Journal of Environmental Education* was discussed with the managing editor, and San Antonio (1990) where 'research' was made an explicit item on the agenda, my initial experiences were that of tensions being brought out among those involved in environmental education research. In fact there was talk of different camps associated with prominent researchers and of contesting research paradigms (although some preferred 'alternative' instead of contesting). The one major research journal in the field at that time, *Journal of Environmental Education*, was considered by some as an extension of the empirical-analytical camp with a heavy bias towards causal modeling and quantitative research. Other camps called for changes in that journal and for new journals in the field that would be more inclusive of other kinds of research, or would have there own bias towards other research paradigms. Those early years can be described as years of contestation.

Albuquerque to me was a kind of homecoming to NAAEE, having moved back—after four years at Ann Arbor, Michigan—to my native Netherlands in 1992 and being unable to attend many of the NAAEE conferences since, or, in having to be selective, choosing to attend the AERA annual meeting instead. The home I found when I was asked to co-organize and kick-off the NAAEE research seminar in Albuquerque together with Justin [Dillon] was quite different from the home I had left in the 1990s.

[Dillon] Prior to 2005, I had only attended NAAEE once (Boston, 2002). The lack of a strong research theme there meant that I had subsequently prioritized other annual conferences in North America (AERA, and NARST—the

National Association of Research in Science Teaching). Michael's invitation to Arjen and myself to share the platform as keynote speakers in Albuquerque was attractive at several levels. Arjen and I had known each other for about 15 years and had worked together on various publications and conferences, so I knew that the experience would be both interesting and entertaining. The success of the 2004 research symposium, as evidenced by the numbers attending and by comments to me from colleagues such as William Scott, led me to believe that attending the symposium would be professionally beneficial.

[Dillon & Wals] What then did we find? What had changed? Clearly, the topic of research in environmental education was no longer a marginal one that—as it was in the eighties and early nineties—attracted only the elite of well-established environmental education researchers that dominated the field. No longer was there the persuasive, defensive arguing between different camps where everybody had something to say, and nobody was listening. The participants of San Antonio were still there, but so were many graduate students and new environmental education research faculty, who were more critical and independent. And, even more encouraging, well-established faculty were more receptive to alternative perspectives than had been the case 10 or so years ago. The seemingly unstoppable trucks of positivism and their 'rival' paradigms appeared to have halted and hybrid forms had emerged.

What is clear is that as the goals and priorities of inquiry shifted and diversified (see Reid & Scott, 2006b), new research programmes and journals representing a range of research approaches and including different kinds of knowing and understanding were establishing themselves, as were the scholars, researchers and students aligned to them, from Canada and all parts of the world, and all were beginning to have more significant profiles at the NAAEE events.

In New Mexico we witnessed convergence: signs of an integrative movement where dialogue prevails, where conflict is seen as healthy dissonance, where there is room for respectful disagreement, and adequate space for pluralism. As a result, we were able to ask a number of potentially 'inconvenient questions' that inspired debate, as opposed to uncomfortable silence. Some of these questions were: Why are we here? What drives us? Whose questions are we addressing? Which principles, values and premises do we challenge as academics? Which ones do we, willingly or unwillingly, reinforce? Who stands to gain the most from our research? And, finally: How does the rhetoric of our research methods and epistemological vantage points relate to our professional lives?

However, with this movement of convergence and the emergence of hybrid methodologies, in reflecting on the symposium we also observed a danger of inappropriately blurring methods, methodologies and ideologies in environmental education research. This led to a paper (Dillon & Wals, 2006) in the special 10th anniversary edition of *Environmental Education Research* which aimed to encourage researchers to make explicit their ontologies, epistemologies and axiologies in the belief that this would help researchers to read research papers more critically and effectively.

[Brody] Following the symposium in Albuquerque, I reflected on our goals for the symposium. We had doubled the participation and presenters. Maybe more importantly the symposium was seen as a great benefit and resource for the NAAEE organization and the main conference. Researchers were coming to the symposium from across the world *and* staying for the main conference. There were fresh conversations in the halls and the symposium sessions. The old mixed with the new. The 'icing on the cake' was that the symposium brought both intellectual and financial resources to NAAEE as more researchers began attending the annual conference.

In asking Wals and Dillon to be keynotes I returned to the idea of 'book-ends' to support the presentations and papers. In this case, I was interested in 'holding things together' from a more critical point of view. Having worked with, listened to and discussed differing perspectives with Wals and Dillon for approximately 10 years at the AERA EEE SIG, I was aware of and appreciated their critical perspectives on environmental education research. I hoped they would push the participants to consider alternative ways of thinking about our work. They, as well as those attending, did not let me down. From the moment in the opening when Wals referred to research as 'mining' to the closing when Dillon displayed an overhead transparency referring to 'scholarship before research', I knew we were on the right trail with good guides. To this day, their questioning of methods and methodologies continues to help me frame critical perspectives of research and evaluation of learning in science and environmental education informal settings (see Brody *et al.*, 2007).

Another outcome of 2005 was that the NAAEE Research Commission became more aware of the implications of the symposium. At the annual business meeting of the Commission, there was lengthy discussion of the event. In general there were very positive reactions from the Commission. Discussion also centred on the gender orientation of the keynote speakers for the previous two years. I agreed to organize the symposium for one more year and to have female keynote speakers in St Paul, Minnesota. To share responsibility, open dialogue, and help develop new leadership, I asked Ron Meyers and Marianne Krasny to work with me to organize the next symposium.

Section 4: Year 3 (2006)—Continuing the Dialogue: Moving into the Future, St Paul

[Meyers] This story of the symposium highlights the desire of many to be a part of a closer community of colleagues, one supportive of open, critical dialogue amongst a plurality of methods and methodologies, supportive of discussions of research in progress, and of young researchers. I sought this too. For Michael Brody, Marianne Krasny and I, as co-chairs of the 2006 symposium, our goal was to continue the development of the symposium along these lines, increase the number of participants, and ensure that it was a viable, sustainable event. It was not clear that we could maintain the intimacy of the first two symposia, given that we expected about 80 participants, so we made an effort to maintain informality, flexibility, and encourage openness and critical, yet respectful dialogue. The review process for papers, posters and roundtables sought to be inclusive and provide means for all to share their work in an appropriate venue. Another goal was to increase diversity of the symposium. While the participants were diverse along gender lines, all prior organizers and keynotes had been white males. Attendance at the third symposium was remarkable, with 114 registrants and a high degree of excitement in the air over what was developing: a significant gathering of environmental education researchers, graduate students, and practitioners interested in research, engaging in the sorts of dialogue envisioned. Participants had a wide range of choices of presentations and roundtables to attend, with researchers sharing work in progress, work completed, and in a daring presentation, a well-attended panel of leading researchers who shared their 'screw-ups'. Nearly all participants continued on to the NAAEE annual conference, many of whom may not have otherwise done so, and one third of attendees were graduate students. Thus, the primary goals were achieved, primarily due to the strong support of NAAEE staff and Michael Brody's continued leadership of the event.

The increase in size did, however, decrease the intimacy of the event, though the informality and the collegiality maintained a friendliness and relaxed atmosphere. The *technics* (attention to the embodied experience, and how our physical environment affects experience) tell a strong story: we had grown to such numbers that we no longer were able to individually recognize and introduce each person at opening events, or to sit in a circle for full group dialogue. How we would maintain the personal and dialogical nature of the event became a concern in looking forward to year four.

[Russell] Like many of my fellow researchers have noted above, I had become disenchanted with NAAEE conferences. I had attended a number as a doctoral student but because I had found only a sprinkling of presentations to be of direct relevance to my own work, I had stopped attending in favour or other conferences in education or environmental studies. Like many environmental education researchers, I have interdisciplinary interests, so there are always a number of conferences that interest me in a given year. NAAEE simply could not compete for my time and money. Obviously, I was not alone in this decision, given one of Michael's stated goals in creating the symposium was to draw researchers back to NAAEE.

For me, the turning point was Year 3. While I had attended the conference in Alaska and taken much out of the invitational seminar described above, I remained disappointed with the wider NAAEE conference. Again, I found only a sprinkling of presentations of relevance to my own work. I thus chose not to go to Biloxi or Albuquerque in favour of other conferences.

Later, I heard that the symposia were growing increasingly interesting and my curiosity was piqued. I also heard that Justin had voiced concern about gender representation in his keynote in Albuquerque. This problem was not a new one. I had raised the issue of representation in Alaska when I read the list of eight or so researchers who were leading a pre-conference workshop on research issues in the field, all of whom were white males. As often happens when I open my mouth in such settings, I was then promptly asked to fill the (gender) gap. So, I quickly found myself giving an impromptu lecture on feminist environmental education research, asserting that the label 'feminist' remained vital (I used one of my favourite lines, 'I'll be post-feminist in post-patriarchy'). My assertions did not go unchallenged by some of my male colleagues, and a number of workshop participants later told me that witnessing the ensuing discussion about feminist research was fascinating to them.

Given this history, I was delighted to be asked to be one of the keynotes in St Paul. As the 'closer', I chose to briefly summarize some of the themes that had emerged in the symposium while acknowledging the challenge of accounting for the encouraging diversity of presentations and remaining coherent. I then chose to focus on the 'So what?' question. Why do we, as environmental education researchers, do what we do? Can environmental education research be imagined as a form of activism? Are our decisions about methodology, method, legitimation and representation congruent with our overarching purpose(s)? Assuming that most of us want to make a difference, how do we ensure that our research reaches our target audiences, be they policy makers, administrators, teachers, fellow researchers, or that amorphous beast known as 'the public'? This led me to a brief discussion of some of the challenges of working in divergent discourse communities. Further, aware of the significant number of graduate students and new researchers in the audience, I chose then to speak briefly about academic publishing and asked the various editors of the journals represented in the audience to stand up, hoping that might facilitate networking and mentoring throughout the conference.

[Krasny] Both Frances Kuo and Martha Monroe posed important challenges to environmental education researchers at the 2006 meeting in St Paul. In her plenary session keynote presentation to the conference, Kuo presented 10 years of research on the impacts of green spaces and plants on people. This impressive body of inquiry, which was conducted with less than $1 million in funding, has had major policy impacts. In Chicago, for example, the Mayor's investment in substantive greening efforts was inspired in part by this work. In conjunction with the work of Richard Louv and the popularization of the idea of Nature Deficit Disorder, the research of Frances Kuo and her colleagues is now leading to major initiatives among multiple Government agencies, with the purpose of getting more youth out into nature. In fact, some research suggests that exposure to nature has more of an impact on environmental attitudes and behaviours than does environmental education per se.

The challenge Kuo presented to the environmental education community was, why can't we, as a community of researchers, join together to answer critical questions about the impacts of environmental education on youth? The implication being that once we are able to do this, we will have a major impact on policy in support of environmental education, just as Kuo and her colleagues have used research on exposure to influence city planning policies. The methods Kuo proposes are large-scale quantitative studies, which often are not the methods participants in the NAAEE research symposium promote. However, I find many elements of Kuo's challenge to the environmental education community provocative and worthy of our consideration. At the same time, I wonder if the reason environmental education research fails to consistently document the kinds of impacts that Kuo has found with her green-exposure research is that education research deals with so many intervening and intertwined variables. Still we cannot ignore Kuo's arguments. She has shown that using quantitative means to document impacts can influence policy-makers, and likely in ways the environmental education community has not yet successfully achieved.

In her keynote address to the research symposium, Monroe also challenged us to think more broadly about how environmental education research can impact policy. She suggested that we conduct research to answer questions about how environmental education can support larger education US policy, such as *No Child Left Behind* (school achievement). If policy-makers are concerned about school achievement, and we can show that environmental education helps students achieve in their core subjects, then environmental education will gain ground in education policy.[2]

Section 5: Conclusions

[Krasny] So where does this leave us? On the one hand, we have environmental education researchers who participate in the NAAEE research symposium calling for exploring innovative ideas, theory, and practice. On the other hand, we have people in prominent positions in environmental education and in the overall greening movement asking us to look at the policy implications of our work, which may entail conducting large-scale quantitative studies. In some cases, such studies may be designed to support policies we don't feel completely comfortable with, such as those calling for school achievement as measured strictly by standardized testing. Or they may employ methods that environmental education researchers feel are unethical, morally wrong, and/ or lead to superficial understandings.

Perhaps future research symposia can be designed to explore these seemingly contradictory trends. We might also explore other types of policies where environmental education might have an impact. For example, Monroe has looked at impacts of extension education programmes on social capital in rural communities. Similarly, at Cornell, we are exploring the role environmental education programmes that integrate civic action, science, and local knowledge play in supporting policies focusing on resilient and sustainable cities.

Is there room for more in-depth discussions of such issues than what is possible in the panel and other presentation formats we have used in past NAAEE research symposia? Should we be drawing in a group of environmental education researchers who have different perspectives, including researchers who have knowledge of the policy process? What perspectives might our international colleagues who have been engaged in policy, or colleagues working in the formal school system and in community-based organizations, have on these issues?

[Meyers] Since 2003, when the 7th Invitational Seminar on Research and Development in Health and Environmental Education was held in Anchorage, Alaska, immediately preceding the NAAEE annual conference, the idea of having an NAAEE Research Commission hosted symposium grew to a successful, established event. In the opening of the paper, Brody identified seven challenges to the field of environmental education that a research symposium could address. To the first of the seven challenges, the authors have shown how the creation of the forum has helped demarginalize environmental education research within the NAAEE community. However, we did not address how environmental education research is marginalized in the broader communities of research and practice in the educational, environmental, and sustainability arenas. We should seek to continue our efforts to extend the impacts of our work throughout not only the environmental education community, but also more broadly as well. The second issue, the need to bring environmental education research and environmental education practice closer together has been addressed by the symposium, as researchers and practitioners gather together at the event and annual conference to engage with what it means to research the field. Regarding our third issue, the symposium also provides a means to continue to reflect upon what it means to be an *environmental education* researcher, and as we gather together, continue to build our professional perspectives on this. To the fourth issue, the symposium has expressly sought to include and give a voice to early-career professionals, particularly graduate students. As an early career faculty member, I found the symposium to be a wonderful professional and academic development opportunity, thanks to the efforts of established researchers to be inclusive and cultivate the growth of early career professionals. The symposium has not only reached out to graduate students through their faculty, but has developed acceptance criteria for papers that gives extra weight to works in progress, so they can be discussed at the symposium. Poster sessions are a common way to include graduate students, and so we have put extra attention to this, increasing posters from three in 2006 to 40 accepted for 2007. In addition, graduate students have been included in the planning of the symposium, and a special venue, the Graduate Student breakfast, planned for the 2007 symposium, was created in response to their requests to have a forum for their professional development. The fifth issue, of the need for discourse about both process and outcomes, has increased through the venues provided in the symposia, and is expected to continue.

The symposium was designed to address the seven issues by creating social learning contexts to help develop professional identities and create more meaningful dialogue. These goals have been advanced, as evidenced through the contributions of these authors and a vibrant event that draws a diverse set of formerly isolated researchers together on an annual basis to share their work and professional lives together, addressing the sixth challenge.

The seventh challenge, the need to accommodate differences in beliefs about methodologies and theoretical frameworks, has received much attention, yet progress is quite mixed. We have created a unique venue where these differences have coexisted, but continue to seek to engage them productively. The contributions of my coauthors reflect their interest in this challenge. It is my observation that it reflects a challenging issue of our day, one that is neither easy to characterize nor productively engage. Since it is Brody's seventh point, and is one that I believe is our most significant challenge, a larger discussion of this critical issue follows.

The perspectives shared here suggest that a unique event was intentionally created where a range of research methodologies are welcome to be present. Clearly though, the symposium *in toto* includes not only the perspectives of the hundreds of people who have participated, but amorphous and ill-defined groupings of those individuals into various communities of inquirers who gather at the research symposium. These are many, overlapping, and not well mapped. It appears to me that we have not yet found the tools to map our commonalities and differences productively to better understand in what ways our individual beliefs differ or are held in common, and so in what ways we do group together roughly with and across similar types of beliefs. Thus, it is extraordinarily challenging to summarize what common perspectives may have emerged. My interests, and those of many who have attended the symposia, are in the philosophy of our research and inquiry, particularly in our methods and methodologies, and the epistemological and phenomenological dimensions of our communities.

It is my observation that over the last 10 years of attending the NAAEE annual conference, and the first three of the research symposium, that there has emerged a new norm among researchers that a greater diversity of research methods are needed to more fully and adequately conduct inquiry in environmental education. It is no longer considered good practice in much of the environmental education research to claim that a quantitative inquiry is sufficient to stand as good work if it cannot answer normative questions of 'so what'; that is, what 'good' is the research seeking, who benefits from it, what are the values that are advanced or diminished through the research? While an earlier generation of researchers in environmental education did often ask these questions, with John Disinger a noteworthy proponent, the challenge of the 'posts' to environmental education research has brought the importance of asking these questions more to the fore.

This goal of more explicit discussion of the normative dimensions of research, is, however, only part of a broader change in environmental education research that has been sought. This is the challenge to absolutism, which has unfortunately been characterized as a challenge to positivism.

It has been unfortunate because it is unclear what set of beliefs actually comprises positivism in a definitionally rigorous way, and who in our community actually holds these beliefs despite the extensive body of work on this topic. Connell's (1997) observations about the (inadvertent, no doubt) misrepresentation of views still seem to hold true. Perhaps the most interesting development over the three years of the symposia has been the moderation of the strong contestation of all empirical methods as positivism, felt strongly by this researcher at the 7th Invitational Seminar, less so at the inaugural symposium, and now more akin to an uneasy truce. In order to have our goal of creating a community of researchers and inquirers that have diverse methodologies, epistemologies and research methods, we have challenged ourselves to be inclusive in our community of views with which we may strongly disagree, yet we are challenged to engage in meaningful dialogue about those differences. I credit the leadership of the symposia for what is truly a remarkable achievement to create this much inclusion. Moreover, rather than cultivating a social norm of merely tolerating disagreeable views, I have sensed the encouragement and emergence of openness to productive dialogue amongst those with perceived methodological chasms. The symposia as an event includes all the socialization that occurs around the formal agenda, of course, and in it a rich, curious, open dialogue has been encouraged. We sometimes attempt to puzzle through what our colleagues of good will (but obviously mistaken views) really believe, probing, testing, through what is emerging as a moment of openness to genuine dialogue. This is priceless, of course. However, we need better tools to see if there really is a chasm—a difference of beliefs, to map its contours, to assess where we are on common ground but not recognize it because of semantic differences, to better assess where our differences might be explored and touched and recreated into new, more productive understandings. In addition, we must continue to identify where the differences and mistakes are such that continue to be contested. A potentially productive start might be to carefully define what we mean by the term absolutists and explore whom amongst us is an absolutist, and why, and how it makes a difference in our work, for this may or may not be the largest schism. How might we find ways to identify which questions are ripe for such dialogue? This is where I hope we might go in the symposia, and see the potential for open-space technology, or other creative sessions, to enable us to identify and explore those questions which most move us in our research, and bring us apart and together into amorphously defined communities of like-minded methodologists.

Interestingly, the symposia have seen many presentations where researchers are using mixed methods, and acceptance, even encouragement, that they do so. Some see mixed methods as necessary, with others continuing to contest the use of quantitative methods of inquiry. While the degree of contestation has lessened, it is my sense that we have not yet had a comprehensive paper or presentation that shows how we can use diverse qualitative and quantitative research methods within a theoretically integrated environmental education research methodology, or an epistemology acceptable to the 'posts' community. However, the embracing of mixed methods because it is the new social norm can be theoretically problematic if it includes unwitting and/or incompatible use of mixed methodologies, as Dillon and Wals (2006) argue. That the symposia provides a forum to raise this type of question in a review of its history shows how far debates about research at NAAEE have come in the last three years.

At the beginning of the paper, I suggested that this article had three purposes: to review the formation of the NAAEE annual research symposium; to illustrate the broadening of what counts as research in the field; and, to jointly develop this account amongst authors with diverse epistemological and methodological views, advancing our sense of being in a community of scholars with a shared mission of increasing our ability to understand and critically evaluate each others' work. The first and second were accomplished through the generosity of my fellow authors as we collaborated to develop this account. It is my hope that my fellow authors also feel that we have made some small progress on the third goal, perhaps more in how we have built trust through engaging in a shared project. In the spirit of this collaborative account, I asked my fellow authors for their perspectives on our progress concerning the third goal. Justin Dillon's response was characteristic, suggesting that he already had a sense of being in a community of scholars with a shared mission. While Marianne Krasny agreed with Justin that the paper did not accomplish this in a major way, she observed that, 'each collaborative activity may contribute toward building such a community of scholars. This trust will hopefully support more critical dialogue amongst ourselves, contributing to the development of a deeper understanding of each other's perspectives'. She also noted that she did gain a better understanding of the other authors, and build trust. Likewise, Martha Monroe wrote, 'I think whenever people interact that they gain and grow—whether it is cooking a meal together . . . or writing a paper. Was it significant? Don't know. But my next "hello" to [them] . . . is likely to include a bit more of a sense of familiarity than it did before'. That some small progress was made for some in increasing our sense of community reflects the challenge of such a goal and the importance of ongoing efforts for this. My hope is that a deeper understanding of methodological and epistemological views will be facilitated as part of the ongoing process, one that never ends.

Looking Forward

[Meyers] We have successfully created a vibrant, inclusive forum in the symposium. While it was modeled on the invitational seminars and the EEE SIG, it differs in several significant ways. It is an annual rather than near bi-annual event (the seminar), and is scheduled immediately prior to the annual conference, making it easier for people to attend logistically and financially. The symposium, like the EEE SIG, is open to all, and thus larger (the seminar is capped at about 50, the symposium had over 100 in 2006, the EEE SIG has approx. 160 members, 2007). The invitational event is as it is named, with the seminar and EEE SIG both having an annual competitive review of papers primarily near completion, and, importantly, with a stronger graduate student presence. All events have strong international participation, with the invitational more so. The symposium is obviously much shorter (the seminar and EEE SIG are typically over five days, the symposium two).

Whilst we have created a unique forum that includes respectful dialogue, it is my assessment that at the symposium we have moved only slightly beyond a detente between those who follow (either consciously or not) the traditional Popperian and the 'posts' approaches to environmental education research, even as new epistemo-logical stances are emerging. However, like the EEE SIG and the invitational seminar, we have begun to engage in authentic dialogue between researchers regarding the importance of identifying the value in the methods and methodologies used by diverse research approaches.

In this historical overview, the authors have observed that there were 'camps' of researchers, one that seems to be a useful way of noting how we saw ourselves. Yet what the actual differences in beliefs about epistemology/philosophy of science, methodology and ontology are not clear, as I argue here—just that we believed that there were significant differences, there were differences, and we acted as if there were. In this researcher's view, we have only just begun to engage in the type of critical, sustained dialogue necessary to advance our understanding of those differences and the concepts we are using, the findings we share, and where we differ. The symposium and other venues have provided the authors time for dialogue, and most of us are familiar with at least a portion of each other's work, yet I would be challenged to rigorously identify where our beliefs differ and are held more closely. An intentional dialogue is needed in order to develop enough understanding to have a truly informed appraisal of our and many other participants' views, itself a necessary step before we can engage in the meaningful and useful dialogue necessary to create a community of researchers who have some semblance of shared meaning in our communication and efforts to advance the field.

To make progress on this goal, the 2007 research symposium is to be organized to facilitate discussion amongst participants on what research questions we share, using a form of open-space technology and pre-conference engagement with participants, encouraging inquiry into what pressing research questions they have, and how we can better understand each other's epistemological, methodological, and ontological conceptual frames and questions. We trust this evolution of the symposium will help us continue the process of deepening the dialogue necessary to continue our development as an inclusive and critical community of researchers.

Notes

1 Personal communication to Meyers, 2007.
2 Monroe offered two additional perspectives in her keynote presentation. First, that moving toward sustainability will require working with many different types of people, from engineers to architects to ecosystem scientists, who may be committed to changing public opinion but lack the insights and skills needed to be effective in the education/communication arena. Second, Monroe spoke to the importance of integrating education with public participation; she spoke to helping not only youth but also adults participate in environmental decision-making by empowering them with information. She called for research on the interface between public participation and environmental education to help us understand how to do this better.

References

Brody, M., Bangert, A. & Dillon J. (2007) *Assessing the outcomes of informal science learning.* Commissioned paper for the National Academy of Sciences Committee on Learning Science in Informal Settings, Washington, DC, The National Academies Board on Science Education, Learning Science in Informal Environments, The National Academy of Sciences.

Connell, S. (1997) Empirical-analytical methodological research in environmental education: response to a negative trend in methodological and ideological discussions, *Environmental Education Research,* 3(2), 117–132.

Dillon, J. & Wals, A. (2006) On the dangers of blurring methods, methodologies and ideologies in environmental education research, *Environmental Education Research,* 12(3/4), 549–558.

Hart, P. (2005) Editorial: transitions in thought and practice: links, divergences and contradictions in postcritical inquiry, *Environmental Education Research,* 11(4), 391–400.

Hart, P., Barrett, M. J., Schnack, K., Dyment, J., Taylor, J. & Clark, C. (2004) Reflections on the 7th Invitational Seminar on Research and Development in Health and Environmental Education: Anchorage, Alaska, USA, 2003, *Environmental Education Research,* 10(4), 563–574.

Iozzi, L. (Ed.) (1984) A summary of research in environmental education, 1971–1982. The second report of the National Commission on Environmental Education Research, *Monographs in Environmental Education and Environmental Studies, Volume II.* Columbus, OH, ERIC/SMEAC.

Kuhn, T. (1970) *The structure of scientific revolutions* (Chicago, University of Chicago Press).

Meyers, R. B. (2006) Environmental learning: reflections of practice, research and theory, *Environmental Education Research,* 12(3/4), 459–470.

Mrazek, R. (1993) *Alternative paradigms in environmental education research* (Troy, OH, NAAEE).

NAAEE (2005) *National Project for Excellence in Environmental Education* (Washington, DC, North American Association for Environmental Education). Available online at: http://naaee.org/pages/npeee/index.html (accessed 14 February 2007).

Reid, A. & Scott, W. (2006a) Researching education and the environment: an introduction, *Environmental Education Research,* 12(3/4), 239–246.

Reid, A. & Scott, W. (2006b) Researching education and the environment: retrospect and prospect, *Environmental Education Research,* 12(3/4), 571–587.

Russell, C. (2006) Working across and with methodological difference in environmental education research, *Environmental Education Research,* 12(3/4), 403–412.

4 The Messy Process of Research

Dilemmas, Process, and Critique

Charlotte Clark, Michael Brody, Justin Dillon,
Paul Hart, and Joe Heimlich

Clark, C., Brody, M., Dillon, J., Hart, P., & Heimlich, J. (2007). The messy process of research: dilemmas, process, and critique. *Canadian Journal of Environmental Education*, 12, 110–126.

Introduction

The error spectrum in publishable research, ranging from slight "undetect-able" flaw to insurmountable foible, is known to exist, but is seldom acknowledged in print unless such an admission is followed by the victorious tale of how the blemish was erased, overcome, or adeptly sidestepped on the way to a research publication. This is not a new tradition; Figure 4.1 contains a 50-year-old tongue-in-cheek research glossary (from metallurgical research) of such sidestepping prose.

We believe that too little attention is given to acknowledging and (especially) disseminating the "stubbed toes" that happen on the research path. In this article, we seek not to sidestep but rather to highlight, and perhaps even celebrate, the unabashedly messy business that is quality research. To do so, we tell stories from our own experiences that illuminate our belief that research happens not at a failure/success binary/dichotomy, but rather along a path where many points can't (and shouldn't) be described as failure or success.

The variety of experience level among authors, from highly published tenured professor to unpublished graduate student, is meant to show that this is true for researchers at every experience level, and that a key is to engage and persist. Writing often tidies up the loose and ugly ends of research, and yet the process of dealing with those parts is often one of the most productive for a project. Our hope is that stories of our flaws and foibles will generate thought and provoke discussion. We wish for others to be heartened, and to see their own difficulties as both more useful to themselves, and perhaps instructive to be shared with others.

The impetus for this paper was provided by a session held at the Third Annual Research Symposium in October 2006, prior to the annual conference of the North American Association for Environmental Education in St. Paul, Minnesota. The session, entitled "What Happens When Research

Sidestepping Research Prose	
Common words and phrases	**Their real meanings**
It has long been known that . . .	*I haven't bothered to look up the original reference.*
. . . of great theoretical and practical importance	*. . . interesting to me*
While it has not been possible to provide definite answers to these questions . . .	*The experiments didn't work out, but I figured I could at least get a publication out of it.*
Three of the samples were chosen for detailed study . . .	*The results on the others didn't make sense and were ignored.*
Typical results are shown . . .	*The best results are shown.*
Presumably at longer times . . .	*I didn't take time to find out.*
These results will be reported at a later date . . .	*I might possibly get around to this sometime.*
It is suggested that . . . It is believed that . . . It may be that . . .	*I think . . .*
It is generally believed that . . .	*A couple of other guys think so, too.*
It might be argued that . . .	*I have such a good answer to this objection that I shall now raise it.*
It is clear that much additional work will be required before a complete understanding . . .	*I don't understand it . . .*
It is to be hoped that this work will stimulate further work in the field . . .	*This paper isn't very good, but neither are any of the others in this miserable subject.*

Figure 4.1 Sidestepping research prose (Graham, 1957).

Goes Bad?" featured a panel composed of Justin Dillon, Paul Hart, Joe Heimlich, and Michael Brody, and was moderated by Charlotte Clark.

As we did at that conference session, we will parse our stories into four themes:

- evolving research questions,
- methodology or methods surprises,
- problematic answers, and
- publication dilemmas.

Evolving Research Questions

In qualitative work, we expect our research questions to evolve as we work and to be preconceived as emergent, rather than preordinate. However, the back story of that evolution in research work is seldom articulated, at least in print (however, see Lather & Smithies, 1997; Russell, 2003). We provide two such back stories here. In the first, the question evolved through external influence through a journal review process; in the second, the question evolved through internal inquiry.

Welcoming External Influence on Research Questions (Michael and Justin)

Michael: This tale emerged from a dynamic editorial process resulting in a career-defining article that led to further development of learning theory related to informal settings, and challenged me to grow professionally by leading me to greater insight and understanding that I otherwise would have overlooked (Brody, Tomkiewicz, & Graves, 2002). The text is presented as a reflective dialogue on the editorial process and subsequent evolving world-view, research questions, research concept, and published paper.

The research contract asked the authors to ascertain if a new Midway Geyser Basin Visitor Guide would have any effect on visitor outcomes; that is, would casual visitors learn anything as a result of using the brochure at the site. I went into this research project with an existing understanding of learning derived particularly from my previous work on misconceptions about the environment, and influenced primarily by Ausubel and Novak (Ausubel, Novak, & Hanesian, 1978). This led me to ask what people already know about Midway Geyser Basin and the associated life forms. I asked what science content visitors brought to the event, how that content might change, and what views and science misconceptions people might have about

- the National Park Service,
- geothermal features at the park, and
- associated microorganisms (especially biopiracy of microorganisms for the biomedical industry (Brody et al., 2002)).

Justin: Michael's first draft was sent to two reviewers of the *International Journal of Science Education* that Bill Scott (co-editor) and I had asked to help with the special edition. The first review, undertaken by a leading researcher in science learning in informal contexts and completed in September 2000, concluded by stating that the paper:

> raises interesting questions, but needs to be more firmly grounded in relevant non-school based research and recent research on alternative conceptions. It also needs a more thorough defense of the research design and an introduction that sufficiently supports the study conducted.

The second review, undertaken by a leading science education researcher, noted that the literature was somewhat out-of-date and suggested other studies should be used:

> Specifically, there are now strongly recognized alternative theories about learners' conceptions. These alternative theories are grounded in research in cognitive science and learners' epistemologies. The more

contemporary research (e.g., Strike & Posner, 1992; Pintrich, Marx, & Boyle; diSessa, Minstrell) is showing that the conditions of the learning environment and the motivation for learning are as important as the initial conceptual schemes held by the students. Simply stated, the conceptual ecology is much more complex than originally proposed by Novak, Ausubel, Champagne, Klopfer, Osborne, etc. Some of the new research maintains that the initial conceptual schemes are not important at all. Here I am referring to the idea of "knowledge-in-pieces" put forth in a theoretical framework by Andrea diSessa and in a practical framework by Jim Minstrell. This research has shown that individuals' conceptual frameworks change with regard to the conditions of the task. This perspective challenges some of the conclusions drawn in the present study. For example, why would we expect a senior citizen with no formal education in geology or evolutionary theory to have a knowledge base about an emerging scientific theory about the origins of life on Earth? This expectation on the part of the author(s) is a serious shortcoming of the present paper because it is a shortcoming of the theories about alternative conceptions and conceptual change teaching.

Both reviewers also noted that there were some issues concerning the sample size and the claims the authors were making about the generalizability of their findings. So, Michael and colleagues were asked to make a major revision to the paper. Although the review process might be seen as a gate-keeping exercise by editors and/or reviewers, editors do (or should) expect authors to defend their submissions if they think the reviewers have missed the point or are mistaken. So, Bill and I waited for Michael's response.

Michael: The comments by the reviewers were the most insightful and telling of any that I had received in my entire career, but they were also the most difficult to accept and, in the end, accommodate. To say that the authors went back to the "drawing board" is an understatement.

I read new literature, attended conference presentations by people cited as important to consider, and constructed concept maps to integrate the new ideas into a more robust conceptual framework. These efforts led to a reformulation of the basic research question, moving from the traditional outcomes-based approach to the more complex task of understanding the experiences of visitors to Midway Geyser Basin. The questions evolved in this way:

- *Old:* What are park visitors' understandings about geothermal features and associated microorganisms?
- *New:* What are the experiences of visitors to Midway Geyser Basin?

As my attention increasingly focused on the experience of visitors, the new theories and principles (both of concept and methodology) the editors introduced to me began to make more and more sense. Figure 4.2 depicts my evolution by showing components of the old (standard text) and the new (italicized text). For example, in the conceptual framework, I had to integrate disparate ideas from various domains into a comprehensive view of learning in this situation—not an easy puzzle to solve!

From the authors' point of view, the reconceptualization of the underlying philosophy, theory, principles, and concepts was the hardest part of the rewriting. Once this had been clarified, it was clear that the data could be

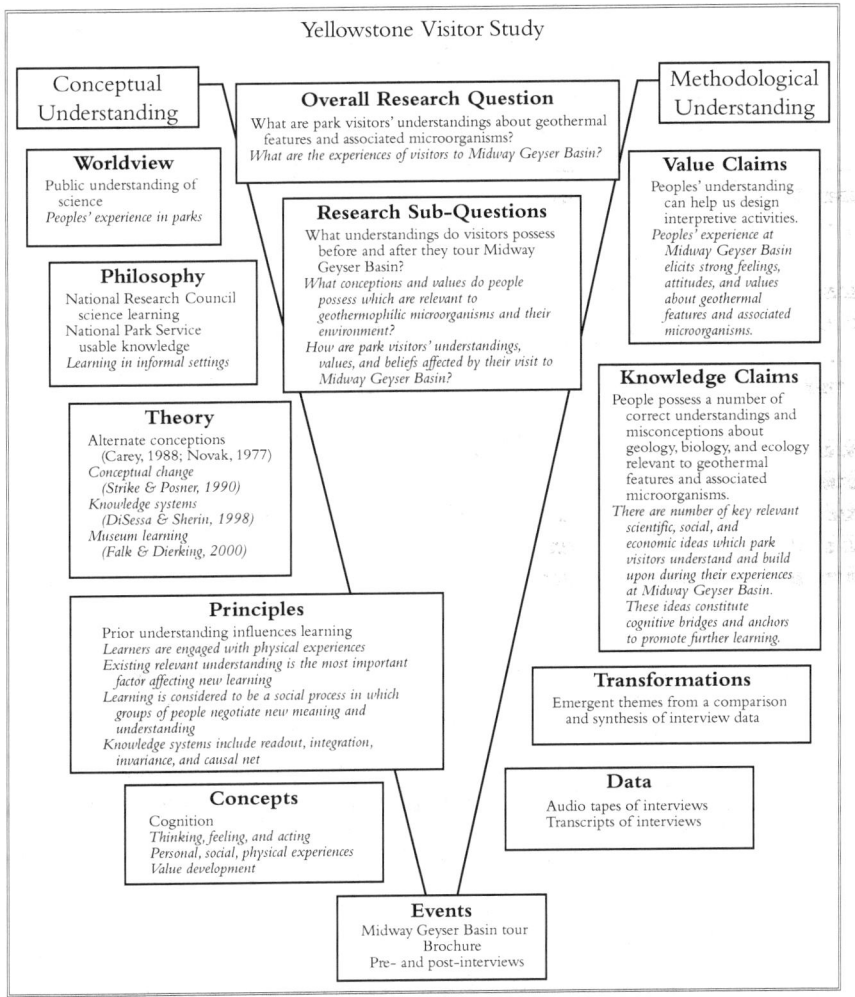

Figure 4.2 Yellowstone visitor study (Brody et al., 2002). Plain text indicates original ideas and framework of study; italicized text indicates new understandings.

reinterpreted in more meaningful ways. This led to different knowledge and value claims, which are depicted in the boxes on the right in Figure 4.2. Fortunately, there were some aspects of the research and final paper that did not change, such as research events, actual data, and transformative emergent themes. A conservative estimate of the amount of change that took place in this paper would be about 75%. Thankfully, the editors allowed approximately six months for the authors to resubmit the paper.

Justin: The final paper, published in 2002, is far stronger than the version that was originally submitted. From an editor's point of view, Michael and colleagues were exemplary in the manner in which they responded to the reviewers' comments.

Michael: The process of producing the final Yellowstone paper will remain a watershed event in my career. On one hand, it was always humbling to be led down a new trail by expert guides—that is, a trail where a part of your brain scolds you, saying "you should have traveled this before." On the other hand, this article is about quieting that scolding voice, and recognizing how empowering it is to know that you can trust your colleagues to give you insights that, with patience and hard work, can result in a career-defining publication.

Welcoming Internal Influence on Research Questions (Paul)

Good research takes time to engage both theory and methodology intensively. Although many university programs work against this longer-term study through fee structures and scholarship award limitations, thoughtful inquiry is not so much a matter of learning the theory as learning to conceive oneself in terms of theory. In this case, a doctoral student's interest shifted over a period of about three years, resulting in new research questions, a significantly different dissertation product, and deep changes to the student's own personal belief system (Barrett, 2005).

Her original interest arose directly from her experience as a secondary teacher who coordinated and taught alternative, semester-long outdoor/environmental education programs. Early in her doctoral program she expressed a narrative interest in understanding what sustains teachers who work somewhat outside the regular high school program. She was curious about what motivates teachers (much like her) and where the energy, persistence, and patience had originated. Her original research questions, as she presents them, are as follows:

- What does it mean to be a teacher of an intensive interdisciplinary outdoor/environmental education program?
- Why do teachers of integrated programs choose to teach the way they do?
- How did they get there?

By mid-course, she began to engage intensively in both poststructural theory and fieldwork. The combination of conversations with one teacher in particular, together with deeper thoughts about how teachers' identities and sense of agency are produced, how teachings are discursively framed, and how life histories are culturally produced, led to a significant shift in research questions. Her new questions were positioned as a critical examination of how environmental education that espouses change can be understood and enacted within the conservative tendencies of the educational system. Thus, stories of experience are not simple tellings, and her evolved questions ask:

- How do discourses of teaching, learning, and "nature" work to constrain and enable environmental educators?
- How have teachers' identities been produced by discourse?
- How can life history research within a feminist poststructural framework be useful in addressing these two questions?

Beyond these evolving questions, she has also constructed a critically reflective re-engagement with her work as questions themselves:

- What assumptions are embedded in the research questions (e.g., about self, agency, knowledge)?
- What kind of answers will the questions produce?

These last two questions, together with St. Pierre's (St. Pierre, 2000; St. Pierre & Pillow, 2000) admonition that poststructuralism needs to turn back on itself to investigate its own sets of assumptions, opened the study to another layer of examination. The study (which is represented in hypertext at www.porosity.ca) has now been extended to investigate ways in which research itself, through its often unacknowledged anthropocentric framing, might be limiting (successful) enactment of outdoor/environmental education.

Methodology or Methods Surprises

Optimally, published works should discuss the methodology, methods, and instruments used, but this does not always happen; methodologies may be the most oft-confused and neglected component (Dillon & Wals, 2006). Published works whose aim is to discuss the perils and pitfalls of these choices are also rare, but not completely absent (Raven, 2006). This section provides three stories of surprises discovered in the midst of research involving methodologies and methods. One describes the unfortunate discovery that a "well-tested" instrument was useless, a second tells of the researcher's realization that the chosen method did not fit the question after all, and a third admits to a study where the frame of the study changed so often after data collection had begun that every type of external validity error was violated.

The Useless "Well-Tested" Instrument (Joe)

We are always seeking ways to ensure that our studies are well-grounded and rigorous. One long-practiced approach is to find an instrument measuring a component of what you hope to be studying, and to use it. Of course, there are the obvious challenges of ensuring that what is being measured is actually the domain of what you want to know, but often there are more hidden challenges as well.

For example, a graduate student was planning to use an instrument as part of a larger study. This instrument was comprised of several scales, had been used in at least 12 studies, and had good reliability measures in each of the studies (ranging from around a 0.69 to a 0.82). However, as we began the necessary work to translate the instrument, we found that several items could be considered positive or negative, other items were neutrally stated, and some were clearly directionally framed. What a shock that this instrument had been replicated and used in other studies, but had construction errors! The reliability measures were strong because there was a clear inter-item relationship, but the validity of the questioning structure would be questionable, as would any analysis on the findings.

Although this original instrument was flawed for our use, sometimes going back to the original instrument is far more valuable than creating a new one or even using a more contemporary adaptation. For example, we've found far stronger reliability and distributions in populations using some instruments developed by psychologists measuring a particular construct, as compared to environmental education instruments measuring the same construct. As a second example, I have found some basic state and trait measures that are better than those that have been adapted specifically for environmental education and science.

In sum, caution is urged both when modifying existing instruments, where errors can be amplified through repetition without critical reflection, and in using existing ones without change, where application to a new field must be carefully considered.

Evolving Epistemology, Methodology, and Methods (Paul)

In this case, a sticking point in data analysis led a veteran teacher to not only a change in methods or even methodology, but in fact an epistemological shift in what counted as legitimate data (not necessarily in that order). The setting for this story is the secondary school physics classroom (not the environmental education classroom), but the story and its outcome are well-applied to the environmental education field.

The teacher in this case developed his doctoral work out of a persistent conundrum in his work with students on vector mathematics (Wessel, 1998). The fact that many high school students experience difficulty in the transition from concrete scenario (e.g., two soccer players each kick a soccer ball simultaneously from different angles) to abstract math symbols and graphical representations, led him to question how adolescents "learn" to conceptualize complex concepts. He came to investigate students' lived experiences in physics problem-solving by creating classroom conditions where students could verbalize their learning experiences interactively with peers and the researcher-as-teacher in an attempt to make their reasoning processes more explicit. Although difficult, he was able to collect rich narrative text through many days of problem-solving.

So while context question and research process were clear enough, the sticking point in this study concerned the method of analysis. What to do with mounds of narrative text is a relatively common dilemma in many emergent interpretive research studies. In this case, given a background in scientific research, the student's first instinct was to organize the data by dwelling inside the text and looking for regularities as patterns or themes. He elected to code the transcripts of his data using an elaborate colour-coding scheme—attempting to make sense of the interactive (reasoning) talk. In my discussions with him during this process, questions arose about what counts as legitimate data and data analysis in various forms of inquiry (quantitative and qualitative). Despite rhetoric to the contrary, it was difficult for this teacher/researcher of science to make the interpretive turn—from viewing data in terms of categories to viewing it differently, perhaps holistically, so that the actual content of the conversations could be brought into sharper relief.

After about three months of coding, he abandoned that analytic process for a form of conversation analysis within a more interpretive frame. I believe the shift was analogous to his physics students' insight into the transition from concrete to abstract thought in solution of their vector problems. A highly independent person with a mind of his own, who would not be convinced of method except through his own experience, produced a thesis rich in conversation analysis. This case, for me, represents an example of distinctive shifts in methodological thinking brought on by a sticking point in analysis that had ramifications far beyond method—in fact, in order for a shift in method of analysis to occur, an epistemological shift in what counted as legitimate data was prerequisite. Such shifts in thinking and data analysis methods are important to consider for questions that are at the concrete/abstract border, many of which exist in environmental education, such as the study of instruction on global warming.

The Case of the Disappearing Frame (Joe)

In a national study, several organizations provided names from which a random sample would be generated. Each group was part of a blocked sample of these organizations and was selected to represent that type of organization based on size of members on their lists. Initially, each group was asked to provide the number of names on their list. The total number was determined and a proportional sample number was assigned to each group. The result was an initial frame, or list of accessible respondents, of around 900 names.

Because the study was under a strict timeline and the volunteer-led groups did not respond in a timely manner, the study began as soon as the first frame was received. This first list, when received, included duplicate names and multiple ways of addressing the same person for some businesses, organizations, and individuals. The total *N* for the study was adjusted, the proportions rerun, the sample drawn, and surveys sent. The second frame was larger than the first, but also had many frame errors. And so it went through all 10 frames. Each time a list arrived it was edited, corrected, and the numbers recalculated. One list of 600 names was actually under 300 discrete names, and another list had over 60 names that were corporate sponsors rather than members.

Clearly, we had a problem. We (the researchers) were in a perpetual state of panic each time a new list of names arrived and differed from the earlier promised list. This created a multitude of emergent threats to validity, among them:

- differential selection–experimental variable interaction,
- selection–maturation interaction effects,
- maturation,
- differential selection of subjects, and
- instability (see, for example, Smith, 1980).

The study could have been discarded, but as most of the threats related to interaction effects and differentiation among the populations, all respondents were considered as one pool. The sample was changed from a blocked design to a single population with no blocks. In doing this, we were able to eliminate two of the major threats. To address the instability threat, significance measures were done using ANOVAs in order to look within and between groups, with the groups being defined by demographics rather than the block design. Finally, the threat of maturation was addressed by looking at differences between late respondents from one frame and comparing them to early respondents of the next frame.

The final result: a national study of a random selection of individuals from 10 organizations that could be considered representative of the larger population. Statistics were nonparametric or, in the case of the ANOVAs, parametric but for the purpose of more clearly describing the population, rather than generalizing. Although the study was published in peer-reviewed journals without a hint of the frame error, the real learning for me was the challenge of addressing the compounding problems.

Problematic Answers

In theory, studies ask questions or state hypotheses without knowing the answer(s) that will be generated. In reality, most students, researchers, supervisors, funders, and sponsors have an opinion or wish as to what an answer might be. Quality study design aims to enable that wish to either be subsumed (as much as possible) under statistical methods, or to be acknowledged and articulated to the extent that it could impact the study findings. Nonetheless, stories of some problematic answers may be useful and interesting to others in the field. These stories illustrate the jeopardy of discovering things a research sponsor doesn't want to know, of learning that the funder (or implementer) is not interested in what you have found, of losing access to some answers mid-stream, and of being caught in political mire.

Finding Answers the Funder Doesn't Want to Hear (Joe)

There is an ethical dilemma when doing evaluation work because the work belongs to the organization paying for the evaluation, rather than the evaluator. This can lead to real problems when someone agrees they want an evaluation to "find out what's working and what can be changed to improve the program" upfront, but when the report is completed decides that they want everything to look good for the funder. On numerous occasions, this shift has led to the organization asking to remove any of the findings that are even slightly critical, and *all* recommendations for improvement. Two alternate approaches could be tried:

- provide the report and allow the organization to pull from the report those things they choose to share, but do not allow alteration of the report itself; or
- offer to present the findings verbally to the funder.

The former creates additional tension around the findings and can lead to resistance to good evaluation. The latter takes additional time, but demonstrates to the program people how to manage findings with funders. It also helps the funders see why a program is truly worthy of additional or continued support, as it is continually improving.

In one study, a board member did not want membership of the organization studied as part of the larger organizational review, and blocked progress on the study. Repeated discussions and messages to stop gathering data led the study team to realize that this individual feared members would not be in favour of some pet projects (changes) of this board member. To satisfy the member and remain ethically true to the research (which needed the input of members), members were not surveyed, but a random selection from the membership roster was interviewed by telephone. This kept the member component of the study separate from the study of staff, board, administration, and the community at large. Not surprisingly, some of the members interviewed liked the changes, and some didn't. However, major discontent was expressed by members over the way the board and administrators were handling the changes. Members felt that too many important things happening were being kept from them, as exemplified by the reluctance to include members in the study itself! Although this observation was not included in the final report, the findings were presented and one bullet point noted the dissatisfaction of the membership. In this way, we feel we were able to include the concept without being confrontational and/or risking that key stakeholders would reject the entire document because they didn't like one piece.

Murky Participant Permission Process (Charlotte)

In a study of a group of people planning to build (and live in) a neighbourhood, I was granted initial permission for the research by the community through a consensus decision process (and a permission form signed subsequently by community leaders). This permission permitted me to attend community meetings, have access to neighbourhood documents, and participate in the neighbourhood Internet list serve. It also allowed me to invite individuals to participate in interviews (I obtained separate written permission on an individual basis). About half the households in the community agreed to (and participated in) an individual interview. Notwithstanding the fact that this study was exempted from the review of the Institutional Review Board at my institution, I have made regular and thorough attempts throughout the years to re-ascertain permission from the community as a whole, and with individuals (especially newcomers) in the community.

After many years of data collection and analysis of these data, I found that some of the most interesting stories and findings feature members of the neighbourhood who did not elect to participate in an individual interview. This produced a dilemma for the research—some findings of high potential interest to the study existed in a murky access situation. These individuals had agreed, through a group consensus decision, to being observed at meetings and to having their emails read that they posted to the list serve. However, I believe that their lack of permission for an individual interview might indicate a discomfort with a more public illumination of their particular participation. Therefore, if I want to highlight those stories and to remain comfortable that permission has been granted, I must decide how, when, and whether to approach these individuals to obtain a second level of permission. A risk exists that a participant may decline permission, which could have implications for many levels of data. Therefore, the decision is actually whether to put some of the data at risk of becoming unusable (with unknown "domino-style" implications), or to use other data (featuring those participants who have given individual interviews) to reveal similar or different findings. In the one case that inspired this anecdote, I asked for and was granted permission to tell the story ("I didn't agree to an interview because I was just busy," she said.) Nonetheless, I wonder whether in the end, I will believe that my initial and ongoing permission process was not thorough enough, or that my data and findings were better and richer as a result of the less invasive approach.

Publication Dilemmas

Many experienced and aspiring academics can tell stories of trying to guess unsuccessfully what is in the editor's head or, when a piece is rejected from a journal, knowing when to work to change the article to fit, and when (and how) to submit the piece to a different journal where it might fit better.

Resubmitting to the Same Journal (Justin)

Different journals operate different review procedures. The *International Journal of Science Education,* for example, has seen the number of submissions rise, particularly with the introduction of an electronic system (Manuscript Central). The electronic system obviates the need for posting printed papers across the globe which, as well as having environmental benefits, also has concomitant time savings. The journal now publishes 15 editions each year and editors can afford to be robust in attempting to raise the quality of the journal by sending papers to two or more experienced researchers. It is extremely unusual for a paper to be accepted without any revisions. In my experience, almost all papers are either rejected (with an encouragement to resubmit) or they are sent back for major revisions. The message is: persist, listen to the reviewers' comments, and be prepared for a lot of extra work after you think you've finished a paper!

Resubmitting to a Different Journal (Michael)

Sometimes, a paper may not need rewriting as much as it may need resubmitting to a different journal. The reason can be a lack of fit with the type of work the journal publishes, or it may be a more amorphous or invisible reason. In the case of the former, an author should try to identify a journal's theme before submission and, where a misfit occurs, hope that comments from the journal will steer the article in a better direction. Two stories of the latter (amorphous or invisible reasons for rejection) follow.

In one case, an editor rejected an article of mine out-of-hand, and without review, because it was based on a well-known author's perspective. Specifically, I had taken the author's perspective on local, social, cultural, and political aspects of a watershed and its potential role in guiding ecological governance and sustainability, and had related it to place-based curriculum development in environmental education. Apparently, the journal editors believed the application of the perspective to a different curricular situation was not unique enough to warrant publication. I chose not to revise but to resubmit to a different publication, where the article was published without any editorial changes.

In a second case, I realized that several papers from early in my career could be combined into a synthesis paper focusing on student understanding of ecological crises. The new paper addressed science and natural resource management concepts across several ecological phenomena and provided insight across disciplines. As with the example above, the first submission was again rejected out-of-hand, and was not sent out for review by contributing editors. The only reason given was that the research involved work over five years with the assistance of approximately 30 graduate students. Again, a subsequent submission of the same paper to an international journal resulted in publication with little revision.

These two experiences have helped me develop a "thick skin" when it comes to reviews that seem to simply dismiss my work. My current perspective is that editors have power that they may choose to wield in a variety of ways, some not as fair or just as others. From my experience, it appears second opinions are just as good in publishing as they are in medical diagnoses.

Conclusion

This article is about encouragement. We identify with the challenges researchers face in exploring their questions and interests in the face of new theoretical perspectives—interpretive, critical, and postmodern. More traditional preordinate frames have been supplanted. In their place, new kinds of questions require diverse theoretical perspectives that often cross disciplinary boundaries. Indeed, we have gone beyond the days when a cookbook approach to methods was sufficient. Researchers now employ a range of methods across a wide spectrum of methodologies that advocate emergent methods more consciously responsive to changing conditions in the field, as well as sensitive to theoretical issues of ethics, power, and authority. The issues, methodology, and methods needs cannot always be anticipated in design.

The authors have attempted to provide insights from work amongst diverse categories and frames of inquiry that are meant to encourage fellow students to persist and problem-solve mid-process. We encourage an emergent trend for authors to write about, and publish, their own struggles in a way that anticipates critique. We hope that increasing instances of researchers telling stories of mucking around the methodologies and ethics of inquiry can be the beginning of a more process-oriented, and perhaps more nuanced conversation that opens active discussion and scrutiny amongst critical friends.

References

Ausubel, D. P., Novak, J. D., & Hanesian, H. (1978). *Educational psychology: A cognitive view.* New York: Holt, Rinehart, and Winston.

Barrett, M. J. (2005). Making (some) sense of feminist poststructuralism in environmental education research and practice. *Canadian Journal of Environmental Education, 10,* 62–78.

Brody, M., Tomkiewicz, W., & Graves, J. (2002). Park visitors' understandings, values and beliefs related to their experience at Midway Geyser Basin, Yellowstone National Park, USA. *International Journal of Science Education, 24*(11), 1119–1141.

Carey, S. (1988). Reorganization of knowledge in the course of acquisition. In S. Strauss (Ed.), *Ontogeny, phylogeny and historical development* (pp. 1–27). Norwood, NJ: Ablex.

Dillon, J. & Wals, A. E. J. (2006). On the danger of blurring methods, methodologies and ideologies in environmental education research. *Environmental Education Research, 12*(3–4), 549–558.

diSessa, A. & Sherin, B. (1998). What changes in conceptual change? *International Journal of Science Education, 20,* 1155–1191.

Falk, J. H. & Dierking, L. D. (2000). *Learning from museums: Visitor experiences and the making of meaning.* Walnut Creek, CA: Altamira Press.

Graham, C. D., Jr. (1957). A glossary for research reports. *Metal Progress, 71*(5), 75–76.

Lather, P. & Smithies, C. (1997). *Troubling the angels: Women living with HIV/AIDS.* Boulder, CO: Westview Press.

Novak, J. D. (1977). *A theory of education.* Ithaca, NY: Cornell University Press.

Raven, G. (2006). Methodological reflexivity: Towards evolving methodological frameworks through critical and reflexive deliberations. *Environmental Education Research, 12*(3–4), 559–569.

Russell, C. L. (2003). Minding the gap between methodological desires and practices. In D. Hodson (Ed.), *OISE Papers in STSE Education, Volume 4* (pp. 125–134). Toronto: University of Toronto Press.

Smith, P. L. (1980, April). *Some approaches to determining the stability of estimated variance components.* Paper presented at the American Educational Research Association, Boston, MA.

St. Pierre, E. (2000). Poststructural feminism in education: An overview. *Qualitative Studies in Education, 13*(5), 477–515.

St. Pierre, E. & Pillow, W. (2000). Introduction: Inquiry among the ruins. In E. St. Pierre & W. Pillow (Eds.), *Working the ruins: Feminist poststructural theory and methods in education* (pp. 1–24). New York: Routledge.

Strike, K. A., & Posner, G. J. (1992). A revisionist theory of conceptual change. In R. Duschl & R. J. Hamilton (Eds.), *Philosophy of science, cognitive psychology, and educational theory and practice* (pp. 147–176). Albany, NY: SUNY Press.

Wessel, W. E. (1998). *Knowledge construction in high school physics: A study of student/teacher interaction.* Unpublished doctoral dissertation, University of Regina.

5 Broadening Views of Learning

Developing Educators for the
21st Century Through an
International Research Partnership
at the Exploratorium and King's
College London

Bronwyn Bevan and Justin Dillon

Bevan, B., & Dillon, J. (2010). Broadening views of learning: developing educators for the 21st century through an international research partnership at the Exploratorium and King's College London. *The New Educator*, 6, 167–180.

Introduction

Two relatively recent studies spotlight the importance of making science more interesting and appealing to children. The first is the *Relevance of Science Education (ROSE)* study, an international survey of students, aged 15, from over three dozen countries (Sjøberg and Schreiner, 2006). This study, which is ongoing, notes that students from industrialized nations are less interested in school science and markedly less interested in science careers than their counterparts from developing nations. The second study is a retrospective data analysis, conducted in the US, that found that, independent of grades, standardized test scores, or family background, 8th grade students who indicated an interest in future science careers were much more likely to go on to major in science in college than were students who had indicated other types of career interests (Tai et al., 2008). These two studies together make a strong case for the need to make science education more appealing and engaging to children, especially in the K–8 years. In light of the growing importance of science and science-related fields in almost all economic sectors, providing engaging science education for all children, even in the most under-resourced schools, becomes a critical issue of access, equity, and social justice.

Traditionally, the focus of any improvement in science education would be school—often high school and in some cases elementary school. Many countries have tinkered with or made wholesale changes to their science curriculum and/or their examination and assessment systems with the aim of changing pedagogy in order to raise attainment. Australia, for example, is implementing a National Curriculum just as England celebrates the 21st birthday of its own version. In the US, task groups are working on new versions of the National Science Education Standards. No one, it seems, is happy with the standards of education of their young people.

More recently, however, the attention of education policy makers has turned to the learning affordances provided by the informal science sector. Science on TV, particularly the natural sciences, has maintained high levels of interest among the viewing public. Millions of people worldwide pay to travel to museums and science centers, often spending hours looking at exhibits and trying hands-on science. Whatever the reasons for their visits, it seems that science can be popular, interesting, and memorable. Not unreasonably, some people have wondered publicly and privately if the formal science education sector could learn something from the so-called informal sector.

Of late, researchers and policy makers from around the world have increasingly called for greater attention to be paid to the educational potential of out-of-school settings, citing the many benefits, and indeed, the necessity, of learning in contexts other than the classroom. For example, the policy statement published by the Informal Science Education Ad Hoc Committee for the National Association for Research in Science Teaching argued that learning "derives from real-world experiences within a diversity of appropriate physical and social contexts" (Dierking et al., 2006). The National Education Standards in the US recognize that science museums and science centers, in particular, "can contribute greatly to the understanding of science and encourage students to further their interests outside of school" (NRC, 1996, p. 45). Many school districts and informal science institutions are forming collaborations to support new curricular programs for students and new training programs for teachers (for example, see Bevan et al., 2010).

But despite a significant number of efforts and experiments to bridge informal and formal settings, most such programs fail to institutionalize as funding dwindles or leaders change. Thus, while the potential power of integrating the resources of the informal sector to support more engaging K–12 science education seems apparent to many, there remain deep-seated institutional divides that appear to draw educators and educational systems back into their organizational boxes when the explicit call or funding for collaboration fades.

This paper describes efforts to develop educators, in both formal and informal settings, who possess conceptions of teaching and learning that not only span but, perhaps, depend on bridging institutional and contextual boundaries. These educators are imbued with broad conceptions of science learning: understanding that it involves not just conceptual knowledge and process skills, but also familiarity with evidence-based ways of knowing in science, as well as how science is used in everyday life and how scientists go about their work. These educators understand that such views of science grow from and also support students' emerging science interests and identities, which are developed not just in school but also over time and across multiple settings.

We argue that it is critical for the new educator, in both formal and informal settings, to understand and work with the learner as the person who in real life is constantly moving across different institutional fields or contextual boundaries. Educators need to understand that learners' experiences with science are neither confined to the classroom nor to the museum. In this view, it is essential that the new educator understands how the learner's interests, identities, and capacities are developed and relate across settings. To do this, educators must understand the possibilities, programs, and experiences provided by other institutions and to seek to build on and expand them.

At the heart of this article is the experience of three institutions separated by thousands of miles but united in a desire to challenge the formal/informal divide. The Exploratorium, a museum in San Francisco, King's College London, and the University of California Santa Cruz worked together to create a center funded by the National Science Foundation to produce young scholars, new educators, and new knowledge about the relationships between teaching and learning in school and non-school settings. Founded in 2002, the Center for Informal Learning and Schools (CILS) has to date worked with more than two dozen doctoral students in science education, developmental psychology, and also the natural sciences; offered about a dozen postdoctoral research fellowships; provided multi-year professional development programs to about 100 informal educators; hosted several conferences for a mix of more than 250 formal and informal educators, researchers, and policymakers. These training programs were largely structured around partnerships between academic and informal science institutions, both in London and in California, that provided practicum sites for CILS participants, and that ensured a two-way dialog across the organizational boundaries. In the next section we highlight hybrid programs at two of the CILS institutions.

King's: Designing a Hybrid Program in an Academic Setting

King's College London has a long history of teacher education, dating back to the late 1800s. When the Exploratorium was first developing a plan and proposal to fund CILS, King's was an ideal partner for many reasons: because of its wealth of experience in science education, because of its pioneering work in informal science education, and because of its location in the heart of a city with internationally leading informal science institutions, such as the London Science Museum, the Natural History Museum, and the London Zoo. King's Science and Technology Education Group was also an international leader in environmental education, and already blended many aspects of learning—working with teachers to develop their capacity to use non-classroom resources in their teaching repertoires. Despite this leading work, at the time that CILS was founded, the group had few sustained contacts with the informal science institutions in London. CILS offered King's an opportunity to rethink some of its education programs and to expand its boundary-crossing work in environmental education to include the development of educators and educational strategies that encompassed a broader range of cultural resources.

Studying Connections

At the time that CILS began, many of the major London science museums were beginning to commit to significant educational research and evaluation efforts intended to study the reach and impact of their programs on the visiting public. These organizations were highly receptive to working with King's faculty and the CILS team of postdocs and doctoral students, including serving as sites for dissertation research. King's responded to this interest by setting up a reading group that included graduate and postgraduate students as well as informal educators. The reading group's quarterly meetings rotate around various institutions including King's, the British Museum, the Natural History Museum, and the Victoria and Albert Museum. Potentially interesting papers are circulated to around 40 group members and recent topics have included a consideration of outcome-based evaluation and possible alternative ways of thinking about or arriving at outcomes. What CILS has learned through these collaborative discussions is that education staff want to engage critically with a diverse range of literature and can evaluate the relevance and generalizability of studies from the point of view of professionals who have daily contact with museum visitors. The traditional "researcher/practitioner" divide has been blurred by the fact that many of the CILS doctoral and postdoctoral students have themselves worked in museums in the US and elsewhere. The success of the reading group is indicated by the fact that it continues to this day. Indeed, the number of participating organizations grows year by year. Museum staff are able to develop ownership of the reading group by taking turns to host the reading group and by making their own suggestions for reading. Sometimes discussions have focused on new exhibitions at the partner institutions such as the Darwin exhibition at the Natural History Museum.

CILS students and faculty at King's have also engaged in research projects such as the design of new galleries in the Science Museum and an evaluation of the public engagement offer of the Natural History Museum's Darwin Center. These projects have led to a cementing of existing networks between academic and museum professionals and a widening circle of involvement. The projects have enabled doctoral students to broaden their experience of research methods and contexts. Staff in departments other than the Department of Education and Professional Studies have taken part in some of these projects, in particular the Work, Interaction, and Technology Group and the Materials Library, further broadening the students' understanding of a range of methods and methodologies. Involvement with projects beyond their own studies has increased the pressure on the doctoral students who have been under pressure to finish their doctoral studies within 4 years. Another issue is that some of the studies have focused more on evaluation than on research, which again has distracted students from their core work.

Nevertheless, as a result of such institutional structures and relationships, CILS students at King's have developed a set of studies that look at issues that are of both practical and theoretical importance. Issues include designing teacher development programs based in museums (Wever-Frerichs, 2008); developing a framework for designing field trips that support a stronger integration of the field trip experience into classroom learning goals, both cognitive and affective (DeWitt, 2007); and understanding how museum exhibits afford opportunities for play and learning (Meisner, 2008). Taken together, these studies have found that although informal science learning settings hold great potential for sparking interest and engaging learning, there are many barriers that impede the use of informal resources to transform teaching and learning in the classroom. For example, in the study of teacher development programs based in the museum, although teachers have been inspired and have developed their inquiry-based teaching skills, limitations as to the extent to which there is time for inquiry in the classroom have contained the potential for the professional development programs to deeply affect classroom learning. A further outcome of the collaboration has been that King's has developed a strength in researching public engagement with science in a range of contexts—an area of growing interest and opportunity for funding.

Exploratorium: Building Bridges in a Museum Setting

The Exploratorium was founded in 1969 as one of the first interactive science museums in the world. Many of the earliest exhibits were museum-sized versions of tabletop classroom curricula developed in the 1960s science education reform movement, such as the Science Curriculum Improvement Study (SCIS) or Elementary Science Studies (ESS). From the beginning, the founder of the museum envisioned the museum as "an adjunctive institution," one that could both supplement and complement science that people had learned or were learning in schools. A core faculty of scientists and classroom educators came together to form two initiatives designed for classroom teachers: The School in the Exploratorium led elementary professional development workshops that provided teachers with firsthand experiences doing and learning science through tabletop inquiries integrating art and science to explore topics such as optics, sound, and electricity. Teachers were provided activity notebooks and curriculum kits to provide the same experiences for their students. The Teacher Institute provided middle and high school teachers with summer institutes and academic-year workshops designed to bolster their science content knowledge and advance their hands-on pedagogical strategies for the classroom. Over time, with significant funding from the National Science Foundation, both of these programs developed and evolved to offer new programs aimed at systemic support of the educational infrastructure. Today, the Institute for Inquiry provides professional development programs for elementary school staff developers from districts across the country. The Teacher Institute provides a two-year novice teacher induction program for teachers from throughout the Bay Area. Both programs have significant evidence of their impact.

Developing "School Sense"

Most informal science centers in the US, UK, and elsewhere offer programs, such as field trips to classroom teachers, that seek to bridge informal and formal settings. Some also develop curricular resources (whether in print or on the web), and some offer teacher programs, mostly designed to prepare teachers to make best use of the field trips. However, a growing number of informal science institutions and educators seek to play a more significant and sustained role in supporting high quality and engaging science in the classroom. These informal educators seek to create, develop, or expand teacher development programs that offer teachers opportunities for content learning, inquiry-based pedagogical strategies, classroom resources, and collegial communities of like-minded educators, thus building the capacity of the formal system to provide engaging, conceptually rich science to school-aged children. This effort is seen as critical to expanding public engagement with science.

In response to this interest and need, CILS created the Informal Learning Collaborative (ILC) program, a professional development program for informal educators who lead teacher development programs. Based at the Exploratorium, over the course of the last five years, the ILC has worked with more than 100 informal educators, representing about 60 institutions and communities around the US and in the UK. The central strategy of the program has been to build a community of informal educators who are conversant in the policy contexts of schools, the inquiry resources of informal science institutions, and the design of professional development programs.

Informal educators, from zoos, aquaria, botanical gardens, and science museums, have participated in the ILC for a period of three to four years, attending workshops every 6 months or so. The 5-day workshops have addressed topics such as Inquiry-Based Science, Theories of Learning, Educational Policies and Assessments, and Professional Development Design. Faculty from King's and UC Santa Cruz have been invited to address particular themes or topics, along with experts from within the Exploratorium and from other informal and formal institutions. ILC participants have also attended the annual Bay Area Institute where they have presented and participated in sessions designed for the broad CILS community of faculty, postdocs, doctoral students, informal educators, and approximately another 100 "friends of CILS" who represent policy agencies, universities, research groups, and informal settings. An independent evaluation, conducted by Inverness Research Associates, has found that participation in the ILC has led to an increase in both the quantity and quality of programs the participating institutions offered to classroom educators.

Bridging Settings through Theories of Learning

The ILC started with an understanding that learning involves not just know-ing, and not just knowing and doing, but rather knowing, doing, being, and becoming (see Herrenkohl and Mertl, 2010). All participants, who enrolled in the program in cohorts of about 30 educators, participated in a 5-day inquiry into light and shadows. This inquiry served as a touchstone for the program experience; it provided participants firsthand experience with the ways in which learners' interests, questions, fears, doubts, learning partners, and prior conceptual knowledge contributed to the ways in which they persisted in their inquiries to gain new insight into the science of light and shadow. Throughout the program, participants were able to reflect back on the shadow inquiry to consider the ways in which their interests and identities positioned them to engage with and in science. Interest and identity became key constructs when thinking about how to support learning across formal and informal settings and boundaries.

The second workshop focused on learning theory. An informal "fireside chat" with Barbara Rogoff, CILS faculty at UCSC, and George Hein, Profes-sor Emeritus at Lesley University, reviewed theories of learning from Dewey to Vygotsky to Bruner and beyond. Again, the goal was to move beyond constructivist ideas of knowing and doing, to include issues of being and becoming: to understand the learner in the processes of teaching and learn-ing, a learner who is constantly crossing borders dividing different institu-tional or formal and informal settings. Discussions focused on how theories of the learner and learning underpinned teaching approaches. Participants observed learners on the museum floor as well as in videos of classroom teaching from around the world, with a focus on beginning to articulate their own operating theories of learning and to consider how that shaped their program designs.

These theories of learning were brought to bear in examination of school policies in the third workshop and of professional development design in the fourth workshop. Both workshops attended to the ways in which informal programs could be designed to support teachers in today's culturally and economically diverse school classrooms.

In general, CILS and the ILC aim to bridge formal and informal learn-ing as well as research and practice. One early measure of success was that at the first Bay Area Institute (BAI), which gathered together all CILS participants from universities and informal science institutions, there was a mini revolt of the informal practitioners who felt that the language and questions of the academics were not applicable to the real questions that they faced on a daily basis. After two days of frustration, a number of the informal educators showed up on the last morning with a long scroll of questions with which they needed help. The divide felt very deep. But a year later, at the second BAI, as an external evaluator noted, the conversa-tions and presentations were so fluid that it was difficult to know who was a researcher and who was a practitioner. A common language was beginning to emerge (supported by the inquiry and learning theory workshops for the practitioners and by a cohort of new CILS graduate students who were keenly attuned to the needs of informal educators because many of them had been in that role only months before), and a shared vision about the purpose and need to bridge formal and informal learning was becoming established—despite the concrete differences in role, questions, and professional practices that exist between academic researchers and informal educators.

Developing a New Educator

Through the work in London, San Francisco, and also in Santa Cruz, faculty and students in CILS came to see the formal/informal dichotomy as highly problematic. Indeed, CILS participants became increasingly frustrated with the term "informal learning." The more time we spent looking at "informal learning," the more apparent it became that we were looking at "learning in informal environments" and trying to understand how it supported, expanded, or was different than "learning in formal [and other] environments." There was not some fundamental difference in the learning, or even the teaching, but rather in the organizational or institutional framework that affected goals, curricular resources, assessments, and perhaps outcomes.

This work through CILS and the institutional partnerships with the Exploratorium and local London informal science institutions has begun to impact the highly regarded environmental education program that has long existed at King's. King's recently recruited a Chinese PhD student who is studying the pedagogy of botanical garden educators and is a key partner in 'Inquire,' a newly funded European Commission project involving the development of inquiry-based learning in 11 botanical gardens in Europe and beyond.

At King's, preservice courses focus primarily on preparing teacher candidates to teach in high schools. Over the years, the course has broadened to reflect the increasing awareness of the outdoor classroom, which is a clear commitment in official UK educational policy documents. However, while government documents and research findings point to the benefits of learning outside the classroom, it would appear that such opportunities are rarely taken up in practice. Indeed, some research points to a decline in the provision and condition of outdoor learning. A study in London found that that there are relatively few planned opportunities for learning outside the classroom in science for students at middle and high school levels and that where such provision does occasionally occur, it tends to focus on particular areas of the science curriculum, such as Biology and Ecology.

Underpinning King's preservice course is a belief that all students deserve to benefit from a range of opportunities (not just in Biology and Ecology) and potentially gain the knowledge, skills, and experiences provided in out-of-school settings. However, we also acknowledge the challenges faced by teachers in providing such opportunities. Furthermore, we note that there are many possible reasons why resources beyond the classroom are not being used. For example, the issues of health and safety, risk management, and cost are amongst the most significant factors in limiting out-of-school learning. Reviewing the literature on learning outside the classroom, Rickinson et al. (2004) also highlighted teachers' confidence and expertise in teaching and learning outdoors; requirements of school and university curricula and timetables; difficulties due to shortages of time, resources, and support; and more generally the susceptibility of outdoor education to the wider changes in the education sector and beyond. Despite the challenges facing teachers, we believe that the learning opportunities afforded by contexts other than the classroom are such that the disparity in provision has serious implications for issues of equity. There is evidence that access to outdoor classroom opportunities—and the advantages that are thereby bestowed—are skewed towards the independent (private) school sector.

Based on our experiences in CILS, and in recognition of the issues raised above, we have increasingly focused on providing opportunities for preservice students to benefit from and to appreciate the opportunities provided for learning beyond the classroom. All science education teacher candidates must, during their one-year course, take part in a range of formal/informal activities including a visit to the Center of the Cell (London's only true science center). Supplementing the preservice course is a book published by the Open University Press, *Becoming a Teacher,* which includes two chapters written by CILS staff that focus on learning in and out of the classroom (see, Dillon and Maguire, 2007).

King's has continued to recruit a number of highly motivated and high achieving doctoral students partly as a result of its profile in the informal sector. Recently, King's and the Natural History Museum announced a PhD studentship to undertake a PhD in an aspect of public engagement at the Natural History Museum's new Darwin Center. This collaborative studentship is unusual in the field of education and indicates the strength of the partnership and the mutual value placed on developing the links between museum educators and university researchers.

CILS work with informal educators also continues to focus on connecting informal and formal settings and opportunities. Much of this work is focused on mapping the landscape of current collaborations and analyzing the ways in which they are configured and supported. Published reports include a field landscape study (Phillips et al., 2006), a white paper created for the Center for the Advancement of Informal Science Education (Bevan et al., 2010), and a study to document features of informal learning activities to understand how they differ, reinforce, and relate to science learning in other settings (described in the next section). CILS participants continue to organize and convene conference sessions in the US and the EU that focus on relating research and practice and learning across settings.

Sustaining the Momentum to Bridge across Institutional Fields

There is a well-known gap between research findings and their application in/to practice (Davis, 2007; Dolan, 2007). The gulf that divides education in formal and informal contexts is less well recognized. Because of the almost ubiquitous use of field trips, many think little of the issue; it seems to work well enough, at least to a point. Yet, more substantial partnerships—ones that fundamentally change the nature of the science curriculum such that informal settings and resources are fully integrated into school subjects, or ones that provide teachers or school systems with new infrastructure—usually fade away when seed grants expire.

One reason for this phenomenon, we believe, is the lack of evidence of the impact of such partnerships, especially the impact on core infrastructure at either schools or museums. Even when individual leaders have a strong vision, it is difficult for them to secure institutional resources without such evidence. But to develop such evidence, it is necessary to understand the ways in which the different affordances of both schools and informal settings come together to create new possibilities for learning, and the further development of students' (and teachers') science interests and identities.

In 2008, with funding from The Noyce Foundation and The Institute for Museum and Library Services, CILS began a project with 13 different informal science institutions and youth development programs to begin to document key design principles underpinning high-quality out-of-school-time (OST) science learning activities. The study involves analysis of some six dozen videos of children engaged in science learning activities across the participating sites. The study, which is still underway, is examining the design of the environment, activity, and facilitation of the learning activities. Not surprisingly, we are finding that features of the informal learning environment are markedly different from many typical classrooms. For example, high-quality informal environments are often designed to inspire and model ideas (through, for example, a strategic level of materials and "mess" that represents an archaeology of ideas, as well as the creation of physical thresholds within a space that allow for different modes and levels of engagement). They also support learner initiative and autonomy (through placement of and access to materials and tools) and allow for cross-pollination of ideas as well as collaboration (through the organization of space to allow for fluid sight lines).

Features of the informal science learning activities themselves include elements such as positioning science as a means to achieve a desired purpose, rather than as an end unto itself; the creation of multiple pathways to account for varying levels of prior knowledge and experience; the use of materials and phenomena that invite inquiry and exploration; the establishment of connections with relevant real-world problems or settings. These features share much in common with high-quality school science activities.

Interestingly, in our study of facilitation strategies we see almost no difference between high-quality teaching in formal and informal settings. Our analysis focuses on how informal educators spark, sustain, and develop student engagement in the activities. Strategies include modeling or engaging in parallel play to spark engagement; providing just-in-time tools or ideas to scaffold learners past frustrating or premature stopping points to sustain their participation; using analogs as well as reflection strategies to help deepen student participation. However, in the context of environments that support autonomy and cross-pollination, and different aged learners working on different projects and at different paces, facilitation may become more logistically complex even while it is less high-stakes.

As we begin to understand critical features that underpin different kinds of informal learning settings—from science museum to nature center to youth development programs held in school cafeterias—we can begin to analyze the ways in which these programs support the development of science understanding, interest, and identities through engaging learners in science concepts and processes. The value of efforts like these, in addition to providing knowledge for informal educators to strengthen their work, is to begin to develop models for how informal and formal educators and institutions can work together to support the developing interests and capacities of teachers and students to engage in science. Documenting the ways in which the different settings provide different opportunities, and how these opportunities relate to outcomes that are valued by both sets of institutional actors, is critical to sustain collaborations across settings. Otherwise, it is too easy for educators and institutions to maintain business as usual. Looking to the learner, as the person who in real life is constantly moving across different institutional fields or boundaries, and understanding how her or his interests, identities, and capacities are developed and relate across settings, is critical to the creation of the new educator in both formal and informal settings.

Conclusion

In exploring the terrain of bridging formal and informal settings, the CILS program seeks to develop a new breed of educational researcher and practitioner who approaches science education with broad perspectives on learning, and who seeks to design and support science learning by drawing on a variety of resources and settings, spanning multiple timeframes and institutional settings. The result of this effort has been a large number of informal educators, representing over 100 informal science institutions, who have not only a vision of but also a growing fluency with how to design programs that span formal and informal settings. For example, ILC graduates have designed teacher preparation programs in collaboration with local universities, have designed youth development programs spanning school coursework and summer field experiences, and have begun to work with state systemic efforts to support teacher development. Many of the doctoral students have gone on to positions of leadership within informal science institutions, prepared with deep study of learning research and theory, now leading educational and public programs that allow them to draw on broad conceptions of learning to more strategically position their institutions within the broader educational landscape. Many of the postdoctoral students have taken on academic positions where their research continues to examine the ways in which children draw on their cultural and community resources to engage in science learning in and out of the classroom.

Such views of science—as emerging from learning across settings and timeframes—are important and have long standing in the research literature. CILS adds an institutional overlay to this question: How can educational institutions and organizations organize and position themselves in ways to provide maximally engaging and effective science learning opportunities for children and their teachers? In efforts to improve science education for all children, it is essential that communities examine the full range of educational resources. Informal settings have been shown to be effective at exciting interest and curiosity. They provide students with views about how science is situated in the everyday world, including how science professionals engage with their communities. They thus bring resources to the sustained effort of schooling that can be critical in making science more appealing and addressing the dwindling levels of interest and participation that typically set in during the middle school years.

But to truly envision and enact such a program that spans resources, we need more educators who have firsthand experiences in seeing the power and thinking through the institutional constraints of such programs. Professional preparation programs, of both formal and informal educators, need to take more consistent approaches, drawing on the research as well as theories of learning, to develop understandings of the following:

- how science learning develops across time and settings;
- the ways in which different institutional settings provide particular learning affordances within a broader educational landscape that also includes the home and community resources that children access;
- how institutional settings can strategically connect, interweaving resources, expertise, and times of exposure;
- how different methods of assessment and evaluation can be used in different settings to support an understanding of progress towards shared goals for student or teacher learning.

We close with a recommendation, based on what we have learned so far through CILS, that leaders of training programs for both formal and informal educators need to develop more opportunities for conversations and concrete collaborative projects, that span the goals and expertise of both formal and informal educators. Taking such broad views of learning will help to strengthen science programs to make them more engaging and meaningful to learners, and thus ensure greater equity and access to science learning for children in schools.

Acknowledgment

This paper was supported in part through a National Science Foundation grant (ESI-0119787). Any opinions, findings, and conclusions or recommendations expressed in this material are those of the authors and do not necessarily reflect the views of the National Science Foundation.

References

Bevan, B., et al. (2010). *Making science matter: Collaborations between informal science education organizations and schools: A CAISE inquiry group report.* Washington D.C.: Center for Advancement of Informal Science Education (CAISE).

Davis, S. H. (2007). Bridging the gap between research and practice: What's good, what's bad, and how can one be sure? *The Phi Delta Kappan, 88*(8), 568–578.

DeWitt, J, (2007). Supporting teachers on science-focused school trips: Towards an integrated framework of theory and practice. http://cils.exploratorium.edu/cils/resource.php?resourceID=1273. Unpublished dissertation. London: King's College London.

Dillon, J., & Maguire, M. (eds.) (2007). *Becoming a teacher* (3rd ed.). Milton Keynes: Open University Press.

Dolan, E. L. (2007). Grappling with the literature of education research and practice. *CBE Life Sciences Education, 6*(4), 289–296.

Herrenkohl, L. R., & Mertl, V (2010). *How students come to be, know, and do: A case for a broad view of learning.* New York: Cambridge University Press.

Meisner, R. (2007). Encounters with exhibits: A study of children's activity at interactive exhibits in three museums. http://cils.exploratorium.edu/cils/resource.php?resourceID=1303. Unpublished dissertation. London: King's College London.

National Research Council. (1996). *National science education standards.* Washington, DC: National Academy Press.

Phillips, M., Finkelstein, D., & Wever-Frerichs, S. (2007). School site to museum floor: How informal science institutions work with schools. *International Journal of Science Education, 29*(12), 1489–1507.

Rickinson, M., Dillon, J., Teamey, K., Morris, M., Choi, M. Y., Sanders, D. and Benefield, P. (2004). *A review of research on outdoor learning,* Preston Montford, Shropshire: Field Studies Council.

Sjøberg, S., and Schreiner, C. (2006). How do students perceive science and technology? *Science in School, 1,* 66–69.

Tai, R. H., Liu, C. Q., Maltese, A. V., & Fan, X. (2006). Planning early for careers in science. *Science, 312*(5777), 1143–1144.

Wever-Frerichs, S. (2008). The role of museums in the on-going professional development of teachers. http://cils.exploratorium.edu/cils/resource.php?resourceID=1302. Unpublished dissertation. London: King's College London.

Section 2

On Methodological Issues

This section establishes the research paradigms within which my work has developed over time. Way back at the start of my academic career, a young Dutch environmental educator visited the UK for a conference organised by a small NGO, Caretakers of the Environment. Arjen Wals became a friend and colleague and we have written together a number of times as well as co-editing the *International Handbook of Research on Environmental Education* (Stevenson *et al.,* 2013). Our first serious collaboration, 'On the dangers of blurring methods, methodologies and ideologies in environmental education research', emerged from a joint session that we led at the Annual Conference of the National Association for Environmental Education (NAAEE). I've used this chapter a number of times with doctoral students and I often recommend to new researchers that, partly for career purposes, they should develop a line of scholarship that focuses on research methodologies and methods as well as developing their interest in particular topics and issues.

Another long-standing friend and colleague is Alan Reid, now at Monash University, Australia, but once at the University of Bath. Alan and I wrote about 'Issues in case study methodology in investigating environmental and sustainability issues in higher education: towards a problem-based approach?' This paper was published in *Environmental Education Research* which has become the leading journal in its field due to the tireless efforts of Chris Oulton and Bill Scott, the founding editors, and Alan himself, who took over as sole editor some years ago.

6 On the Dangers of Blurring Methods, Methodologies and Ideologies in Environmental Education Research

Justin Dillon and Arjen E. J. Wals

Dillon, J., & Wals, A. (2006). On the dangers of blurring methods, methodologies and ideologies in environmental education research. *Environmental Education Research*, 12(3/4), 549–558.

Introduction

When deciding what to cook, there are certain questions to be borne in mind including: who the meal is for; what food is available; and what cooking utensils and equipment one has. Other, more individual factors, include personal tastes, diet and philosophies. As a vegetarian, for example, Justin is not going to be cooking meat; Arjen who does occasionally eat meat, might. One of us likes spicy food, the other does not. Likewise, as people over the age of 40 'watching our weight', we both try to avoid consuming too many 'calories'. So, our end product, the meal, depends on a range of factors over which we have no control (what's available); a range of factors over which we have some control (depending on our personal choice); and a range of beliefs and preferences that reduce our choices. Usually research is more complicated but, in essence, this analogy serves to remind us that the final product—the substance of our conclusions—also depends on factors over which we have no control, factors over which we have some control, and a range of beliefs and preferences that reduce our choices in designing, conducting and reporting research. What people think of our meal/research conclusions depends largely on them and their preferences and persuasions: there's nothing intrinsic to either the cooking/research process that guarantees what someone will think about its end product in terms of quality, suitability, choices or compromises.

In this contribution, we would like to caution against blurring methods, methodologies and ideologies in research. We do this by drawing on two earlier articles in *Environmental Education Research* that focused on this issue as well but from quite different vantage points: Hart's (2000) paper in which he problematizes the generating of generic guidelines for designing and judging different strands of research, and Connell's (1997) paper in which she suggests that a plurality of methodologies and multi-paradigm research appear more fruitful, provided that the mixing is done with a thorough understanding of the social, political and philosophical contexts of the research.

From our own experience of reviewing published work (see Dillon *et al.*, 2003; Corcoran *et al.*, 2004; Rickinson *et al.*, 2004) and articles submitted to environmental education research journals, we note that many authors do not pay much attention to some aspects of research that some—including the authors of our two focus articles, Hart and Connell—value as essential. We shall discuss a series of confusions that seem particularly persistent, and conclude by calling for both a more cautious use of language, and a better examination and articulation of methodology in research. Such examination would need to include a discussion of the ontological, epistemological and axiological underpinnings of a methodology.

Methods and Methodology

> One of the most confusing developments in educational research over the past quarter-century has been the proliferation of epistemologies—beliefs about what counts as knowledge in the field of education, what is evidence of a claim and what counts as a warrant for that evidence.
>
> (Pallas, 2001, p. 6)

We start with a recurring confusion present in the mixing up of methods and methodology. Methodological considerations involve examining positionings and tensions in research ontologies, epistemologies and axiologies. Ontology looks at what we're dealing with (the *what*)—the nature of reality—are we 'researching', for instance, people's knowledge, attitudes, the words people use, the number of books in a school library, or how long people spend looking at an exhibit? Epistemology refers to how we make knowledge (the *how*)—for example, do we look for patterns and themes in what people say in answer to our questions, do we give people tests, or do we watch what people do and infer their thoughts from their actions? Axiology relates to ethical considerations and our own philosophical viewpoints (the *why*)—such as, do we take a positivistic stance, use feminist epistemologies, involve participants as researchers?

Less-experienced researchers might be more open to consider and reconsider their methodological approaches than more established academics. As Pallas (2001) has noted, this issue is particularly crucial to less-experienced researchers, given that as participants in a complex field, we are all faced with an increasing diversity of positions and philosophies in one's reading and at conferences, and while we are invited to engage them in our work situations and networks through ongoing debate and dialogue, we must all demonstrate sound reason and judgement in choosing which to attend to.

If environmental education researchers want to develop a community of reflective scholars, we need to look more closely at the whole range of available epistemologies, ontologies and axiologies, and appreciate them not in some relativistic sense because they are there, but because they offer different perspectives. By appreciating them all, we do not mean using them all. To continue the food analogy, when at an 'all you can eat' buffet we need to show some restraint and combine with some care in order to have a decent meal. One of Hart's (2000) points, with which we wholeheartedly agree, is that the unwitting or thoughtless mixing of methodologies can be an indicator of a poor understanding of what research entails.

Part of the confusion over methodologies and methods has arisen because of commonly articulated misunderstandings in the literature around the use of the terms qualitative and quantitative. We regard methods as the tools or instruments used for data *generation* or *collection* (for example, questionnaires, audio-taped interviews, focus groups or texts from chatroom exchanges), or data *analysis* (frequency counts, thematic coding, inferences, and so forth). Note here that we are separating methods of production from the processes to which data are subjected in methods of analysis. Examples of this confusion can be found in published environmental education research papers. Chatzifotiou, for instance, after introducing the semi-structured interview as her key *methodology*, writes 'the *overarching method* was *qualitative* because every teacher possesses a distinct idea of what environmental education is and what practice entails' (Chatzifotiou, 2005, p. 507; our emphasis). These are not research paradigms or approaches, although it is common to find the phrases 'qualitative/quantitative research/method' in papers. Yet, a questionnaire can include open-ended and closed questions—the issue of whether it is qualitative or quantitative is meaningless. If we choose to count the number of people who choose 'very strongly' as a response to a question then we are analysing the data in a quantitative manner; if we code people's responses to open-ended questions then we are analysing the data in a qualitative way. But we could alternatively count the words in responses to open-ended questions and present the data in a quantitative manner. So, the notion that questionnaires are exclusively a quantitative research method represents a misunderstanding of basic ideas, while embracing a qualitative—quantitative binary is not very helpful in improving that understanding.

Methodology and Ideology

Perhaps it is potentially more fruitful to move to the level of methodologies and their corresponding ideological partners. Returning to the papers of Connell and Hart, Connell's purpose was 'to contribute to methodological discourse about research approaches to environmental education . . . as a way of opening opportunities for diverse pathways of research in environmental education' (Connell, 1997, p. 117). Hart's stated purpose was to 'overcome some confusion about qualitative studies within environmental education research' (Hart, 2000, p. 37). Both authors have much in common but see the world through different lenses. At one level, they both observe a distinctive shift in educational research and a growing acceptance of alternative methodological approaches within it. However, the subtle differences in wording between the two authors seem to indicate significantly different philosophical stances. Whereas Hart sees research itself as shifting, Connell sees a new 'alternative' approach to research coming into existence.

Either way, it is clear from looking back over 10 years of *Environmental Education Research* that the field continues to be a broad though sometimes divided 'church' with its own complement of agnostics and atheists. Following some groundbreaking debate initiated in the early 1990s (Mrazek, 1993; Robottom & Hart, 1993), many environmental education scholars have become familiar with the existence of a range of different views about doing research. The emergence, since then, of new research journals that are more open to different approaches to, and ways of thinking about, research than was the case with existing outlets (for example, until then the *Journal of Environmental Education*), illustrates the shifts observed by both Hart and Connell. Similarly, at a recent pre-conference research symposium at the 2005 annual conference of the North American Association for Environmental Education, it was evident that more and more people have come to appreciate the important differences in perspectives between the many research approaches that we now find.

One value in republishing these two focus articles is to illustrate that the arguments that the writers made are still valid. Hart argues that the debates about methodologies have helped researchers to identify and state their positions and assumptions, which, in turn, has helped other researchers. Hart's underlying premise is that researchers should 'appreciate the independent and interrelated roles of ontology, epistemology and methodology in rendering more transparent these issues' (Hart, 2000, p. 38). We agree with Hart's deprecation of the 'academic sectarianism . . . as reflected in superficial dialogues about method and simplistic debates about essentially related schools of thought' (p. 38), and his objection to reducing (i.e., non-positivistic) research to a single mode of inquiry. For him:

> . . . the range of qualitative approaches, from ethnography and phenomenology, through various participatory forms of inquiry as well as feminist and postmodern approaches, is too complex and varied in philosophy and history to be represented as a unified or even a coherent field.
>
> (Hart, 2000, p. 38)

We broadly support his conclusions, but not his notion that ethnography and phenomenology can be adequately termed 'qualitative approaches'. They may well be, but do not *have* to be.

Robottom and Hart (1993) have also argued that different ontologies and associated epistemologies have very definite implications for methodologies and methods which must not be ignored by field researchers if high quality research is to be carried out. The point, for Hart, is:

> . . . not to accommodate or reconcile multiple paradigms of educational thought but to recognize them as unique, historically situated forms of insight which require reconciliation not at the paradigm level but at the level of meta-paradigms—for example, whether people can agree on the relationship of education to the goals and ideals of democracy and social justice.
>
> (Hart, 2000, p. 38)

Hart refers to the paradigmatic and ideological underpinnings of research which require critical reflection and explication as they have serious consequences for the methodologies available and the ability to blend methodologies. One could argue that it is the questions posed that determine the methodology and that it should not be the other way around. But, the kinds of questions we ask, the purpose for asking them in the first place, how we ask them, to whom we ask them (and whom we exclude), how we value people's responses, how we relate to those who partake in a study, who is to benefit from the study, and so on, are worldview-laden. In Table 6.1 we present three purposefully simplified representations of approaches to environmental education research to illustrate three different ways of conceptualising research. There are, of course, other ways of doing this but we present these three to show the need for articulating purpose, roles, relationship with those who are part of the research, and, more generally, assumptions about the role of science in society.

Table 6.1 Three simplified representations of environmental education research

	Research as evidence	*Research as co-learning*	*Research as activism*
Modus of understanding	Empirical analytical	Hermeneutic–interpretive Holistic–descriptive	Socially-critical
Locus of impact	Universal	Trans-contextual	Contextual-transformative
Key research competencies	Good tester, designer and modeller	Good listener, interpreter and storyteller	Good ally, critical friend, advocate
Main researcher modes	Passively-detached Neutral expert objective	Actively-detached Passively-engaged Explicitly-biased	Actively-committed Explicitly-partisan
Role of the researched	Source of data	Active informant Co-learner	Change agent Co-learner
Desired outcomes include	Explanatory models, Tests of hypotheses Definitive answers	Improved understandings Thick descriptions Increased (self) awareness	Transformation (Systemic) change

Pluralism in Educational Research: Some Considerations

Both authors of the focus articles address issues pertinent to the purpose of research and to ways in which researchers engage with problems. Whereas Hart regretted the 'academic sectarianism' reflected in 'superficial dialogues about method and simplistic debates' (p. 38), Connell objects to Robottom and Hart's 'antagonistic attitude' (Connell, 1997, p. 118) towards an empirical–analytical methodology. Connell summarises three arguments that she says characterise the common descriptions of the empirical–analytical methodology and the associated post-positivist ideology. Having set out the arguments, Connell then asks if they are justified. Just as Hart points to the different paradigms within, in his terms, qualitative research, one of Connell's key points is that there are two paradigms within the empirical–analytical methodology— traditional positivist and post-positivist. Although it can be argued that even though post-positivist approaches are more aware of bias, subjectivity, assumptions with regards to the nature and value-ladenness of scientific knowledge, they still embrace traditional empirical–analytical values and assumptions (Phillips & Burbules, 2000). In the light of Hart's appeal, above, that researchers value ontology, epistemology and methodology, it is worth noting that Connell uses ontology, epistemology and methodology as lenses to examine the different paradigms. But this is clearly a problem if ontology and epistemology are both key aspects *of* methodology.

What both Hart and Connell do particularly well is to argue that different research paradigms deserve respect and need to be better understood by more members of the research community than currently seems the case. But whereas Hart argues that we are 'between stories' in educational research and that, for the present, 'we are left to cobble together our stories to help us and others understand how and what we did . . .' (2000, p. 44), Connell, presciently, notes with affirmation the call by other researchers for an 'associated community of researchers . . . where not all researchers do all kinds of research but all do what they do well . . . ' (1997, p. 30).

Connell concludes her piece by suggesting that the emerging binaries in the literature may 'distract researchers from considering more enlightening complementary and cooperative dialogues about research methodologies in environmental education' (p. 129). However, a look at the contents of *Environmental Education Research* from 1998–2005, in particular, its special issues, would indicate that her understandable worry was unfounded. Connell also points out the growing number of researchers promoting multi-paradigm research, a trend that seems common across social science research. Our question is whether this can be done when the ideological underpinnings of various research methodologies are so different.

Writing more recently on a related topic, Johnson and Onwuegbuzie seem to have no trouble blending methodologies which, incidentally, they refer to as methods. They argue that 'mixed methods' research is a 'research paradigm whose time has come' (2004, p. 14), pointing out that the qualitative/quantitative debate has being going on for more than a century ('ardent dispute' is how they describe it). Arguing against the purists on both sides of the argument who advocate the 'incompatibility thesis' between the qualitative and quantitative (see Howe, 1988), Johnson and Onwuegbuzie claim that what they recognise as the 'qualitative and quantitative research paradigms, including their associated methods' (*ibid.*) can and should be mixed. In this way, the authors posit mixed methods research as a third paradigm. Johnson and Onwuegbuzie contend that 'researchers and research methodologists need to be asking when each research methodology is most helpful and when and how they should be mixed or combined in their research studies' (2004, p. 15). They argue that 'many (or most?) qualitative and quantitative researchers (that is, post-positivists) have now reached basic agreement on several major points of earlier philosophical disagreement such as the theory-laden perception or the theory-ladenness of facts' and 'the social nature of the research enterprise' (2004, p. 16).

At this point, we note that Martyn Hammersley provides a useful perspective on divergence and convergence in social science research methodology. He asks whether the amount of pluralism in educational research ought to be applauded or whether it poses barriers to progress in educational research itself and to its playing a worthwhile role in policy and practice (Hammersley, 2004). He distinguishes two camps in answering the question. One side, which in all likelihood includes many of the mixed methods proponents, argues that there is a need for methodological consensus building in order to be able to construct a cumulative body of knowledge that can more effectively inform policy and practice. Others argue that attempts to do so will lead to a reimposition of a positivist paradigm, and that this will undercut much valuable research (Hammersley, 2004). Hammersley suggests that the 'evidence-based' trends in health research but also in educational research embody such a reimposition. In answering the question he finds himself somewhere in the middle:

> I don't think that any simple bringing-together of all the various methodological trends that have developed, can be engineered, or even brought about through enlightenment. Indeed, some of these trends seem to me so obviously antithetical to the very nature of research that they cannot be tolerated. . . . Nevertheless, there is a need for more effort on the part of social and educational researchers to build bridges between different kinds of work and to see how they might be usefully combined or integrated. . . . [However] complete consensus would not be a good thing; some difference and diversity is always necessary in order to stimulate further development.
>
> (Hammersley, 2004, pp. 8–9)

Johnson and Onwuegbuzie argue that 'mixed methods research should . . . use a method and philosophy that attempt to fit together the insights provided by qualitative and quantitative research into a workable solution' (p. 16). They go on to advocate the use of pragmatism as the philosophical partner of mixed methods research. By this they mean that the answer to the question 'how should research approaches be mixed?' is, pragmatically, 'in ways that offer the best opportunities for answering important research questions'. This answer begs the question of who decides what the 'best opportunities' are? Mixed methods, they argue, is 'a movement that moves past the paradigm wars by offering a logical and practical alternative'. The authors argue that what is 'most fundamental is the research question' and that 'research methods should follow research questions in a way that offers the best chance to obtain useful answers' (pp. 17–18). While, in his classic book, *Against methods*, Paul Feyerabend takes this even further by suggesting that science should be an anarchistic enterprise and that theoretical anarchism is more humanitarian and more likely to encourage progress than its law-and-order alternatives (Feyerabend, 1975).

Paul Hart is not arguing for anarchy or even the liberal blending of methods and/or methodologies tailored to each specific research question. Instead he stresses the need for a 'well-articulated methodological strategy':

> Qualitative researchers need more than a methods toolkit; they need a well-articulated methodological strategy, capable of arguing epistemological and ontological positions and responding within a particular system of inquiry.
>
> (Hart, 2000, p. 42)

What is particularly interesting to note here is that Hart recognises that researchers operate within particular systems of inquiry. He suggests that researchers articulate and argue their strategies and positions from *within* their specific vantage point. We agree but at the same time note that many researchers don't, or are perhaps not too worried about this issue. Having looked closely at the nine research articles published in the last available issue of *Environmental Education Research* at the time of writing this contribution (Volume 11(5)), we note that there a wide range of methodologies and methods used by the various researchers. The volume illustrates that the landscape of research methods and methodologies employed in environmental education has become far more diverse. Nonetheless, some authors do not articulate their methodology. In fact, none of the authors who could be positioned in a more empirical–analytical end of the spectrum do this; rather they limit themselves to presenting their methods (Table 6.2).

A good discussion of methodology requires the articulating and arguing of one's vantage point, the recognition of other vantage points, and clarifying

Table 6.2 Overview of self-reported methodologies and methods used in EER 11(5)

Paper	Methodology	Methods
Jurin and Hutchinson Chatzifotiou	Grounded-theory *Not specified*	Ecological autobiographies Semi-structured interview questions 'qualitative'
Mueller-Worster and Abrams	Phenomenology and grounded theory	Focused life-history interviews Details of experience interviews Multiple observations Informal interviews Document analysis
Moore	Feminist epistemology Participatory action research	Active participation Interviews Document analysis Observation Critical friend
Ekborg	*Not specified*	Questionnaires Semi-structured interviews
Tal	*Not specified*	Pre- and post short answer questionnaires Analysis of students' work Semi-structured interviews
Brody	Learning in nature theory	Case study Content analysis (of notes, writings and drawings)
Summers, Childs and Corney	*Not specified*	Questionnaires (partly open, partly closed)
Banks, Elser and Saltz	*Not specified*	Pre- and post-surveys (Likert-scale) *Post hoc* survey (Likert-scale) Open questions

why one vantage point was preferred to others. Admittedly, this will be no easy task for, as Hart states:

> . . . the range of qualitative approaches, from ethnography and phenomenology, through various participatory forms of inquiry as well as feminist and postmodern approaches, is too complex and varied in philosophy and history to be represented as a unified or even a coherent field.
>
> (Hart, 2000, p. 38)

As noted earlier we are uncomfortable with the qualitative–quantitative divide, but we imagine the same points can be made at the quantitative end of the spectrum.

Like Hart, Connell too seems to have problems with the pragmatic, needs-based, contingency approach, as advocated by Johnson and Onwuegbuzie. Although happy with the idea of 'methodologies . . . selected to meet clearly identified research needs' (1997, p. 130), she sees those needs as being 'balanced with a clear understanding of the social, political and philosophical contexts in which the techniques are located'.

Looking Ahead

So, are we any nearer to overcoming some of the 'confusion about qualitative studies within environmental education research' (Hart, 2000, p. 37)? *Does* mixed methods research, for example, offer us a new methodological paradigm—a third way to do research? The answer to these questions, we suspect, is 'it depends'. As Pallas (2001, p. 7) argues, meaning 'arises from two complementary purposes', participation and reification. In essence, participation is the process that communities (social, cultural or geographical) use to share and develop ideas and meaning (see also Lave & Wenger, 1991). Reification is the social consolidation of that meaning. So, in an academic research group, participation takes place during presentations, seminar discussions, shared reading, notifying each other of useful references or online resources. Reification occurs through the production of papers, websites and posters, and so forth, which are engaged with through participation.

If educational researchers cannot understand and engage with one another, both within and across at least some educational communities, progress in research and its potential contribution to policy and practice will be hampered. To prevent a recurring pattern of epistemological single-mindedness, educational researchers will need to engage with multiple epistemological perspectives to the point that members of different communities of educational research practice can understand one another, despite, or perhaps through, their differences (Pallas, 2001, p. 7).

In order to move forward we need to start seeing ourselves as members of a community of reflective scholars, and not just as aggregates of individuals, or as competing camps, or as a pluralistic field of multiple unconnected research paradigms without common interests. In order to move in that direction, we need to consider the view that environmental education research, as a member of the broad family of educational research:

> . . . [has] empirical, interpretive, and normative dimensions; that there now are and can in the future be developed public and rational ways to make warranted knowledge claims in each of these dimensions; . . . and that we should open-mindedly presume the goodwill and intelligence of each other as researchers until proven otherwise.
>
> (Soltis, 2004, p. 9)

Soltis makes another important point we wholeheartedly support which is that open-mindedness is not the same as empty mindedness. It is not the same as tolerance of all views, but rather it requires a sincere attempt to consider the value of other views and their underlying claims. It does not release us from our duty as critical scholars to exercise judgment.

So, the challenge for new researchers (and for those who advise, supervise and support them) in doing research, is to appreciate the value of a broader range of methodologies or at least not dismiss some out of hand. At the same time, however, we argue that researchers will need to resist being drawn into mixing methodologies however pragmatically useful or convenient this may seem. Whatever methodologies are chosen, researchers will need to consider the ontological, epistemological and axiological ramifications of such an approach. At the same time inquiry should be driven by questions, not by preferred methods or even methodologies.

References

Connell, S. (1997) Empirical-analytical methodological research in environmental education: response to a negative trend in methodological and ideological discussions, *Environmental Education Research,* 3(2), 117–133.

Corcoran, P. B., Walker, K. E. & Wals, A. E. J. (2004) Case studies, make-your-case studies, and case stories: A critique of case-study methodology in sustainability in higher education, *Environmental Education Research,* 10(1), 7–21.

Chatzifotiou, A. (2005) Implementing environmental education in the primary schools of northern Greece, *Environmental Education Research,* 11(5), 504–523.

Dillon, J., Rickinson, M., Sanders, D., Teamey, K. & Benefield, P. (2003) *Improving the understanding of food, farming and land management amongst school-age children: a literature review. Research report 422* (London, Department for Education and Skills).

Feyerabend, P. (1975) *Against method* (Atlantic Highlands, NJ, Humanities Press).

Hammersley, M. (2004) Social science research methodology today: diversity or anarchy? Available online at: www.arts.monash.edu.au/lcl/newmedia_in_langlearn/assests/sources_hammersleypaper.pdf (accessed 26 January 2006).

Hart, P. (2000) Requisite variety: the problem with generic guidelines for diverse genres of inquiry, *Environmental Education Research,* 6(1), 37–46.

Howe, K. R. (1988) Against the quantitative-qualitative incompatibility theses or dogmas die hard, *Educational Researcher,* 17(8), 10–16.

Johnson, R. B. & Onwuegbuzie, A. J. (2004) Mixed methods research: a research paradigm whose time has come, *Educational Researcher,* 33(7), 14–26.

Lave, J. & Wenger, E. (1991) *Situated learning: legitimate peripheral participation* (Cambridge, Cambridge University Press).

Mrazek, R. (1993) Alternative paradigms in environmental education research, in: R. Mrazek (Ed.) *Monographs in environmental education and environmental studies, VIII* (Troy, OH, North American Association for Environmental Education).

Pallas, A. M. (2001) Preparing education doctoral students for epistemological diversity, *Educational Researcher,* 30(5), 6–11.

Phillips, D. C. & Burbules, N. C. (2000) *Postpositivism and educational research* (New York, Rowman & Littlefield).

Rickinson, M., Dillon, J., Teamey, K., Morris, M., Choi, M.Y., Sanders, D. & Benefield, P. (2004) *A review of research on outdoor learning* (Preston Montford, Field Studies Council).

Robottom, I. & Hart, P. (1993) *Research in environmental education: engaging the debate* (Geelong, Deakin University Press).

Soltis, J. F. (1984) On the nature of educational research, *Educational Researcher,* 13(10), 5–10.

7 Issues in Case-Study Methodology in Investigating Environmental and Sustainability Issues in Higher Education

Towards a Problem-Based Approach?

Justin Dillon and Alan Reid

Dillon, J., & Reid, A. (2004). Issues in case study methodology in investigating environmental and sustainability issues in higher education: towards a problem-based approach? *Environmental Education Research*, 10(1), 23–37.

Introduction

'Case studies' are frequently used in environmental education to illustrate applications of theories in practice as well as being used as potentially generalizable exemplars in research reports and advisory texts for teachers. They serve several functions including exhortation, valorization and exemplification. In the research methods curriculum, the term case study has been conceptualized in radically different ways—as methodology, method, approach, and as being none of these things, simply a limiting case of research where n = 1 (Wellington, 2000; Bassey, 2003).

In this paper we critically examine the use of 'case study' as a term within the context of environmental education research. We do this partly as a response to the paper by Corcoran *et al.* (2002), which claims, inter alia, that 'Case study methodology . . . is the ideal research tool to investigate sustainability in higher education' (p. 4). We begin by exploring what others have written about case study and then state our own position, before critically examining Corcoran *et al.*'s call for 'a framework for critical case study research in sustainability in higher education'. Finally, we discuss the use of problem-based methodology as a conceptual tool that might more adequately serve the problem-oriented function currently assigned to observers and participants in critical case-study methodology.

Cases, Case Studies and Confusion

Brown and Dowling (1998), in *Doing research/reading research,* point out that 'professional educational practice and academic educational research . . . are distinct fields of activity' (p. 165). In recognizing that the two fields stand in dialogic relation to each other in what is 'potentially a productive relationship' (p. 165), the authors note that 'failure to recognise the distinctive natures of the two fields will result in the one being unduly subordinated to the principles of the other' (p. 165). Later we will refer back to this argument as we critically examine the ideas put forward by Corcoran *et al.* (2002).

Building their case, Brown and Dowling argue that 'failure to regard research as a distinctive activity [will lead to] the plundering of research for the techniques which will facilitate the genesis of the all-singing/all dancing practitioner-researcher' (1998, p. 165). They warn that higher education curricula and research methods textbooks may end up comprising 'a chapter or seminar on surveys, one on case studies, another on interviews, another on participant observation' (p. 165). 'One of the dangers,' they continue, 'is that an unprincipled organization of the diversity of methodological approaches to research may lead to the fetishizing of methods. Case study . . . is, indeed, a case in point' (p. 165).

Brown and Dowling draw on a quote from Bob Stake that they are unable to verify but which seems authentic (Stake, in Brown & Dowling, 1998, p. 166):

> Case studies are special because they have a different focus. The case study focus is on a single actor, a single institution, a single enterprise, maybe a classroom, usually under natural conditions so as to understand it—that bounded system in its natural habitat.

This perspective they dismiss as:

> a mythologizing and a romanticizing of the world in general. The 'natural' world is presented as thinkable in terms of a collection of mutually independent (bounded) systems which are nevertheless transparently knowable to us.
>
> (p. 166)

The selective quoting does a disservice to Stake, or at least it does if his more recent writing is taken into account, as we will show later. Nevertheless, their reading is partly supported in that one of Stake's more recent general writings (Stake, 1995) is a chapter entitled 'Case studies' in the second edition of Denzin and Lincoln's *Handbook of qualitative research* (Stake, 2000). However, also in that chapter, Stake writes that '[c]ase study is not a methodological choice, but a choice of subject to be studied' (p. 236). At the heart of case-study research, then, is the question *what is this a case of?* Stake continues, 'As a form of research, case study is defined by interest in individual cases, not by the methods of inquiry used' (p. 236). This to us is a firm rebuttal of some of the insinuations made by Brown and Darling. Their other criticisms are addressed, at least partly, by Stake's explanation that:

It is common to recognize that certain features are within the system, within the boundaries of the case, and other features outside. Some are significant as context.

(pp. 236–237)

Stake asks what can be learned from the single case. This, to us, is a critical question and one that is fundamental to our criticism of the points made by Corcoran *et al.* (2002) below. Stake quotes Louis Smith in arguing that the case is a 'bounded system' (Smith, 1978) but at the time that Stake was writing, 'The concept of case remain[ed] subject to debate and the term study [was] ambiguous (Kemmis, 1980)' (Stake, 2000, p. 237). For Stake, a case study is both the process of learning about the case and the product of our learning (p. 237), each one entailing a respectful attention to context, boundary-drawing (i.e. conceptualization and definition of the case), and the concerns of stakeholders. Stake also points out that Lawrence Stenhouse advocated calling the product of the learning a 'case record' (Stenhouse, 1984) but that the term 'case study' has become widely accepted. (While moving beyond the scope of Stake's writing, the significance of case records and case reports in making possible the public evaluation of the selections, presentations and interpretations of case data in case studies had occupied Stenhouse for some time by then [e.g. 1978, 1980]. For example, in a presidential address to the British Educational Research Association [1980, pp. 4–5] he argued:

> The problem of field research in case study is to gather evidence in such a way as to make it accessible to subsequent critical assessment, to internal and external criticism and triangulation . . . It would be useful to our field . . . to assess critically the process of case study in respect of its collection of evidence or data, its problems of verification and cumulation and its responsibility to address educators.)

Leaving these matters of evidentiary warrant to one side for the moment, we would agree with Stake insofar that the term 'case study' is widely used and that there is debate about the concept of case. We would also agree that case study is not a methodological choice but, as we will explain later, we are cautious about Stake's claim that case study can be conceptualized as a form of research. Indeed it has become (and maybe always was) an inadequate and confusing term, as we now show.

Types of Case Study—Stake's View

Stake described three 'types of study' (2000, p. 237): *intrinsic, instrumental* and *collective* case studies. Stake places studies into the categories based on his interpretation of the purpose and design of the attending research activity (cf. Bassey, 1999; Wellington, 2000 for other typologies).

An Intrinsic Case Study

> is undertaken because one wants better understanding of this particular case. It is not undertaken primarily because the case represents other cases or because it illustrates a particular trait or problem, but because, in all its particularity and ordinariness, this case itself is of interest.
>
> (p. 237)

(Arguably, readers of this type of case study should be able to identify the warrant and provenance of claims to identifying and investigating the unique or typical qualities of the case study.)

An *instrumental case study* involves studying a case 'to provide insight into an issue or refinement of theory' (p. 237). This is a familiar function: a case study of the particular is drawn upon to address and/or elucidate more concisely some essential, underlying principle, issue or point the author seeks to highlight, even to the extent that general features or lessons may be drawn. It is about deepening understanding and knowledge of the issue at focus. Stake explains that:

> The case is of secondary interest: it plays a supportive role, facilitating our understanding of something else . . . The case may be seen as typical of other cases or not . . . there is no line distinguishing intrinsic case study from instrumental; rather, a zone of combined purpose separates them.
>
> (p. 237)

The final type of study, *collective case study,* occurs when 'researchers . . . study a number of cases jointly in order to inquire into the phenomenon, population, or general condition'. Stake qualifies this further by explaining that it is not the study of a collective but an instrumental study extended to several cases, e.g. an issue is studied in several situations. The cases, he argues, 'are chosen because it is believed that understanding them will lead to better understanding, perhaps better theorizing, about a still larger collection of cases' (p. 237). Elsewhere, Stake (1983) links this exploration of similarities and differences amongst cases to *naturalistic generalization,* that which might provide illumination by engaging the vicarious experience and tacit knowledge of the case reader in relating to the case study. While Eisner (1991) likens this to the *thematics of connoisseurship:* the concrete universals of the case are evaluated, the lessons are learned, the moral of the story is disclosed and, as such, are of likely interest to others outside the case study.

We would reiterate at this point Brown and Dowling's claim that 'professional educational practice and academic educational research . . . are distinct fields of activity' and that the two fields 'stand in dialogic relation to each other' (1998, p. 165). Nothing that Stake posits seems to challenge that claim; although we do note that Robinson, in advocating a problem-based methodology, is highly critical of such distinctions being made on artificial or academic grounds, matters we will return to towards the end of this paper.

Case Studies in Corcoran et al.?

In the context of environmental education research, particularly with respect to environmental and sustainability issues in higher education—the subject of the Corcoran *et al.* paper (2002)—one might illustrate the three categories by reference to three imaginary pieces of research:

> *The intrinsic case study.* A study of the implementation of a sustainable energy policy in a university—*undertaken because it is intrinsically interesting.*
>
> *The instrumental case study.* A study of the implementation of a sustainable energy policy in a university—*undertaken because it is might enable the researchers to examine particular data collection strategies.*
>
> *The collective case study.* A study of the implementation of a sustainable energy policy in a university—*undertaken because it might lead to better understanding about a larger number of universities.*

It is worth noting Stake's caveat that 'authors and reports seldom fit neatly into such categories, and I see these three as heuristic more than functional' (2000, p. 238). Stake also commented that there are other kinds of cases study. As with the drift in our comments on the instrumental category, particularly when questions as to application of a case study to other settings are in mind, they include the 'teaching case study . . . used to illustrate a point, a condition, a category, something important for instruction (Kennedy, 1979)' (p. 238). Stake's point in identifying three types of case study was to 'emphasize variation in the concern for, and methodological orientation in, the case' (p. 238).

The Use of Case Studies

It is the use to which case studies are put that is the main focus of this section of the paper. Stake recognizes that 'the end result [of a case study] regularly presents something unique (Stouffer, 1941)' (2000, p. 238). The uniqueness, he argues 'is likely to be pervasive, extending to:

1 the nature of the case;
2 its historical background;
3 the physical setting;
4 other contexts, including economic, political, legal, and aesthetic;
5 other cases through which this case is recognized;
6 those informants through whom the case can be known.'

Stake notes that the uniqueness 'is not universally loved', and that, 'case study methodology has suffered somewhat because it has sometimes been presented by people who have a lesser regard for study of the particular' (p. 238). 'Many social scientists,' he continues, 'have emphasized case study as typification of other cases, as exploration leading up to generalization-producing studies, or as an early step in theory building' (p. 238).

Stake goes on to discuss how learning occurs from a case study. But there is little in his explanation that appears to be different from learning from other genres, whether they are ethnographies, group interviews or questionnaire surveys. In all of these methods, the researcher has made a selection in terms of what data to collect, how to interpret the data, and how to present the data to the reader. Generally, the way in which data are collected is more explicit in 'questionnaire research' than in 'case-study research', but the fundamental principle is the same.

In focusing on the purpose of the researcher, and on the choice of tactics, there is a danger that we ignore the key role that the reader plays in turning the research report into new knowledge. In this respect case studies are no different from any other method of data collection. But case studies are often selected and used by third parties for purposes that might not be those that the researcher intended. One might ask, given the unique nature of the case study, 'Why do people bother to read them?' We believe the answer is, often, because those who publish case studies overtly or covertly intend them to be used to indicate more generalizability than Stake suggests is possible. Whereas Stake, in a Popperian way, suggests that cases can be very effective in falsifying theories, case studies may be used by the lazy, the intellectually bankrupt, and the morally corrupt to support strategies, theories and policies that are, by any other tests, ineffective, redundant or still unproven and inadequately tested (see also Swann, 2003).

The Significance of a Case: Empty, Underdetermined or Indeterminate . . . ?

Although Stake's typography is interesting and his caution about generalization well taken, the question has to be asked, is there anything particularly special about only studying one case? Since case-study research can involve quantitative as well as qualitative data collection, and given that no system in life is completely independent of its context, do we actually need the term case study?

Taking a cue from Galileo, one could begin to argue that we need a term for the study of two cases, three cases, and so on. We could have 'Twin Case Study' research or 'Bi-case study', clusters of cases or 'case-study family' research. In such circumstances, we propose that authors, editors and reviewers reject the term as representing a form of research; rather, we advocate that it is a long-winded way of indicating that the level of generalizability of the findings is such that readers without an intrinsic interest should skip the article.

A Critique of Corcoran *et al.* (2002)

In 'Case studies, make-your-case studies, and case stories: a critique of case study methodology in sustainability in higher education', Corcoran *et al.*

note that 'a range of case studies have emerged' (2002, p. 2) as higher education institutions have begun to implement sustainable policies and practices.

Corcoran *et al.*'s view is that 'Case study methodology . . . is the ideal research tool to investigate sustainability in higher education' (p. 4), arguing that it is 'a common and appropriate tool' (p. 3) and that 'the case study approach allows the researcher to '"go deep", to learn what works and what does not' (p. 4). They note that, according to Yin, case studies enable the researcher to 'reveal the multiplicity of factors [which] have interacted to produce the unique character of the entity that is the subject of study' (Yin, 1989, p. 82). Drawing on the same source, they argue that case study:

> is a method of learning about a complex instance through description and contextual analysis. The result is a description and theorizing about why the instance occurred as it did, and what may be important to explore *in similar situations.*
>
> (Corcoran *et al.*, 2002, pp. 3–4, emphasis added)

This seems to us to be too particular a reading of the purpose of case-study approaches (cf. Stake's categorization above). Moreover, a contradiction in their argument occurs when they claim that sustainability in higher education is an appropriate field for case-study methodology because it 'is complex: *there are no two institutions alike and within institutions, no two schools alike'* (p. 3, emphasis added). We would also argue that it is true to say that no two people would give the same description of any one institution or school either, so it is hard to say definitively exactly what an institution is like.

Another possible contradiction in their argument is that having argued that case-study methodology is 'a common and appropriate' tool, they then claim (as we have done above) that some confusion about case study exists, pointing to the 'flexible and adaptive nature of the typology (Winegardner, undated)'. They note that:

> There is an imprecise understanding of case study and, according to Merriam (1998), it is often misused as a 'catch-all' research category for anything that is not a survey or experiment. Indeed, 'a case study can accommodate a variety of research designs, data collection techniques, epistemological orientations and disciplinary perspectives, each with its own standards of scholarship' (Winegardner). No matter what the researcher's epistemology the case study is an appropriate strategy for answering questions about how or why (Robson, 1993).
>
> (Corcoran *et al.*, 2002, p. 4)

So what is 'common and appropriate' about case-study methodology? Or are they themselves misusing it, as, to use Merriam's term, a 'catch-all' research category?

Having pointed to Winegardner's broad-minded interpretation of case study, they set about arguing for a much tighter definition of the term in the context of sustainability in higher education. The problem here is that they note that a 'case study can be defined by its special features and these features are not mutually exclusive' (p. 4), adding that 'Merriam (1998) describes the special features as particularistic, descriptive, and heuristic':

> Particularistic case studies focus on a special event, situation or program. The descriptive refers to the end product, which means inclusion of as many variables and analysis of their interaction over time. Heuristic means that case studies enhance the reader's understanding of a phenomenon in such a way that the study extends the reader's experience.
>
> (Corcoran *et al.*, 2002, p. 4)

In an attempt to clarify their understanding of case-study research, they note that it 'has many differences depending on the purpose of the study, the size of the study, the people involved, the theories developed and the theories tested' (p. 4), adding that case-study methods 'vary according to the researcher's purpose in conducting the case' (p. 5). But, having elaborated at length that, in their and others' opinions, case-study methodology has many epistemologies and many purposes, they attack 'case study research [*sic*]' in sustainability in higher education because it 'has not lived up to its potential for improving practice in institutions' and that it 'has been descriptive but not transformative' (p. 1). It would seem to us that this is an unfair criticism given the plurality of interpretations, the singularity and the complexity of the subject matter—all of which they themselves have identified.

Specifically, the problem that Corcoran *et al.* identify is that:

> The research does not problematise practice, instead it sets up a dichotomy of practice. Stories of successes are reported but the data supporting these successes are not available for public critique. Such success stories may mask the problems experienced by the institution in implementing sustainability.
>
> (p. 1)

But why *should* it 'problematise practice', particularly if the case study is an intrinsic one? Why *should* research do anything? Who is to say? It would appear to us that different readers of research will take different things from research depending on their intrinsic interest, their ability to read research and the situations that they know and understand (or wish to know and understand). Their point is weakened further by their use of Yin's work that, like Stake, stresses the variation in the form and use of case study.

For example, drawing on Yin (1993), Corcoran *et al.* elaborate 'three forms of case study: exploratory, explanatory, and descriptive':

> In exploratory case studies, fieldwork, and data collection may be undertaken prior to definition of the research questions and hypotheses. This type of study has been considered as a prelude to some social research. Explanatory cases are suitable for doing causal studies. In very complex and multivariate cases, the analysis can make use of pattern-matching techniques. Descriptive cases require that the investigator begin with a descriptive theory, or in other words, they form hypotheses of cause-effect relationships.
>
> (p. 4)

Corcoran *et al.* note that Bassey (1999) 'defines a range of purposes for educational case studies that include theory-seeking and theory-testing case study, story-telling and picture drawing case study and evaluative case study' (2002, p. 4). For Bassey, 'Case studies may involve description, explanation, evaluation and prediction' (p. 4). We would agree—but Corcoran *et al.* have previously criticized case-study work that 'has been descriptive but not transformative' (p. 1). Again we would refer to Brown and Dowling's point that 'professional educational practice and academic educational research . . . are distinct fields of activity' (1998, p. 165)—research does not have to be transformative. It is the use of research 'evidence' that is or is not transformative, not the research itself.

Critical Case Study

Having pointed to a wide typology of case-study methods, Corcoran *et al.* (2002) argue for a specific type of case study—'critical case study'. They elaborate their criticisms by explaining that case-study research in sustainability in higher education 'rarely includes information on the theoretical approach to the methodology or on the methods used to gather the data' (p. 1). They go on to outline 'a framework for critical case-study research in sustainability in higher education' (p. 1).

They note that in many cases the research is carried out externally by 'an outside evaluator or critical friend who sets out to critique the practices of an institution' and who 'may provide feedback to the institution as a whole in the form of a report on the success or otherwise of the implementation of sustainability in the institution' (p. 5). When the case study is carried out by an internal researcher, 'the aim is to engage in a self-study of their own practices. While this form of case study has the disadvantage of not providing critical external feedback it is a valuable tool in improving practices' (p. 5). They note that 'qualitative case study methodology is of course not the only research tool used' in the area, adding that 'the other most common tool is quantitative studies using indicators of sustainability' (p. 3).

Drawing on the work of Feagin *et al.* (1991), Corcoran *et al.* note that wherever the researcher comes from, they 'are striving for a holistic understanding of cultural systems of action', which involves them in considering 'not just the voice of individual actors, but also of the relevant groups of actors and the interaction between them' (p. 5):

> Indeed, case study research is a study of practice. It is a study of all the players, or practitioners, involved directly, or indirectly, in the innovation. Further, it is a study of the practitioners' actions and the theories they hold about their actions. Improvement in practice occurs when practitioners confront their existing theories (own and others) and in so doing engage in theory building to bring about change.
>
> (p. 5)

Corcoran *et al.* ask three related questions:

> How do we move from impressionist 'feel good' case studies towards case studies that have educative value and stimulate institutional learning? How can case studies become more critical and reflective of practice at one's own institution and at the same time be of interest to practitioners elsewhere? How can we strike some kind of balance between the call for universal models and prescriptions and the need for contextual relevance and meaning?
>
> (p. 9)

Towards Piety?

In setting out their model for case-study methodology, Corcoran *et al.* (2002) argue that:

> there should be a clear purpose for the case study and that the study adequately addressed this purpose. The purpose of the case may be to improve one's own practice or to improve the practice of others.
>
> (p. 9)

> It is important that all the actors, representing potentially diverging interests, be involved in the study and their role in the innovation be explained.
>
> (p. 9)

> it is important that a study addresses how the learning from one institution can become transformative beyond the context in which the case was developed.
>
> (p. 9)

A study should challenge the reader and/or set challenges for the writer.

Corcoran *et al.* deduce from the fact that the case studies were published that 'the authors felt that there was something of value in their case study that extends the context itself' (p. 13). They add that 'It is not unthinkable that many of the cases presented were . . . reconstructed, to appeal to audiences outside the institutional context of the case itself' (p. 13).

On another tack, Corcoran *et al.* ask 'whether the case studies . . . were sufficiently, if at all, validated by all actors involved in the case or only by the outside reviewers serving on the editorial board of the scientific journal' (p. 13). 'This', they argue, begs the question, 'Whose case-study is being reported?' (p. 13).

Corcoran *et al.* see a tension 'between the instrumental use of case studies and the more emancipatory use of case studies' (p. 13). The former, they argue, 'often leads to prescriptive guidelines, criteria or standards that others, at worse, should copy or implement or, at best, should adapt to their own circumstances':

> Such case studies tend to be more of an impressionist and convincing nature and are to appeal to a large audience and multiple stakeholders and interest groups. Such case studies can easily be hijacked to serve a political agenda.
>
> (p. 13)

Whereas, the:

> more emancipatory use of case studies is more process-oriented and may provide ideas, suggestions or imagery that might sensitise outsiders to issues they may have overlooked or not considered, particularly with regards to the process of institutional change. This tension also relates to the notion of a case study as a fixed snapshot of reality as interpreted or reconstructed by a single individual, versus the notion of a case study as an ongoing, collaborative learning process characterized by a higher level of critical reflection.
>
> (pp. 13–14)

Corcoran *et al.* identify a final tension 'between "good" and "bad" practice', arguing that:

> this really is a false dichotomy when considering that in all practice there are lessons to be learned and when recognizing that to label something 'good' or 'bad' is a normative business, highly dependent on one's vantage point.
>
> (p. 14)

In closing, they posit 'a need for distinction between a case as such, a case study and case study research [*sic*]' (p. 14).

This attempt at elucidation seems to make matters more rather than less confusing. Rather than continue to unpick what is becoming a threadbare argument, we will look at another approach that we think has potential, albeit limited, for case-study research in environmental education.

Robinson's Problem-Based Methodology

The final strand of our paper explores the possibility that aspects of Robinson's problem-based methodology (PBM) may provide an appropriate basis for critical approaches to research aimed at investigating environmental and sustainability issues within higher education.

Walker (1995), an author of the Corcoran *et al.* (2002) paper, has both defended and critiqued this approach in her doctoral work, as has one of the authors of this paper (Reid, 1998) and Walker continues to draw on Robinson's (1993) approach in her ongoing work (e.g. Gough *et al.*, 2000). In this commentary, and in light of our earlier discussion, we outline key features of this approach and explore whether it might contribute to Corcoran *et al.*'s vision of the promise of 'critical case study research', or might actually be part of the problem, so to speak.

According to Robinson (1993), the purpose of PBM is:

> to contribute to the understanding and improvement of problems of practice. In brief, PBM involves the reconstruction of theories of action which are operative in the problem situation, the evaluation of such theories, including the assessment of their possible causal role in the problem, and, where necessary, the development, implementation and evaluation of the alternative theory of action. Ideally, these stages of inquiry are embedded in a 'critical dialogue' between researcher and practitioner; that is a conversation that is simultaneously critical and collaborative.
>
> (p. 15)

This approach to enquiring about real-world practice sits well with Schön's attempt to promote reflective practice by engaging practitioners with the research process. It is an attempt to bridge the distinct realms of activity we have identified earlier, and an approach that puts transformation of research (and research evidence) at its centre.

Schön (1991) has characterized the traditional, hierarchical model of professional knowledge in the following way:

> Researchers are supposed to provide the basic and applied science from which to derive techniques for diagnosing and solving the problems of

practice. Practitioners are supposed to furnish researchers with problems for study and with tests of the utility of results.

(p. 26)

Yet, Schön argues:

> In real-world practice, problems do not present themselves to the practitioner as givens. They must be constructed from the materials of problematic situations which are puzzling, troubling, and uncertain . . . Problem-setting is a process in which, interactively, we name the things to which we will attend and frame the context in which we will attend to them.

(p. 40)

According to Robinson (1993), uncertainty and controversy in the defining and delimiting of the case study are likely to be among the authenticating characteristics of the case, but that the features which are often 'controlled out' by the researcher may be the very phenomena 'that are most significant to practitioners' decisions about how to act' (p. 14). To remove these complexities and ambiguities in a case study may result in a 'loss of meaning for the practitioner'. In this light, case-study methodology should not pretend to provide a neatly packaged relationship between the problem, an innovation and the solution. Indeed, from Robinson's perspective, we are reminded to be suspicious of any findings that seem to be too tidy, even when what is offered is intellectual or theoretical security, because they may well fail to encompass the untidy complexities experienced in inquiring into an environmental or sustainability issue in higher education.

As Robinson (1993) notes, users of a problem-based methodology will be interested in the ongoing process of 'resolving educational problems' rather than simply in finding 'a solution', which may restrict the enquirer to a 'single-loop search for a solution that fits the existing constraint structure of the problem' (p. 48). Problems are seen, fundamentally, as an opportunity for ongoing learning, not a fixed, or finalized utterance to end all utterances about the case (see Gough *et al.*, 2000).

Furthermore, Robinson (1993) argues that enquirers using a problem-based methodology will deliberately explore, critique and evaluate existing and alternative practices, that is, the ways in which problems are usually framed are open to reframing and restructuring, and investigated through exploratory action and 'critical dialogue' (p. 262). Such action produces the kinds of phenomena that Schön (1991) describes as 'unexpected changes which give the situation new meanings' (p. 68).

Schön suggests a compelling metaphor for this process when he describes the way in which the 'inquirer . . . shapes the situation, but in conversation with it, so that his [*sic*] own models and appreciations are also shaped by the situation' (pp. 150–151). This idea of a 'reflective conversation', which spirals through stages of appreciation, action, critique and re-appreciation, focuses attention on the perceptions and understandings of the case-study participants and/or its enquirers and authors. Thus, reformulating the problem can help problematize our understandings of the nature of the problem and its representation as much as its outcomes, and even lead to further questions and different problems in a case study.

From Robinson's perspective, then, if there is to be change and improvement in a higher education institution in relation to environmental or sustainability issues, it must engage with the ideas that the participants have about themselves; about one another; and about the processes entailed in identifying and addressing environmental and sustainability issues in higher education. Educational practice, Robinson (1993) suggests, is action informed by beliefs about how to achieve educationally important purposes in particular circumstances. Ignorance of them through blind spots, blank spots or simply hiding them in environmental education research, will not facilitate the processes of reflexivity through dialogue or collaboration.

Closing the Case?

For Robinson (1993, p. 7), 'practice is said to be improved when problems that arise in the pursuit of our goals or the satisfaction of our needs are resolved in ways that enhance our ability to resolve other problems that we experience.' Moreover, changes in local theories of practice are necessary if change is to be non-coercive. If practitioners in the case study begin to think about their work in different ways then their practice will change, as it were, from the inside. This kind of change is qualitatively different from the 'improved' practices that are imposed on practitioners from 'above' or from 'the outside', as we identified earlier on in relation to the uses to which case-study approaches are put. This process of self-reflection in turn entails self-awareness and self-questioning, for all participants, and perhaps an uneasiness with glib case-study findings. Thus the kind of critical dialogue Robinson espouses is characterized by:

- a willingness to ask challenging questions without looking for simplistic answers;
- acceptance and celebration of uncertainty;
- an ongoing process of clarifying meanings and understandings of the case study.

However, in drawing attention to Robinson's (1993) account of educational theory and practice, we are aware that users and readers may choose not—or fail—to probe the adequacy of dichotomous constructions of theory and practice, in much educational research in general, or in Robinson's work in particular. The reduction of educational problems to accounts of theories of action, their adequacy and their criticality, by corralling 'theory' and 'practice' into largely western, modernist meanings, overlooks the 'cultural hegemony and bias' that Scott and Usher (1999) have noted are prevalent in prevailing educational research discourse about case studies (pp. 84–90).

Rather, we argue here, questioning the account of a case study's methodological and methodic consistency and eclecticism is better achieved if epistemological and ontological considerations are 'bracketed in' rather than 'out' of the evaluation of its worth. For example, the guidelines in Table 1 of Corcoran *et al.* (2002) primarily present single-loop paths to solutions where limitations are prescribed and outlined, whereas a PBM-informed approach lays claim to (and should involve the practice of) double-loop solutions, explicitly challenging the use of 'single-loops', e.g. privileging either the solution of the outsider (typically, theorist or academic or evaluator) or the insider (typically, the practitioner or teacher or manager), as in the failure to engage in fruitful critical dialogue between parties.

Touchstone theory, as invoked by Robinson (1993) and Walker (e.g. 1995), has the potential to offer a robust set of considerations for critical case-study development. It seems particularly pertinent when matters of particularity and universality of a case are raised—the location a common criticism of case-study methodologies—as in purporting to provide completeness, correspondence and compatibility in alternative or competing accounts of praxis. Moreover, questions of the extent of partisanship and the role of power present serious challenges to the theorizing and representation of critical case-study-based approaches—in theory and in use—and suggest fruitful lines of poststructuralist and post-colonialist critique of the 'force of the force of argument' in case-study research, e.g. the positioning of terms like variables, inference, rigour, ecological validity, trustworthiness, reflexivity, additivity, verification, predictiveness and generalizability.

To conclude, then, we suggest that approaches like PBM are not taken seriously enough by Corcoran *et al.* (2002) in advocating a critical case-study methodology. Both in terms of its promise and its being part of the problem, PBM may be (in)adequately or (in)consistently applied by researcher and researched. Neither role in the research process is exclusively 'owned' by any of the parties involved in the 'improvement of practice' through case-study methodology, although the poisoned chalice of voluntaristic solutions may appear all too tempting in the rhetoric of case studies for investigating environmental and sustainability issues in higher education.

If, as interpretivists argue, knowledge is contextualized meaning, then issues of authenticity, authority and analytical orientation remain central issues within case-study methodology (Gomm *et al.*, 2000). Or, as Swann (2003, p. 26) suggests in a rather pregnant argument about adjudging the criticality of different approaches to educational research: 'I don't take the view that a project must necessarily be large-scale in order to be scientific. One well-conducted case study has the potential to cast doubt on existing assumptions.'

References

Bassey, M. (1999) *Case study research in educational settings* (Buckingham, Open University Press).

Bassey, M. (2003) Case study research, in: J. Swann & J. Pratt (Eds) *Educational research in practice: making sense of methodology* (London, Continuum).

Brown, A. & Dowling, P. (1998) Doing research/reading research: a mode of interrogation for education (London, Falmer Press).

Corcoran, P. B., Walker, K. E. & Wals, A. J. (2002) Case studies, make-your-case studies, and case stories: a critique of case study methodology in sustainability in higher education, conference paper presented at the *Annual Meeting of the American Educational Research Association*, New Orleans, LA, 2 April.

Eisner, E. W. (1991) Taking a second look: educational connoisseurship revisited, in: M. W. McLaughlin & D. C. Phillips (Eds) *Evaluation and education: at quarter century* (Chicago, IL, University of Chicago Press), 169–187.

Feagin, J., Orum, A. & Sjoberg, G. (Eds) (1991) *A case for case study* (Chapel Hill, NC, University of North Carolina Press).

Gomm, R., Hammersley, M. & Foster, P. (Eds) (2000) *Case study method* (London, Sage).

Gough, S., Walker, K. E. & Scott, W. A. H. (2000) Lifelong learning: towards a theory of practice for formal and non-formal environmental education and training, in: B. B. Jensen, K. Schnack & V. Simovska (Eds) *Critical environmental and health education: research issues and challenges* (Copenhagen, Research Centre for Environmental and Health Education, Danish University of Education), 285–298.

Kemmis, S. (1980) The imagination of the case and the invention of the study, in: H. Simon (Ed.) *Towards a science of the singular* (Norwich, University of East Anglia, Centre for Applied Research in Education), 93–142.

Kennedy, M. M. (1979) Generalizing from single case studies, *Evaluation Quarterly,* 3(4), 661–666.

Merriam, S. B. (1998) *Case study research and case study applications in education* (San Francisco, CA, Jossey-Bass Publishers).

Reid, A. D. (1998) *How does the geography teacher contribute to pupils' environmental education? Unpublished doctoral thesis,* University of Bath.

Robinson, V. M. J. (1993) *Problem-based methodology* (Oxford, Pergamon Press).

Robson, C. (1993) *Real world research* (Oxford, Blackwell).

Schön, D. (1991) *The reflective practitioner: how professionals think in action* (Avebury, Ashgate Publishing Ltd).

Scott, D. & Usher, R. (1999) *Researching education: data, methods and theory in educational enquiry* (London, Cassells).

Smith, L. (1978) An evolving logic of participant observation, educational ethnography and other case studies, in: L. Shulman (Ed.) *Review of research in education* (Ithaca, IL, Peacock), 6, 316–377.

Stake, R. E. (1983) The case study method in social enquiry, in: G. F. Madaus, M. Scriven & D. L. Stufflebeam (Eds) *Evaluation models: viewpoints on educational and human services evaluation* (Boston, MA, Kluwer-Nijhoff), 279–286.

Stake, R. E. (1995) *The art of case study research* (London, Sage).

Stake, R. E. (2000) Case studies, in: N. K. Denzin & Y. S. Lincoln (Eds) *Handbook of qualitative research* (2nd edn) (Thousand Oaks, CA, Sage), 237–247.

Stenhouse, L. (1978) Case study and case records: towards a contemporary history of education, *British Educational Research Journal,* 4(2), 21–39.

Stenhouse, L. (1980) The study of samples and the study of cases, *British Educational Research Journal*, 6(1), 1–7.

Stenhouse, L. (1984) Library access, library use and user education in academic sixth forms: an autobiographical account, in: R. G. Burgess (Ed.) *The research process in educational settings: ten case studies* (London, Falmer Press), 211–234.

Stouffer, S. A. (1941) Notes on the case study and the unique case, *Sociometry*, 4, 349–357.

Swann, J. (2003) A Popperian approach to research on learning and method, in: J. Swann & J. Pratt (Eds) *Educational research in practice: making sense of methodology* (London, Continuum).

Walker, K. E. (1995) *Improving the learning and teaching of environmental education in the primary school classroom: a problem-based approach.* Unpublished doctoral thesis, University of Technology, Sydney.

Wellington, J. (2000) *Educational research: contemporary issues and practical approaches* (London, Continuum).

Winegardner, K. E. (n.d) *The case study method of scholarly research.* Online at: www.tgsa. edu/online/cybrary/case1.html (link no longer active).

Yin, R. K. (1989) *Case study research: design and methods* (Beverly Hills, CA, Sage).

Yin, R. K. (1993) *Applications of case study research* (Beverly Hills, CA, Sage).

Section 3

Developing Theories of Learning, Identity and Culture

Although much of my work is set in out-of-school places, the theoretical framework has often involved a focus on learning, identity and culture. This section contains three papers that show how my thinking developed around the turn of the century. 'Identity and culture: theorising emergent environmentalism' was an invited paper for a special issue of the journal *Environmental Education Research*. It was one of the first papers that I wrote with doctoral students, in this case Elin Kelsey (from Canada) and Ana Maria Duque-Aristizábal (from Colombia). One of the joys of being an academic is working with emerging scholars from around the world and working with Elin and Ana Maria was truly a pleasure.

'Reconceptualising environmental education—taking account of reality' was co-authored with Kelly Teamey another doctoral student (from the US). Kelly and I worked with Bill Scott and Alan Reid on a UK Department for International Development project which looked at mainstreaming environmental education in to aid programmes and projects. While Bill and Alan found themselves in the West Indies, Kelly and I carried out field-work in Pakistan although a pinched nerve in my back severely hampered my contribution.

Finally, in this section, I have included an invited response to Mark Rickinson's seminal review of the literature on learning in environmental education research. 'On learners and learning in environmental education: missing theories, ignored communities', also published in *Environmental Education Research,* is an example of the academic space between science and environmental education that I have come to inhabit.

8   Identity and Culture

Theorising Emergent Environmentalism

Justin Dillon, Elin Kelsey and Ana Maria Duque-Aristizábal

Dillon, J., Kelsey, E., & Duque-Aristizábal, A. M. (1999). Identity and culture: theorising emergent environmentalism. *Environmental Education Research*, 5(4), 395–405.

Introduction

Volume 4(4) of *Environmental Education Research (EER)* was a special issue edited by Thomas Tanner which looked at research on significant life experiences (SLE). Our commentary here is focused on the two articles that made up part II of that edition: New Research. The two articles in this section were 'Significant influences on the development of adults' environmental awareness in the UK, Slovenia and Greece' (Palmer *et al.*, 1998a), and 'An overview of significant influences and formative experiences on the development of adults' environmental awareness in nine countries (Palmer *et al.*, 1998b). These two articles, and a later article in Volume 5(2) of *EER* (Palmer *et al.*, 1999), are all products of the Emergent Environmentalism Research Project (EERP) which is directed by Joy Palmer.

Notwithstanding this central focus, our approach necessarily acknowledges the wider theoretical context of SLE research. In justifying this research, Tanner (1998a) argues that 'it is imperative that we understand how activists (informed, responsible activists) got to be the way they are' (p. 400). Palmer *et al.* have investigated influences and experiences that, it is claimed, have led to environmental concern among their samples of environmental educators. The focus of much of the work of Palmer and her colleagues is the nature of the education that takes place in the formal education context. Tanner identifies one of the goals of Palmer's research as being 'to investigate . . . children's acquisition of environmental subject knowledge and awareness and to monitor development of these as the children proceed through school' (1998a, p. 425).

Palmer and her associates are concerned with:

- looking for patterns in the responses of like-minded individuals who have chosen to belong to specific groups (e.g. members of the UK-based National Association for Environmental Education);

- looking for similarities and differences between people from different countries with different cultural backgrounds and experiences; and
- inferring the relative importance of categories of events in influencing people's subsequent lives and affirmations.

The model that appears to underpin this approach to SLE research begins with children's acquisition and development of knowledge and awareness of environmental issues during school (though these processes can be constrained by incomplete or stereotypical information). Such knowledge and awareness may then be translated into positive actions during a student's formal education, and such actions can be carried over into and developed within adult life. The research is proceeding through comparisons of adult autobiographical statements with the influences that emerge from longitudinal research samples in a range of countries (adapted from Tanner 1998b, p. 425).

This model, though acknowledging a variety of influences on the development of children and adults, hints at a technical rationality view of education: a view which would advocate children being exposed to particular sets of experiences and given certain knowledge which lead to positive attitudes as a means of their developing into an environmentally sensitive and active citizenry. This view is supported by Palmer *et al.*'s (1998b) contention that the outcomes of their research:

> . . . convey crucial messages for environmental educators. They emphasise without doubt the importance of nature and the countryside; those *in*-the-environment experiences that nurture attitudes of appreciation, care and concern for the world that will endure the passing of years.
>
> (p. 443)

Desirable though such studies may be, the model and the philosophy only go so far in illuminating what 'makes' us what we are, and others what they are. One of our major criticisms of the work reported in Volume 4(4) of *Environmental Education Research* is the lack of a convincing conceptual framework to support and explicate both the methodology and data analysis. It is not that there is a lack of theoretical frameworks available, as Haywood and Mac an Ghaill (1997) point out:

> . . . at a time of political retreat and intellectual self-doubt among educationalists, critical social scientists are making available productive conceptual frameworks within which to explore identity formations in educational sites.
>
> (p. 261)

In this article we advocate that the emergent environmentalism research effort needs to be underpinned by materialist and deconstructivist views of identity. We also argue that the validity of the interpretation of this approach to research is open to question.

Identity Theory

Theories of what make us think what we think, and do what we do, abound. To some extent, our identity is defined by others, but an extreme social construction thesis, as Head (1997, p. 8) points out ignores 'the essential psychology'. This is to say, it ignores that we are 'reflexive beings who interpret what others do and say', as opposed to being individuals merely 'shaped by social forces' (Head, 1997, p. 8). One of our criticisms of the work of the Palmer group is that it appears to diminish the role of the individual in interpreting often complex experiences. We agree with Haywood and Mac an Ghaill (1997) when they argue that:

> . . . subjective positions . . . are constituted by a range of narratives that speak identity. . . . As an individual can be located in a range of social relations at one time, the formation of identities through a range of discursive positions is a highly complex, ambivalent and unfinished position.
>
> (pp. 267–268)

Theories of identity and its development offer possibilities to explain and enhance the role of significant life experiences in helping to understand how people get 'to be the way they are' (Tanner, 1998a p. 400). But what is meant by 'identity' and 'identity development'? Erikson's (1974) definition signals the duality of the concept of identity:

> A sense of identity means being at one with oneself as one grows and develops: it also means, at the same time, a sense of affinity with a community's sense of being at one with its future as well as its history or mythology.
>
> (pp. 27–28)

As well as being a personal psychological construct, identity is also associated with a group characteristic, which may be corporate and/or cultural in origin. This duality is a key to understanding how particular experiences might affect individuals in later life. Erikson viewed identity development as being staged, and argued that the epigenesis of personality involves eight stages that are faced at particular phases of life. For example, in adolescence, young people go through a psychosocial crisis of 'Identity versus role confusion'.

> The growing and developing youths, faced with [a] physiological revolution within them, and with the tangible adult tasks ahead of them are

now primarily concerned with what they appear to be in the eyes of others as compared with what they feel they are, and with the question how to connect roles and skills cultivated earlier with occupational prototypes of the day . . .

(Erikson, 1950, p. 235)

Erikson's work on vocational identity developed Freud's notion that happiness comes from *Lieben and Arbeiten* (love and work) focusing less on psychosexual development and more on psychosocial development (Head, 1997, p. 13). Erikson (1968, p. 13) argued that 'One methodological precondition, then, for grasping identity would be a psychoanalysis sophisticated enough to include the environment; the other would be a social psychology which is psychoanalytically sophisticated'. In understanding why people do as they do, it is important to recognise not just the importance of the influence of the environment, but also to focus on the agency of the person concerned. This is something that Palmer and her colleagues would seem to ignore.

Identity and Postmodernism

John Head (1997, pp. 9–10), developing the ideas of Erikson and others, defines identity development as 'the process of making choices which allow one to live effectively as an adult and identity is a functional life script'. Head makes the point (1997, p. 7) 'that children are largely defined by the significant adults in their life, but in adolescence they have to make a series of self-defining choices to allow them to function as autonomous adults'.

> The social class of children is that of their parents. They live where their parents choose to live and go to the school arranged by the parents. In school their actions are largely controlled by teachers. At home, in their leisure time and holidays, their activities are promoted and constrained by the parents. In becoming adults they have to sort out a number of issues. Most obviously they have to start on a career, a process which may start midway through secondary schooling, through the choice of school subjects for study . . . A further requirement is to settle on a desired lifestyle, a term which embraces such abstract notions as beliefs and values, and such practical matters as whether one lives in a city or the country and how leisure time might be spent.
>
> (Head, 1997, p. 7)

As autonomous individuals we seek ways to signal our identity to others—to provide clear-cut signs of who we are and what we stand for. Côté (1996), for example, argues that we are increasingly using consumerism as a means of defining identity. Our choice of clothes, cars and holiday destinations signals something about who we are. But we use other non-consumerist methods too in order to signal our identity to others. We join groups that we think match our values and attitudes and we display that membership through badges, car stickers or other means. We become vegetarians or public transport users or conservationists and signal that to others overtly or covertly. We make statements, not simply through utterances but through the discourse of what we do and what we do not do. A richer understanding of such choices might allow us to be more confident in explaining the significance of events in influencing our 'functional life script' (Head, 1998). Such an enhanced understanding would certainly serve to render more complex the theoretical underpinning of ideas on emergent environmentalism.

However, we are living in a world in flux—socially, culturally, politically and economically, and many of us inhabit a range of cultural sites—our work place, our home, public spaces, and so on. Within each of these we have an identity or multiple identities in response to our perceptions of how others see us. Freud argued that the ego is a 'frontier creature' which is 'menaced by three dangers: from the external world, from the libido of the id, and from the severity of the super-ego'. Freud saw a dynamic between the various aspects of the psyche whereas later psychologists saw more evidence of resolution of life's crises as being desirable and achievable. Given that the world is in flux—with jobs tending to become more transient, major shifts in the views of political parties, and globalisation bringing ostensibly a more diverse though possibly eventually more uniform cultural life—it is possibly more useful to see identities as being transient, more controllable by ourselves and others, and more useful in the bricollage of everyday life [1].

Identity, as Erikson argues (Head, 1997, p. 8), simultaneously describes sameness and difference. This view of identity is problematic if we are trying to define a singular concept of 'identity'. The notions of 'group identity' or 'national identity' indicate that identity can be possessed by groups as well as by individuals. Critics of identity theory might argue that modernist concepts do not capture the complex, multiplicity of life for the individual actor in their own cultural and personal space. We argue that denying the existence and importance of self and the influences on it ignore both the 'biological integrity of the individual' (Head, 1997, p. 20) and the possibility of order in what are non-random, though complex, influences on human existence. Again, such issues would seem to be significant for significant life experience research.

Culture and Identity

Identity, as mental representation, is developed, we argue, through processes that involve personal choice and external influence. Louise Chawla (1998a) echoes this when she writes that:

> The experiences that people remember as significant in motivating their care and concern for the natural world may be characterized as

exchanges between an external and an internal environment: an external environment composed of the qualities of physical surroundings, *and social mediators of the physical world's meaning;* and an internal environment of the child's needs, abilities, emotions, and interests.

(p. 380, emphasis added)

Much of this mediation is through language of some sort. People that share both language and mental representations can be thought of as sharing the same culture. We frequently form our picture of groups by comparing and contrasting what we perceive as their behaviour and motivation with our own. Growing sensitivity to the existence of cultures and sub-cultures has led to increased academic interest in their similarities and differences—in their discourses compared with our own. Our personal experience of interactions with others is qualitative rather than quantitative.

Chawla (1998b, p. 385) argues that the most important strength of 'significant life experience (SLE) research' is that it '*is* qualitative' (emphasis in original). Chawla argues that the qualitative aspect of the research adds an understanding of why people do what they do, and claims that another major strength of SLE research is that it has become a:

. . . cohesive, self-referencing tradition that centers around the common goal of understanding people's own explanations of their environmental feelings and actions.

(Chawla, 1998b, p. 385)

Tanner's commentary on the work of Palmer *et al.*, which is itself primarily a quantitative study, invites the question—to what extent is significant life experience research a coherent body?

What we have done so far is to suggest that identity theory might provide an appropriate conceptual framework through which to mediate the work of the emergent environmentalism research project. However, there are other aspects of this research that we now wish to consider. There are two reasons for doing this. First, in doing so we can illuminate further the utility of a more psychological approach; secondly, we wish to continue the dialogue established by Tanner's sharp and cogent commentary (Tanner, 1998b).

Continuing the Dialogue with the Emergent Environmentalism Research Project

Our specific critique of other aspects of the two articles which report the emergent environmentalism research project begins by summarising and subsequently developing Tanner's (1998b) concerns, and are organised within traditional categories (sampling, data collection, data analysis, background/context, and social, cultural and psychological factors). It is important to stress here that we recognise that such categories are interrelated, and that it would be a mistake to think that improvements in one aspect would not have implications for the others. A key issue in this section is the disempowering nature of a research framework that tends to confuse 'country' with 'culture', and conflate 'national' and 'cultural'. Comments on the categories now follow.

Sampling

Palmer and colleagues claim that the utility of their work is justified in that it provides 'important insights into cultural/national similarities and differences' (Palmer *et al.*, 1998a, p. 429). There seems little evidence to support this claim. The work repeated in Palmer *et al.* (1998a) involves participants from the UK, Slovenia and Greece, while that reported in Palmer *et al.* (1998b) involves research in nine countries: Australia (n = 82), Canada (48), Greece (97), Hong Kong (131), Slovenia (245), South Africa (92), Sri Lanka (203), Uganda (128) and the UK (233). We have considerable concerns about the representativeness of these samples and, because of this, doubt that any meaningful comparisons can be made. For example, to extrapolate from a sample of 233 UK environmental educators to a generalisation about the population would seem unjustifiable.

Tanner himself regrets the fact that neither of these Palmer reports 'describe precisely the methods of selecting subjects and establishing reliability' (Tanner, 1998b, p. 427). Indeed, the issue of sample selection, for example, is hardly mentioned in these articles. We are only told that the 'aims of the study and a questionnaire were given to environmental educators in each of the three countries' (Palmer *et al.*, 1998a, p. 431). It is not clear whether the questionnaires were sent to a stratified random sample of environmental educators or whether it was sent to all environmental educators known to the researchers in each country. The articles do not state how many questionnaires were sent out, only the number of those returned. Further, the titles of both articles state that the research is into adults' environmental awareness, whereas it would seem to have been more accurate to say that the research is focused around a sample of (adult) environmental educators.

It is also unclear from the articles whether, within the United Kingdom for example, the sample includes a cross-section of environmental educators in England, Wales, Scotland and Northern Ireland. The UK is made up of a diverse cultural mix. We would, therefore, have expected to see some reference to the possibility that different cultural and/or ethnic groups are represented by the sampling which would affect the generalisibility of the conclusions. We would argue that a similar point could be made in relation to the cultural mix of the samples from the other countries covered by the studies.

Respondents are identified by gender. In one article (Palmer *et al.*, 1998a, p. 431), a remarkable imbalance (10 men/229 women) is reported for Slovenia, and the researchers argue that it is therefore 'not possible to do any worthwhile analysis between the groups according to sex (*sic*)'. Without knowing the gender breakdown of the people to whom the questionnaire was sent it is impossible to agree or disagree with this assertion. Similarly, there is an issue about the sample of 97 Greeks and 245 Slovenians—whom do they represent, and how were they selected?

Data Collection

Respondents completed a questionnaire which included a request to 'write an autobiographical statement identifying those influences and experiences that led to their environmental concern' (Palmer *et al.*, 1998a, p. 431). Although respondents were actually asked about their 'concern', as has already been noted, the titles of the articles refer to environmental 'awareness'. Outlining the aims of a study in this way might, of course, be construed as leading respondents, and this is a concern which we share with Tanner (1998b, p. 427). There are several assumptions in this method which we argue should at least be discussed by researchers, not the least of which is the question of the extent to which respondents are actually able to identify 'those influences and experiences'—especially when perhaps quite subtle, and both transient and in the distant past. Palmer *et al.* comment that 'Previous work in this field suggests that, generally, when the process of autobiographical memory is unconstrained, and it uses multiple cues, it is both durable and accurate' (1998b, p. 431). We question, however, whether the multiple cues that were used in these questionnaires were adequate to the task—a point which Chawla argues determines whether any recall is 'durable and roughly accurate' (Chawla, 1998a, p. 388).

To check respondents' environmental credentials, the researchers provided a list of seven 'pro-environmental' behaviours from which respondents were asked to select those that they engaged in 'regularly' (Palmer *et al.* 1998a, p. 431). These categories seem poorly operationalised: reading; buying 'green'; outdoor activities; practical conservation; recycling; curriculum development; membership of organisations/attending public meetings. They are not obviously mutually exclusive—outdoor activities, for example, might just include practical conservation. Again, Chawla points out the importance and utility of using comparable, operationalised coding (1998b, p. 393), a point with which we would concur. Tanner (1998b, p. 427) wishes that the authors of the studies 'had more fully defined, explained, and/or justified some of their categories'. We agree with this criticism. It is also not absolutely clear to us where these categories originated. They may have been developed from the original UK study; if that is the case, they do not appear to have been trialled with respondents from other countries.

Data Analysis

The complexities and potential pitfalls of data analysis in this field of research are considerable. We might easily imagine that a respondent could write a statement such as:

> I became interested in the environment when my geography teacher took us outdoors on a field-trip to see the pollution on the river which we had seen on the local TV. When we got back home, some of my friends decided to start a 'Clean River' campaign and I joined in.

It is not clear how such a complex response would be coded. Is this 'primary school' or 'secondary school'? Is it 'pollution'? Is it 'older friends' or 'friends'? The utility of the classification is obviously questionable for such compound but quite believably 'normal' descriptions of events.

Palmer *et al.* state (1998a, p. 442) that 'the most influential factor in developing personal concern for the environment is childhood experiences of nature and the countryside'. The lack of clear operationalisation of the categories of influence makes the analysis of the data problematic. For example, many responses are coded as 'nature/outdoors'; however, it is unclear as to whether that includes quite different activities such as species conservation or fishing. Chawla also points out that such activities may well have a gender bias (for example, more boys than girls tend to go fishing) (1998a, pp. 371–372).

Further, the low incidence of responses in certain categories may have multiple interpretations. For example, the numbers attending tertiary education in the UK are much higher than those in Sri Lanka—so it may not be particularly significant if more respondents in the UK mentioned the impact of tertiary education than did those in Sri Lanka. The analysis of such data is handicapped by a lack of demographic information about respondents, and we are unable to tell whether there are sufficient data available to the researchers to draw these conclusions in a valid way.

Another data analysis issue is the possible inter-country questions that might have been addressed, but were not. For example, 40% of UK respondents are deemed to have been influenced by 'organisations' (p. 441) compared to around 5% of both Greek and Slovenian respondents. But what can this tell us about the roles of organisations in different countries? Are UK organisations more effective, for example, and thus more likely to influence members' views? Or is the income of UK respondents such that it allows membership of these organisations? Currently, it is not possible to distinguish such issues from the ways in which data are reported and analysed, which is a considerable weakness within the reported studies.

Background/Context

We have already indicated our concern about the weak theoretical underpinning of the work of the emergent environmentalism research project. The absence of any references at all in the Palmer *et al.* (1998b) article, and the fact that the later article (Palmer *et al.*, 1999) has only four references (all to work involving Palmer) suggests that the value of the 'self-referencing tradition' of significant life experience research (Chawla, 1998b, p. 385) may be somewhat overstated. In its lack of reference to other studies, the work of

Palmer and her associates appears to be distancing itself from the more main-stream 'significant life experience' camp and possibly even seeking to create a separate niche in the research field. Indeed, Palmer *et al.* seem to take a view of the environment that is at odds with many other researchers. Their view seems to be compartmentalised, apolitical, objective and decontextualised and contrasts strongly with the positions of most other contemporary researchers. See, for example, Robottom and Hart (1993).

Social, Cultural and Psychological Factors

In looking at the respondents whose views Palmer *et al.* report as having researched, and at where these respondents are domiciled, a number of social, cultural and psychological factors spring to mind that might have been used to help interpret data, and possibly lead to further research questions. The five factors which are raised briefly here are, we feel, undervalued by the emergent environmentalism research project in its own consideration of issues which might emerge from the research.

> *Occupation.* All respondents are, in some way or another, educators. One might hypothesise that educators might have had some experience of working with children outdoors either during their training or otherwise.
>
> *Gender.* The majority of respondents are female. Women tend to spend more *time* with their own children in an educative role. One might hypothesise that they are also more likely than men to spend more time outdoors with children.
>
> *Family size.* This tends, in some countries, to be greater than in others, and family size in both Greece and Slovenia is likely, on average, to be larger than that in the UK. One might hypothesise that children in large families are more likely to be influenced by their siblings than those in smaller families.
>
> *Location.* It might be thought that those living in large cities and metropolitan areas would be less aware of seasons, of the human impact on the environment, and of the source of natural resources than those who live in smaller communities such as small towns and villages. One might hypothesise that the percentage of respondents from the UK who have spent most of their lives in such urban areas is greater than those in many of the other countries.
>
> *Climate.* We believe that climate and weather patterns have potentially important impacts on environmental sensitivity. One might hypoth-esise that children in Greece spend more time outdoors than children in the UK, irrespective of where they live.

These brief points must serve to convey our reservations about the con-duct and reporting of research which, by underplaying such social, cultural and phychological factors, has missed an opportunity to link that research to everyday lives and the social contexts of respondents—the very issues which the research sets out, ultimately, to influence.

Summary

We believe that it is important and useful to concern ourselves with the development of an understanding of how individuals find themselves engaged in environmental education. We also believe that it is important that research addresses the role of the individual in making such choices about their vocation, their values and their contribution towards environmental education. Earlier we commented on the utility of identity theory in helping to conceptualise the data from the emergent environmentalism research project. However, as we also mentioned earlier, identity has come to be seen as a cluster of characteristics which bind together groups of individuals. As Head points out, 'social identity develops from an internalisation of the images, albeit stereotyped, of the groups to which one does and does not belong' (1997, p. 35). The interaction between personal identity and social identity is one that explains the 'identity crises' that many people experience, particularly in adolescence. Although not everyone experiences crises as the word is commonly understood, we believe that the dilemmas of a postmodern world have an impact on developing the individual's response to the social discourse in which they find themselves. As a result, we are forced to challenge both the ontological stance implicit in the emergent environmentalism research project as well as its explicit epistemology. In essence, we argue for a greater theoretical underpinning to the research and specifically advocate the use of identity theories to explain personal and social phenomena. We also argue for the use of theoretical frameworks that empower rather than 'capture' sub-groups, and for research that collects and analyses data in a way which opens up opportunities for developing a more sophisticated understanding of factors at play in this complex area.

Acknowledgements

We would like to thank Philip Payne of La Trobe University for his comments and suggestions while he was working with us at King's. We acknowledge a great debt to John Head of King's College London for his invaluable help in developing our understanding of identity and its development.

Note

[1] In this filmscript dialogue, Alice assumes a discourse that she thinks will engage Tom's empathy. She expresses an interest in the environment, shows sympathy with the death of Bambi's mother and agrees with Tom's attempt to turn correlation into causality.

Alice is speaking Tom's language in order to identify her as being the 'same' as Tom in some way.

Alice: How did you get involved with environmental causes? I think it's great.

Tom: You interested in the environment?

Alice: Very much so.

Tom: Actually there's a theory that the environmental movement of our day was sparked by the re-release of Bambi in the late 1950s—that many of the Baby-Boom generation were traumatised [by] the hunters killing Bambi's mother.

Alice: Yes, it was terrible.

Tom: For six-year-olds it was devastating. To this day, no one wants to identify with those hunters.

Alice: I think you're right.

(*The Last Days of Disco*, Stillman, 1998)

References

Chawla, L. (1998a) Significant life experiences revisited: a review of research on sources of environmental sensitivity, *Environmental Education Research,* 4(4), pp. 369–382.

Chawla, L. (1998b) Research methods to investigate significant life experiences: review and recommendation, *Environmental Education Research,* 4(4), pp. 383–397.

Coté, J. E. (1996) Sociological perspectives on identity formation: the culture-identity link and identity capital, *Journal of Adolescence,* 19, 417–428.

Erikson, E. H. (1950) *Childhood and Society* (New York, Norton).

Erikson, E. H. (1968) *Identity, Youth and Crisis* (New York, Norton).

Erikson, E. H. (1974) *Dimensions of a New Identity* (New York, Norton).

Haywood, C. & Mac an Ghaill, M. (1997) Materialism and deconstructivism: education and the epistemology of identity, *Cambridge Journal of Education,* 27(2), pp. 261–272.

Head, J. (1997) *Working with Adolescents: constructing identity* (London, Falmer Press).

Palmer, J. A. & Suggate, J. (1996) Influences and experiences affecting the pro-environmental behaviour of educators, *Environmental Education Research,* 2(1), pp. 109–121.

Palmer, J. A., Suggate, J., Bajd, B. & Tsalili, E. (1998a) Significant influences on the development of adults' environmental awareness in the UK, Slovenia and Greece, *Environmental Education Research,* 4(4), pp. 429–444.

Palmer, J. A., Suggate, J., Bajd, B., Hart, P. K. P., Ho, R. K. P., Ofwono-Orecho, J. K. W., Peries, M., Robottom, I., Tsaliki, E. & Van Staden, C. (1998b) An overview of significant influences and formative experiences on the development of adults' environmental awareness in nine countries, *Environmental Education Research,* 4(4), pp. 445–464.

Palmer, J. A., Suggate, J., Robottom, I. & Hart, P. (1999) Significant life experiences and formative influences on the development of adults' environmental awareness in the UK, Australia and Canada, *Environmental Education Research,* 5(2), pp. 181–200.

Robottom, I. & Hart, P. (1993) *Research in environmental education: engaging the debate* (Geelong, Victoria, Deakin University Press).

Stillman, W. (1998) *The Last Days of Disco* (Director, Whit Stillman, Westerly Pictures, USA).

Tanner, T. (1998a) Choosing the right subjects in significant life experiences research, *Environmental Education Research,* 4(4), pp. 399–417.

Tanner, T. (1998b) Editor's Preface to Part II, *Environmental Education Research,* 4(4), pp. 425–428.

9 Reconceptualising Environmental Education

Taking Account of Reality

Justin Dillon and Kelly Teamey

Dillon, J., & Teamey, K. (2002). Reconceptualising environmental education: taking account of reality. *Canadian Journal of Science, Mathematics and Technology Education*, 2(4), 467–483.

Manchar Lake is the largest freshwater lake in Asia. In recent years, its freshwater intake has been less than the saline and toxic effluents entering the lake through the Main Nara Valley Drain. As a result, the lake has deteriorated to such an extent that its fish stocks are severely depleted and the concentration of some toxic material is more than ten times the safe level. One of the many different communities living on or near the lake, the Mohana, some of Sindh Provinces' oldest inhabitants, is on the verge of extinction. Manchar communities face a wide range of problems including difficulties in obtaining health services, safe drinking water and adequate incomes. Illegal actions by private 'landlords' have further deprived local communities of their economic independence. Fifty percent of all men over 15 and 90 percent of all women over 15 live below the poverty line (US$1 per day). Three quarters of the boys and 84 percent of the girls in the 5–15 age group have never been to school. Poor housing, over population and poverty have ensured the prevalence of TB, anaemia, malnutrition, gastroenteritis, and skin and waterborne diseases. More than 80 percent of the women and children suffer from one disease or another.

—*Extract from* Manchar Lake: Saving Nature, Overcoming Poverty . . .
(Sindh Education Foundation, Pakistan, June 1999)

Introduction

The above description of the lives of the Manchar Lake communities illustrates that, for a major proportion of the world's population, everyday life is extremely difficult. The difficulties can be categorized under a range of headings: environmental, health-related, economic, cultural, political, and so on. People do not experience this ontological separateness; they are faced with interrelated issues that are often complex, out of their control, and, mainly as a result of an inappropriate education, incomprehensible.

In this article, we argue that environmental education has a key role to play in helping people to comprehend the complexity of the issues facing them and, critically, in empowering them to be able to do something, however small, to change the circumstances in which they find themselves.[1] We take a radical stance in arguing for the need for a reconceptualization of the purposes of environmental education if its potential is to be realized.

Fundamentally, we believe that the challenge for environmental education is not just to find the knowledge, skills, values, and pedagogic strategies that are appropriate to individuals and communities but, *simultaneously,* to aim to affect the social context in which those being educated live and work. In our opinion, environmental education cannot be isolated from the social and political context in which it is planned and experienced. The views of curriculum planners, teachers, employers, politicians, and the public at large need to be influenced by environmental education, not just the students'. All those concerned with planning and implementing education need to appreciate that the environment is not something that can be treated independently of human action and that economic development, social inclusion, health, and the environment are inextricably linked.

In order to achieve our goal of a reconceptualized, effective environmental education, we argue, such education needs to be mainstreamed within a range of development policies *and* practices—that is, within the discourses of education and development. If governments, nongovernmental agencies, and aid providers do not incorporate environmental education into their policies across all sectors—not just specifically educational policies—the spiral of decline that characterizes much of the lives of communities across the world will be even more difficult to reverse. Mainstreaming environmental education, however, requires a range of policies and concomitant practices, for example, in environmental management and micro-economics, to support educational policies rather than (as they do now) to ignore or, worse still, to contradict them. The immensity of the challenge of such coordination, particularly in the face of globalization, does not escape us (see Jones, 1998, for more details of the challenges). Full mainstreaming would recognize that it is the policy makers, the donors, the non-governmental agencies, the teachers, the parents, and the employers who need educating about the environment and about links between education and development, in order to begin to modify dominating discourses. Finally, we note that the contributions of science, mathematics, and technology education are, in their own ways, critical to a reconceptualized environmental education and to mainstreaming it within society.

This article is in three major parts. By way of introduction, and in order to provide some historical and conceptual frameworks, we discuss issues of curriculum construction. We look first at the relationship between environmental education and science, mathematics, and technology education. For several reasons, the focus of this section is more on the relationship of environmental education with science education than on its relationship with the other two subjects. Science continues to offer insights into environmental issues, and, in recent years, science education and environmental education have evolved in ways that raise key issues about both areas of knowledge. We believe that sufficient opportunities for changes in the school curriculum are now opening up to make our calls for a radical reconceptualization of environmental education realistic and achievable.

The second part of the article looks more closely at the role of environmental education in the lives of individuals and communities, particularly in countries undergoing substantial development. We draw on work that we have undertaken in Pakistan, and the growing body of literature on environment and development education, to discuss the issue of mainstreaming environmental education within a range of policies and practices. We make a case for a much more radical form of environmental education than currently exists—one that recognizes that the purpose of environmental education should be to influence the lives of all involved with planning and implementing education, and the public at large, not just students. We situate these changes primarily in terms of the school curriculum, but we also recognize the important role played by education and training in the non-school sector.

In the third and final section, we propose specifically that science and environmental education in the West should reflect the day-to-day social, political, economic, and environmental reality of the lives of people and communities in the rest of the world. Unless this happens, we will not be able to face the challenge of the environmental degradation and social injustice that we see all around us.

Environmental Education and Science, Mathematics, and Technology Education (SMTE)

We begin the journey towards a reconceptualization of environmental education by discussing issues of curriculum construction, in particular those affecting science, mathematics, and technology education (SMTE). We believe that if environmental education is to succeed in gaining a stronger foothold in the curriculum, this can be achieved only through integration into these core curriculum subjects.

Our experiences overseas and our contact with visitors from a wide range of countries leads us to think that there are some signs that SMTE is undergoing significant shifts in emphasis. Although this may be a false dawn, glimmers are visible in many parts of the world. For example, writing from a Canadian environmental educator's perspective, Paul Hart points out that 'there is mounting evidence that the curricular ideologies that have dominated science education for decades are undergoing significant challenges' (2002, in press). Hart's optimism results partly from the reported success of SMTE projects under the auspices of an Organisation for Economic Cooperation and Development (OECD) study (Black & Atkin, 1996; Olson, James, & Lang, 1999). In England, the new Advanced Supplementary 'Science for Public Understanding' syllabus encourages 17- and 18-year-old students to examine the nature of science as well as addressing the content of science. For example, a section devoted to 'Risk and Risk Assessment' is meant to educate students so that they should, *inter alia,* understand different ways of expressing the size of a risk; be aware of the range of factors that can influence people's willingness to accept specific risks; and be aware of the contribution of risk assessment to decisions about the management of risk (AQA, 1999).

Hart identifies some of the reasons for the increased pressure on traditional curricular ideologies, which, he argues,

> include problems with the perception of science among young people, problems with international test scores broadly interpreted as a problem of standards, distress about social and community-based issues such as adolescent health and environmental deterioration, and a desire to make science and mathematics more authentic, that is, more genuine and pertinent for students and more like 'real' science, as practised by scientists (see Solomon, 1999).

> (in press)

The Australian scholar Annette Gough, writing from a critical eco-feminist position, concurs with Hart's conclusions and reasoning. The drift from science, Gough argues, can be explained because 'it would seem that while students come to secondary school from primary school interested in studying science, the content of the science curriculum is a strong negative influence and students rapidly lose interest in pursuing further studies in science beyond the compulsory years' (in press). Such pressures also exist outside the English-speaking world. In Japan, for example, economic crisis has led to a serious reappraisal of the role of education:

> After nearly a decade of denial, the Japanese now acknowledge that key economic, political and social structures have gotten out of step with their nation's rapidly evolving needs. This reappraisal has penetrated even to what some consider the bedrock of Japan's harmonious society and what has been until recently a successful economy—the educational system. The current system's weaknesses have been studied extensively since 1970, and a government-led effort to implement reform was launched in the mid-1980s. At that time, however, the problems were not viewed as acute, and the reform movement was stalled by entrenched, conservative political and bureaucratic interests.

> (Choy, 1999)

Other 'tiger economies' are worried about the possibility of extinction and are responding through educational reform. In South Korea, the Ministry of Education Website contains references to its 'Major Projects.' In the section entitled 'Education for the Future' (MoE, 2002), the three goals for education reform are clustered under the heading 'Students' development of their creativity and specializations':

- With the flowering of open education, students' individual differences will be respected and they will be able to make the most of a variety of education materials at their disposal, thereby leading an exciting and satisfying school life.

- With diversified subjects, they can select what attracts their interests most and develop their aptitude and ability while studying with students of similar ability.
- They will have increased opportunities to learn about nature and agriculture in the field study provided in the exchange classes between city and rural schools.

In discussing the impact on the curriculum of these pressures for change, which have already resulted in a substantially increased environmental focus (i.e., links between science and the environment are more explicit), both Hart and Gough see opportunities for a new relationship between science education and environmental education. According to Gough, this 'different relationship [. . .] arises from the need to respond to students' declining interest in science despite their increasing levels of environmental concern' (A. Gough, in press). Gough's analysis can be understood in terms of the parallels and relationships between the following entities: science and science education; environment and environmental education; science and the environment; and science education and environmental education.

Science education's current problems, which are partly mirrored in technology and mathematics education, are interlinked with fundamental debates about the nature of science itself. Western science finds itself in an ironic position—*credited* with explaining a wide range of phenomena (e.g., it can explain the causal mechanisms by which the Manchar Lake, mentioned earlier, has become an environmental disaster) and yet deeply *criticized* by growing numbers of the public and the media (see, e.g., RSA, 2001). Western science, partly as a result of the scale of its own success and partly through the inappropriate expectations and attitudes of some of its major figures, has managed to antagonize a large number of people who, as a result, have voted with their feet. As Noel Gough points out,

> The global reach of European imperialism has given Western science the *appearance* of universal truth and rationality, and many people (regardless of their location) assume that it is a form of knowledge that lacks the cultural fingerprints that seem much more conspicuous in knowledge systems that have retained their ties to specific localities . . . and comparable knowledges of nature produced by other indigenous societies. This occlusion of the cultural determinants of Western science has contributed to what Harding (1993, p. 1) calls an increasingly visible form of 'scientific illiteracy,' namely, 'the Eurocentrism or androcentrism of many scientists, policymakers, and other highly educated citizens that severely limits public understanding of science as a fully social process.'
>
> (N. Gough, in press)

Science education has provided some of the critics and some of the ammunition used against science (see, e.g., Solomon, 1999). Bryan Chapman, in a highly critical attack on the utility of science education in England and else-where, notes that 'the survival of the planet, and the reduction of the obscene disparities of wealth both between and within nations, which characterise today's world and its economic system, are issues demanding education in politics and sociology not science and technology' (Chapman, 1991, p. 59). As Hart points out, this internally driven criticism, from those who teach the subject, raises 'deep questions concerning the epistemological and ontological basis for science education policy and practice.' But, as Hart also writes, 'Many of these questions . . . align with issues of disciplinarity and social value that also define political and curricular struggles in environmental education' (in press). We are cautiously optimistic that radical voices, promoting a more socially realistic environmental education, might be heard more now than they were during the last decade, when national curricula and standards and outcomes-based education—all of which have tended to replicate traditional content organization—were the norm (the exception here is the United States, where standards now seem to be becoming even more restrictive and prescrip-tive). The question to which we now turn is this: Is environmental education ready to offer something to the rest of the curriculum?

Despite tracing its roots to the late 1940s, environmental education gained prominence only during the curriculum struggles of the late 1960s and the 1970s. The delay may not be surprising, given that the process of establish-ing a subject usually involves argument and debate about boundaries and about the meanings of terms (Goodson, 1987). Annette Gough argues that 'environmental education as a formal education movement has its origins in the concerns about environmental degradation and decreasing quality of life expressed by scientists in the 1960s' (in press). Early definitions and thinking about environmental education took links between science and technology almost for granted. For instance, Gough notes that the Tbilisi Declaration, a key document in environmental education's evolution, stated that 'education utilising the findings of science and technology should play a leading role in creating awareness and a better understanding of environmental problems' (UNESCO-UNEP, 1978, p. 24).

In its early days, environmental education provided an opportunity for science knowledge to be applied *for* the environment through both indi-vidual and collective actions (see Jickling and Spork, 1998, for a critique of the idea of being 'for' the environment). Responsibility for changing the planet lay with those capable of thinking globally while acting locally (see N. Gough, 2002, in press, for a criticism of this position). Responsibility for changing individual attitudes and thus, it was hoped, behaviours became, de facto, that of teachers and curriculum designers. Science educators dominated much of this early curriculum design and development, partly through the influence of science, technology, and society (STS) education (see, e.g., Fensham, 1978; Solomon, 1981). For much of the time, what could be classified as environmental education took place within the context of the science curriculum (Gayford, 1986). However, dissatisfaction with the science focus of environmental education and concerns about the dangers of the dominance of science teachers began to be voiced towards the end of the 1970s (Hall, 1977; Lucas, 1980). Dissatisfaction with the science curriculum's general lack of social relevance was also evident around the same time. Dissidents such as Fensham and May (1979) argued that science education had to change to be more in tune with the needs of an environmentally concerned public. Others—perhaps rightly, in hindsight—believed that science education was so entrenched that it would be more effective for the links between environmental education and science and technology education (which were tenuous anyway) to be weakened substantially (see, e.g., Robottom & Hart, 1993).

The dominance of scientific/ecological knowledge was eroded by those who advocated the need for environmental education to develop both skills (e.g., decision making) and values and to adopt a more openly political stance (see, e.g., Fien, 1993). The discourse of environmental education gradually incorporated words and ideas such as 'interdisciplinarity,' 'multidisciplinarity,' and 'holistic' to reflect the synthetic possibilities that environmental education offered—though rarely delivered. More recently, some environmental educators have developed systematic models for the development of 'action competence' (see, e.g., Jensen & Schnack, 1997), while others have promoted 'socially-critical environmental education' as *the* way forward (Fien, 1993; Greenall-Gough & Robottom, 1993; Payne, 1995). As Annette Gough points out, 'These aspects of environmental education did not sit comfortably with conventional representations of science in science education as an objective, rational and value free search for "one true story"' (Harding, 1986, p. 193) (in press). However, they did mirror moves within science education away from the teaching of the separate sciences and towards more integrated science (the term 'balanced science' was used in England and Wales). Our criticism of the socially critical approach is that it has rarely delivered on its promise of radical change in education. Later in this article, we argue for more structural change in education as a whole to support an environmental education that meets the needs of individuals.

During the 1980s and 1990s, educational change focused largely on the widespread implementation of national curricula, standards, and frameworks. Environmental education paid the price for being perceived as multidisciplinary, spending a long period on the margins of curriculum debate—whereas science, mathematics, and technology generally prospered. Some countries, such as India, created space in the curriculum for environmental education as a separate, sometimes optional, subject. Now, however, as science education in some countries stands on the verge of possibly radical change, shifting away from a focus on conceptual development towads a more environmentally and socially relevant curriculum, some environmental educators are actively seeking a revision of the links between science education and environmental education. Gough, for example, articulating a desire for a 'mutualistic' relationship between the two, argues that if we are to achieve sustainable development then science education must have a role in encouraging ecological thinking (instead of being kept at a distance) and environmental education must move on from the insecure relationships that accompany the abstract arguments for it to adopt 'a holistic approach, rooted in a broad interdisciplinary base' (UNESCO 1978, p. 24).

(A. Gough, in press)

We will discuss education and sustainable development in the second part of the article, but at this point we need to note that a *rapprochement* such as Gough suggests involves more than curriculum change. Teachers and their pedagogic practices, founded on their covert or overt beliefs about society, education, and the environment, play a key role. As Hart points out, 'The introduction of socially critical dimensions to issue-based science curricula also raises pedagogical issues which go straight to the heart of teaching as a moral and a political enterprise' (in press). This is not a new problem, but it has new dimensions, as Hart adds:

> Although the notion of STS as an integrating, broadening, more practical, and relevant frame for science education is not a new concept (see Hodson, 1992), the addition of an environmental dimension (i.e., the E in STSE) brings into sharp relief certain epistemological and pedagogical issues involved in changing science curriculum policy and practice,
>
> (in press)

Hart notes that 'the changes implied by environment in the science curriculum are philosophical as well as practical and go far beyond the simple addition of environmental science units to a science curriculum' (2002, in press). It is hard to see enough teachers supporting a radical change in the science curriculum that would challenge their own view of science as a neutral body of facts without professional development that challenged their most fundamental philosophical positions. So, is the enterprise doomed? Hope lies, ironically, in the curriculum pressures exerted by the growing dissatisfaction with science and scientists mentioned above. An education that offers a more complete vision of science as a cultural—and, thus, a human—activity might halt the perceived flight from science that has so perplexed the minds of the great and the good of science and science education in recent years. SMTE that shows how environmental decision-making and management are based on an agglomeration of social, scientific, and technical knowledge, values, and consideration of epistemic questions might capture the minds and hearts of young people in ways that conventional education has so abundantly failed to do. In today's society, students are interested in studying subjects that appear credible—that is, socially relevant as well as academically coherent.

Nevertheless, the scale of the challenge facing those promoting a more socially critical and values-based curriculum are clearly visible. Not only will teachers need a new knowledge base, they will also need to have to hand a range of pedagogic strategies that help their students to appreciate philosophical and epistemological aspects of a range of subjects—not something that occupies much time in pre-service and in-service teacher education courses.

Teachers can only work within the boundaries set for them by the curriculum, assessment, and inspection frameworks within which they operate. However, there are signs that, in some countries at least, teachers will have more flexibility in the future rather than less. As dissatisfaction with lowest-common-denominator international comparisons and national curricula increases, opportunities arise for individual teachers to reclaim their rights to curriculum ownership in order to reflect local conditions, if nothing else. More curriculum and pedagogic autonomy might increase teacher retention in many developed countries, although, in less economically developed areas, the problems caused by the need for many teachers to have two or more jobs will still constitute major handicaps to change. Such autonomy may become easier as scientists, technologists, and mathematicians lose or loosen their grip on the curriculum. Similarly, as governments demand that more students have access to higher education, the school curriculum needs to be more appealing and less exclusive. Environmental education might well be able to offer answers to some of the problems faced by the established subjects and, as a result, might be in a position to be mainstreamed in the curriculum in ways that have not been possible before. However, this cannot take place if politicians and policy makers are unaware of the links between education, environment, economics, and development.

In concluding this section, in which we have pointed out that it might be an appropriate time to promote environmental education in the curriculum, we have indicated that any changes would require changes in teacher education and in assessment policies and practices. What we would wish to point out is that implicit in that statement is a need for environmental education to affect curriculum designers, policy makers, teacher educators, and all those who influence the curriculum and its implementation. No mean task, but one that merits serious consideration. In the next section, we discuss a framework for contemplating the implications of a reconceptualized environmental education in the context of so-called developing countries.

Environmental Education and Development

So far, the focus of our discussion has been on schools and teachers and on the advantages of mainstreaming environmental education within the curriculum. In this section, we wish to take a critical look at the limitations of focusing discussions of environmental education on curriculum and pedagogic issues without taking into account a much broader range of factors. We argue for a framework for reconceptualizing environmental education that takes into account the macro-, meso-, and micro-level influences on the lives of individuals and communities.

We have not, as yet, defined what we mean by 'environmental education.' This is deliberate; the history of environmental education, as we have shown, is one of shifting conceptualizations and definitional disagreement. As we see it, and as we will argue below, monolithic definitions are of little help to people on the ground. However, we do make the assumption that the environment is an integral part of an individual's and a group's identity. Environmental education and educators need to recognize that humans and the environment are separable at some levels of analysis and inseparable at others (see also Dillon, Kelsey, & Duque-Aristizábal, 1999; Payne, 2000). The importance of this assumption will become clearer later.

In making our claim for a need to reconceptualize environmental education and to hint at the relevance of science, mathematics, and technology education, we return to the Manchar Lake, described at the beginning of the paper. In the face of massive and complex 'problems,' a range of 'solutions' have been initiated:

> The Sindh Education Foundation (SEF), a semi-autonomous organization has been working with local communities on participatory processes for developing educational facilities in the area. SEFs vision is to 'empower disadvantaged communities towards social change by creating and facilitating new approaches to learning and education.' SEF has opened and assisted 120 Fellowship Schools, 180 Community Supported Schools, 21 Home Schools and 220 Community Village Schools for girls. It has also arranged for adoption of government schools and activated and trained 2,100 Parent Teachers Associations in government schools. SEF has proposed short-term solutions, including the review of fishing rights by 'landlords,' as well as longer term solutions, such as finding alternatives for the Right Bank Outfall Drain so that the effluent from the southern Punjab and parts of Sindh do not end up in the lake. SEF is engaged in short term social actions including expanding its community school programme, adding components such as: adult literacy; skills training for alternate employment, orientation for better environmental practices, microcredit, improved hygiene and health practices, support for purifying drinking water and sanitation infrastructure.
>
> (Sindh Education Foundation, 1999, n.pag.)

Given the immensity and complexity of this real-life situation, the challenges for environmental educators become clearer. In those parts of the world where environmental education can make a big difference to people's lives within a short time, there are usually few schools or teachers; the curriculum does not reflect the students' needs; education often excludes women or minorities (such as children with AIDS), deliberately or otherwise; and opportunities for employment are frequently non-existent. We believe that in order to conceptualize what environmental education might be able to achieve, it is important for all those who are concerned with education policy and practice to understand the different levels of factors affecting the lives of groups and individuals.

Educational change within nation states is wrapped up in an array of factors ranging from the macro to the micro. For example, at the macro level, government policy might result in more money being spent on military hardware than on schools and teachers. At the meso level, the lives of a region's population might be dominated by a particular industry, such as logging. At the micro level, an individual person might not be able to go to school because he or she is needed to work at home. Adopting policies aimed at changing the school curriculum (specifically, making SMTE education more socially critical), we believe, is unlikely to do more than scratch the surface of the problems faced by millions of people. However, we do argue that environmental education has a major role to play in addressing the fundamental needs of a much greater number of people than is now the case. But this can happen only when links between education and poverty are understood by those making the key decisions.

Politicians and policy makers aim to meet the needs of the world's most disadvantaged people by setting internationally agreed development targets, which, to some, seem rather optimistic and unachievable. Nevertheless, many countries are committed to the following:

- Universal primary education (UPE) by 2015
- Elimination of gender disparity in primary and secondary education by 2005
- National strategies for sustainable development to be under implementation by 2005
- A 50% reduction in the proportion of people living in extreme poverty by 2015

We would argue that environmental education has a role to play in achieving all these targets—or at least making progress towards meeting them. We can only go into limited detail here, but further information can be found in Dillon, Gough, Hindson, Scott, and Teamey (2001), which is based on research carried out for the United Kingdom's Department for International Development. The research involved an extensive trawl of relevant literature, interviews with a range of individuals, and field visits to several countries, including Pakistan (Dillon & Teamey, 2001). We concluded that the process of mainstreaming environmental education into a wide range of social policies (i.e., integrating learning about the environment into programs across a range of policy initiatives) would begin to address many of the problems faced by those in less economically developed countries. By way of explanation, we address each of the four targets described above and show how the issues facing individuals can be examined at the macro, meso, and micro levels. We then indicate, briefly, how environmental education, in its broadest sense, might *begin* to engage with the issues—provided that the focus is not solely on the student's own knowledge, attitudes, and behaviours. By indicating the reality within which environmental education operates, we aim to highlight the issues and complexities that need to be appreciated by policy makers, planners, and those responsible for funding social reform.

Universal Primary Education (UPE)

Attaining UPE is not simply a matter of building enough schools and training enough teachers. In many cases, parents keep their children at home for economic and cultural reasons. At the macro level, this might be because girls are traditionally expected to work at home. At the meso—level, it might be because local industry needs every available pair of hands at some times of the year. At the micro level, it might be because a child's parents think that the school curriculum is irrelevant.

In our experience, environmental education that promotes the development of economically useful skills and knowledge—such as new technologies of farming or conserving potentially useful resources, for example, mangrove swamps—is likely to be seen as valuable by parents and by students, particularly if literacy and numeracy are simultaneously enhanced. The work of the International Union for the Conservation of Nature (IUCN) in Karachi, Pakistan, provides a good example. Environmental education that promotes understanding of a range of world-views is also likely to challenge ingrained cultural practices that result in social injustice.

Gender Disparity

Gender disparity can exist because of macro-level issues such as the dominance of political and religious philosophies that have traditionally segregated societies (particularly in rural areas). At the meso level, girls may be denied access to education because they are forbidden from travelling long distances to school, whereas boys are expected to travel. At the micro level, a girl may be expected by her parents to follow a particular vocational route, unlike her brother. This can be as true in developed countries as in less economically developed ones. A factor that needs to be borne in mind by policy makers is that, in general, the more educated women become, the fewer children they have.

Environmental education that promotes awareness of the links between social practices and cultural values at least provides opportunities for discussion and, in some cases, challenges to existing orthodoxies. Successful environmental educators, in our experience, often work with and through local spiritual and religious leaders rather than trying to work around them. Most religions and many major philosophies promote positive attitudes towards the environment, and local spiritual leaders can be powerful levers to promote environmental concern.

Sustainable Development

'Sustainable development' was defined by the Brundtland Commission as being that 'development that meets the needs of the present without compromising the ability of future generations to meet their own needs' (WCED, 1987). Policy makers soon linked education to sustainable development. One of the key outcomes of the 1992 Earth Summit in Rio de Janeiro was Agenda 21, chapter 36 of which stated that 'education is critical for promoting sustainable development and improving the capacity of people to address sustainable development issues' (UNCED, 1992).[2]

At the macro level, many governments do not have the will, the resources, or the ability to move towards sustainability within the global economic structures. Many believe that their job is to bring the economy into line with those of more economically developed countries; if that requires unsustainable practices to raise the growth rate, then so be it. At the meso level, individual villages may find it impossible to adopt resource or waste management practices that are remotely sustainable. At the micro level, millions of individuals lead totally unsustainable lives and do not have the ability to approach sustainability in any shape or form beyond living at the poverty level.

Environmental education can provide opportunities for individuals to examine their own practices. Such education might rely on teaching strategies such as role-plays, street theatre, or puppetry rather than on traditional knowledge transmission modes (see Fien, 1995). Sustainability is easier to teach in the community than in the school. However, schools have a role to play in explaining the science and technology behind waste and resource management, in support of community-based initiatives. Again, the focus of the environmental education is not just on what happens in the minds of students in classrooms—thought has to go into supporting the environmental education of parents, employers, and community leaders, thereby linking formal initiatives with informal/out-of-school/community initiatives.

Poverty Reduction

Many countries find themselves with levels of debt that dominate their ability to spend money on education. If this is coupled with high levels of spending on military hardware, the situation is exacerbated. At the meso level, such as the Manchar Lake example mentioned above, the practices of local 'landlords' result in families being unable to break out of poverty. At the micro level, a person who has no economically useful skills will find it difficult to raise adequate income to support any reasonable quality of life.

An appropriate environmental education in such situations might involve working with local communities to develop skills of environmental management with a view to increasing income from eco-tourism, for example. If the school curriculum reinforces the messages that the non-formal education promotes, then so much the better. Environmental education might also involve teaching adults about the danger of water-borne diseases, which, in turn, can lead to a decrease in infant sickness, allowing parents more time to farm or work in other employment.

The immensity of the challenge facing environmental educators is only too clear from the following quotation from an interview that we carried out with an academic involved with environmental education in Pakistan:

> We are in a third world developing country—if there *is* a link between environmental education and poverty alleviation—how can we possibly alleviate poverty? It has been there for centuries. What do we do about it? If a man is freezing and his family is freezing, and there is a tree down there—will he preserve it for the environment or chop it down? This is the level where people are many times—harsh realities.
>
> (Academic, Aga Khan University Professional Development
> Centre, Pakistan, February 2001)

The hierarchy of needs implied by these remarks—in which short-term solutions take precedence over long-term consequences—is not the only issue facing environmental educators (see also Gaskell, 2001). Simply knowing that one's way of doing something might be damaging to one's health (for example, using solid fuel indoors to cook), or being taught that erosion is hastened by overgrazing, does not automatically translate into a change in practices. Even if education is relevant, it is unrealistic to expect people to change simply because they know something that they did not know before. Health educators have grappled with this phenomenon for some time, and the links between health and environmental education, in terms of philosophy and pedagogy, have been the subject of some study (see, e.g., Jensen, Schnack, & Simovska, 2000).

The issues and the strategies described above relate primarily to our experience in Pakistan. However, our work with, and experience of, other countries convinces us that many of the issues are pertinent to a much wider audience. The US Academy for Educational Development (AED) has produced a series of over 40 case studies from around the world of what it terms 'examples or real projects that have made a real difference in people's lives and their environment':

> Puppet shows in Guatemala bring smiles to the faces of young and old even as the characters speak of the disastrous consequences of deforestation. Call-in radio shows explain conservation initiatives in Indonesia and build public support for biodiversity in selected conservation areas. Teachers along Mexico's Sonoran coastal wetlands help their students learn about this valuable ecosystem and launch habitat restoration activities. Eco-club members in Jordanian secondary schools explore home water conservation and measure their families' water consumption.
>
> (Monroe, 1999, p. 8)

The key features of the projects described in the collection are identified as follows:

- Devolving power to local communities and increased use of their expertise
- The inclusion of scientific, social, economic, political, and cultural aspects
- The identification and involvement in educational decision making of a variety of stakeholders
- The fostering of an environmental ethic and the provision of assistance to residents in developing decision-making skills
- The incorporation of a gender component
- Flexibility in operation and realism in setting timetables
- The use of appropriate tools for evaluation

This list goes some way towards recognizing that a range of factors are involved in enabling environmental education to be successful. What it does not do, however, is to recognize that any success requires environmental education of local communities and other stakeholders. In the next part of the article, we examine the implications of this reconceptualized environmental education for science education in the West.

Environmental Education, Society, and Sustainable Development

We have argued that the challenge for environmental education is not just to find the knowledge, skills, values, and pedagogic strategies that are appropriate to individuals and communities but, *simultaneously*, to aim to affect the social context in which those being educated live and work. We believe that one cannot isolate environmental education from its social and political context. In our opinion, it is not just students who need influencing; it is curriculum planners, teachers, employers, politicians, and the public at large.

To some extent, non-governmental organizations and development agencies have a key role—coordinating and sustaining appropriate initiatives. Despite some early setbacks, the current trend towards sector-wide approaches (SWAPs) to development—for example, addressing educational change through aid aimed simultaneously at curriculum reform, re-engineering the examination system, and improving teacher education—shows that there is growing recognition of the interconnectedness of different parts of a social sector. Mainstreaming environmental education, however, requires a range of policies, such as strategies for environmental management and micro-economics, to support educational policies rather than, as they do now, to ignore or (worse still) contradict them. Such 'joined-up thinking' is not easy, particularly in the face of globalization—which tends to be driven by the imperative of the market rather than by notions of social justice (see, e.g., Jones, 1998).

A growing body of research might offer policy makers and practitioners some hope of finding strategies that work and that can be sustained (Hart & Nolan, 1999; Rickinson, 2001). An area of research that we feel is particularly useful focuses on the interface between environmental learning in what many refer to as formal and informal education settings. Hart and Nolan state that 'environmental education is a unique field in that it branches out from educational institutions into the non-formal or public domain more so than other areas of educational research' (1999, p. 18). Formal learning, in some instances, is more effective when it is set in the everyday context rather than transposed, however good the intentions, into the classroom.

We hinted earlier at the link between environmental education and identity. Environmental education needs to reflect the inseparability of people and their environment—take them away from their environment and they will think and act differently. There are many who believe that the impact on the actions of individuals is affected more by non-formal experiences (which are more likely to be engaged in voluntarily) than by formal (school-based) experiences. Others, however, are more critical of the non-formal approach and advocate much more community-based approaches that might involve both formal *and* informal approaches (Dillon & Teamey, 2001). Briefly, the argument is this: we believe that environmental education in schools is likely to be more effective if it is supported by out-of-school initiatives. Such reinforcement might come from parents, employers, the media, or through government policies. However, further research is needed into the efficacy of environmental learning in a range of contexts and on the impact of different pedagogic strategies on the knowledge, skills, and values of a range of groups, including, but not exclusively, those in formal education.

Earlier in this article we mentioned a caveat about sustainable education, although so far we have focused almost exclusively on environmental education and avoided using the terms 'education for sustainable development' or 'education for sustainability.' A decade ago, the United Nations Conference on Environment and Development (UNCED) in Rio de Janeiro produced, inter alia, Agenda 21 chapter 36 of which states that 'education is critical for promoting sustainable development and improving the capacity of people to address sustainable development issues' (UNCED, 1992). In recent years, the overlap between environmental education and development education has resulted in a range of initiatives and publications aimed at shifting the emphasis in the curriculum away from environmental studies and towards education for sustainability (see, e.g., Gough & Scott, 2001; Wheeler & Bijur, 2000). We remain to be convinced that this is a useful direction for environmental educators to be heading. Our concern revolves around both the ambiguity of the notion of sustainability in ecological and economic terms and the potential incompatibility of 'sustainable' and 'development'—some resources cannot be sustainably developed (see Sachs, 1997). Even in more economically developed countries, incorporating such a diffuse concept into mainstream education is problematic. Cross, for example, found that, while a sample of Australian teachers took the language of 'sustainable development at "face value,"' they 'were inhibited by a lack of knowledge of the complexities of the issues and how their teaching might contribute' (1998, p. 50). Writing in support of linking education and sustainability, Sterling (2001) argues that education is behind other fields in developing new thinking and practice in response to the challenge of sustainability. We believe that environmental education has more immediate and pressing problems to solve and that focusing too much attention on the sustainability issue might be a distraction.

In Conclusion

At the conclusion of their review of environmental education, Hart and Nolan state that 'much of the research . . . has focused on knowledge-attitude-behaviour connections. Perhaps the future will cause a rethinking of the entire teaching-learning process, maybe through a more ecological approach which parallels Capra's (1996) notion of a new ecological synthesis in science' (1999, p. 42). There is no doubt in our minds that environmental education needs a radical overhaul if it is to justify its place in the curriculum. But the time is right for such changes to take place, and the indications from some parts of the world are that changes are taking place, particularly in science, mathematics, and technology education. What is missing, we believe, is a vision of the immensity of the challenges facing environmental education if it is to respond to changes in SMTE and, simultaneously, have an impact on the lives of the millions of people who are in most need. It is often stated that curriculum change without changes in teacher values and actions is unlikely to succeed (see, e.g., Helsby, 1999), but there is another significant dimension. Educational change in a vacuum—unsupported by broader changes in society—can be only partially effective. The challenge for environmental education is, at the local level, to prove itself useful and relevant to the needs of individuals and communities. The challenge at the national level is to command the support of broader social and political forces—and this is something that development agencies and non-governmental organizations can assist with.

Whatever the future holds, environmental education, by its very nature, will almost certainly be contested, which, as Annette Gough argues, 'keeps it healthy' (1997, p. 165). The role of SMTE in environmental education depends primarily on educators' abilities to offer insights into the major environmental issues affecting life on Earth. Without their insight, the future is bleak. Looking back on the early days of environmental activism, Arthur Lucas writes,

> While many did not expressly say so, it was the environment of the human species that most activists considered worth fighting for. [From a *bio*-centric position, as opposed to an *anthropo*-centric one, we could

argue that the best way to ensure the continuity of life on earth would be to allow the human species to expand so much that it forced itself into extinction by destroying the features of the earth that supported it: then another evolutionary explosion such as the one that followed the extinction of the dinosaurs would occur, with new life forms evolving that exploited the new niches created by the circumstances that drove *Homo sapiens* and many other species to extinction.]

(Lucas, 1995, p. 2)

Acknowledgements

We gratefully acknowledge the helpful comments of three anonymous referees and of Derek Hodson, Editor of *CJSMTE*. We also valued comments from Elin Kelsey, whose own research touched on many of the global issues that are fundamental to our argument.

Notes

1 A caveat here: In Foucault's opinion, becoming 'empowered' means simultaneously becoming disempowered, because people are still objects of power, regardless of how much they understand the power structures that affect them (Usher & Edwards, 1994, pp. 97–98). They become more self-regulatory and learn the limits of their possibilities—these limits are functions of discourse rather than natural factors. Foucault argues that no one becomes empowered; rather, people only participate in discursive positions within power-knowledge formations.

2 A caveat: Although sustainable development has become, especially since the Brundt-land Report (WCED, 1987), a part of the discourse of environmental education, we have a series of reservations about its utility as a concept, some of which we outline below.

References

Assessment and Qualifications Alliance [AQA], (1999). *Science for Public Understanding Advanced Subsidiary 5401 Specification*. Manchester, UK: Author.

Black, P., & Atkin, J. (Eds.). (1996). *Changing the subject: Innovations in science, mathematics and technology education*. London: Routledge/OECD.

Capra, F. (1996). *The web of life: A new scientific understanding of living systems*. New York: Doubleday.

Chapman, B. (1991). The overselling of science education. *School Science Review, 72,* 47–63.

Choy, J. (1999). Japan's educational system heads for reform. *US-Japan Links*. Available: http://www.us-japan.org/edufeature.html (accessed June 9, 2002).

Cross, R. T. (1998). Teachers' views about what to do about sustainable development. *Environmental Education Research, 4,* 41–52.

Dillon, J., Gough, S., Hindson, J., Scott, W. A. H., & Teamey, K. (2001). *Mainstreaming environmental education: A report with recommendations for DFID*. Preston Montford, UK: Field Studies Council.

Dillon, J., Kelsey, E., & Duque-Aristizabal, A. M. (1999). Identity and culture: Theorising emergent environmentalism. *Environmental Education Research, 5,* 395–405.

Dillon, J., & Teamey, K. (2001). *Mainstreaming environmental education in Pakistan.* London: King's College London.

Fensham, P. (1978). Stockholm to Tbilisi: The evolution of environmental education. *Prospects, 8,* 446–455.

Fensham P. J., & May, J. B. (1979). Servant not master: A new role for science in a core of environmental education. *Australian Science Teachers Journal, 25,* 15–24.

Fien, J. (1993). *Education for the environment: Critical curriculum theorising and environmental education.* Geelong, Australia: Deakin University.

Fien, J. (1995). Teaching for a sustainable world: The environmental and development education project for teacher education. *Environmental Education Research, 1,* 21–33.

Gaskell, J. (2001). STS in a time of economic change: What's love got to do with it? *Canadian Journal of Science, Mathematics and Technology Education, 1,* 385–398.

Gayford, C. (1986). Environmental education and the secondary school curriculum. *Journal of Curriculum Studies, 18,* 147–157.

Goodson, I. (1987). *School subjects and curriculum change.* London: Falmer Press.

Gough, A. (1997). *Education and the environment: Policy trends and the problems of marginalisation.* Melbourne: Australian Council for Educational Research.

Gough, A. (in press). Mutualism: A different agenda for environmental and science education. *International Journal of Science Education, 24*(11).

Gough, N. (in press). Thinking/acting locally/globally: Western science and environmental education in a global knowledge economy. *International Journal of Science Education, 24*(11).

Gough, S., & Scott, W. (2001). Curriculum development and sustainable development: practices, institutions and literacies. *Educational Philosophy and Theory, 33,* 137–152.

Greenall-Gough, A., & Robottom, I. (1993). Towards a socially critical environmental education: Water quality studies in a coastal school. *Journal of Curriculum Studies, 25,* 301–316.

Hall, W. (1977). Where next for environmental education? In R. D. Linke (Ed.), *Education and the human environment* (pp. 65–76). Canberra: Curriculum Development Centre.

Harding, S. (1993). Introduction: Eurocentric scientific illiteracy—a challenge for the world community. In S. Harding (Ed.), *The 'racial' economy of science: Toward a democratic future* (pp. 1–22). Bloomington: Indiana University Press.

Hart, P. (in press). Environment in the science curriculum: The politics of change in the Pan-Canadian science curriculum development process. *International Journal of Science Education, 24*(11).

Hart, P., & Nolan, K. (1999). A critical analysis of research in environmental education. *Studies in Science Education, 34,* 1–69.

Helsby, G. (1999). *Changing teachers' work.* Buckingham, UK: Open University Press.

Hodson, D. (1992). In search of a meaningful relationship: An exploration of some issues relating to integration in science and science education. *International Journal of Science Education, 14,* 541–562.

Jensen, B. B., & Schnack, K. (1997). The action competence approach in environmental education. *Environmental Education Research, 3,* 163–178.

Jensen, B. B., Schnack, K., & Simovska, V. (2000). *Critical environmental and health education: Research issues and challenges.* Copenhagen: Research Centre for Environmental and Health Education, The Danish University of Education.

Jickling, B., & Spork, H. (1998). Education for the environment: a critique. *Environmental Education Research, 4,* 309–327.

Jones, P. W. (1998). Globalisation and internationalism: Democratic prospects for world education. *Comparative Education, 34*, 143–155.

Lucas, A. (1980). Science education and environmental education: Pious hopes, self-praise and disciplinary chauvinism. *Studies in Science Education, 7*, 1–26.

Lucas, A. (1995, March). *Beware of slogans.* Paper presented at the opening session of British Council Seminar 'Environmental Education: From Policy to Practice.' King's College London.

Ministry of Education [MoE]. (2002). Education for the Future. Web site of the Ministry of Education, South Korea. Available: http://www.moe.go.kr/eriglish/mproject/index.html (accessed June 9, 2002).

Monroe, M. C. (Ed.). (1999). *What works: A guide to environmental education and communication projects for practitioners and donors.* Gabriola Island, BC: New Society Publishers.

Olson, J., James, E., & Lang, M. (1999). Changing the subject: the challenge of innovation to teacher professionalism in OECD countries. *Journal of Curriculum Studies, 31*, 69–82.

Payne, P. (1995). Ontology and the critical discourse of environmental education. *Australian Journal of Environmental Education, 11*, 83–105.

Payne, P. (2000). Identity and environmental education. *Environmental Education Research, 7*, 67–88.

Rickinson, M. (2001). Learners and learning in environmental education: A review of recent research evidence. *Environmental Education Research, 7*, 207–317.

Robottom, I., & Hart, P. (1993). *Research in environmental education: Engaging the debate.* Geelong, Australia: Deakin University Press.

Royal Society of Arts [RSA]. (2001). Bad chemistry. *RSA Journal, 4*, 22–25.

Sachs, W. (1997). Sustainable development. In M. Redclift & G. Woodgate (Eds.), *The international handbook of environmental sociology* (pp. 71–82). Cheltenham, UK: Edward Elgar.

Sindh Education Foundation [SEF]. (1999). *Manchar Lake: Saving nature, overcoming poverty . . .* Karachi, Pakistan: Author.

Solomon, J. (1981). STS for schoolchildren. *New Scientist,* 8 January, 121–123.

Solomon, J. (1999). O P-ED: Meta-scientific criticisms, curriculum innovation and the propagation of scientific culture. *Journal of Curriculum Studies, 31*, 1–15.

Sterling, S. (2001). *Sustainable education: Re-visioning learning and change.* Totnes, UK: Green Books / Schumacher Society.

United Nations Committee on Environment and Development [UNCED], (1992). *Promoting education and public awareness and training. Agenda 21, Conference on Environment and Development.* New York: Author.

UNESCO–UNEP. (1978). The Tbilisi Declaration. *Connect, 3*, 1–8.

Usher, R., & Edwards, R. (1994). *Postmodernism and education: Different voices, different worlds.* London: Routledge.

Wheeler, K. A., & Bijur, A. P. (Eds.). (2000). *Education for a sustainable future.* New York: Kluwer Academic.

World Commission on Environment and Development [WCED]. (1987). *Our common future: From one earth to one world.* London: Oxford University Press.

10 On Learners and Learning in Environmental Education

Missing Theories, Ignored Communities

Justin Dillon

Dillon, J. (2003). On learners and learning in environmental education: missing theories, ignored communities. *Environmental Education Research*, 9(2), 215–226.

Introduction

Mark Rickinson's review of empirical studies of learners and learning in school environmental education, which constitutes issue 7(3) of *Environmental Education Research* (Rickinson, 2001), 'focuses specifically on the nature and quality of the *evidence* generated by the work in this area' (p. 207, emphasis in original). Rickinson argues that this approach was chosen because of the tendency for 'previous reviews to focus on methodological trends more than research findings' (p. 207). In effect, Rickinson's point is that previous reviews (e.g. Hart & Nolan, 1999) have focused on epistemological issues whereas his own review concentrates on ontological ones.

Rickinson goes on to argue that shortcomings in environmental education theory and research need to be addressed through 'a thorough and grounded understanding of what studies have, and have not, been undertaken on students and learning' (p. 207). I argue in this article that Rickinson's review is only partially successful in achieving what he sets out to do, and that other, major shortcomings in environmental education theory and research have still to be addressed, not least the lack of an appreciation by researchers of what is known about learning. Although Rickinson focuses on the 'nature and quality' of the research evidence, by neglecting to critically examine explicit, or more often, implicit, theories of learning held by the researchers, his review stops short of delivering the thorough and grounded understanding that he promises.

I should begin by acknowledging the major contribution that Rickinson's article makes to the field. The review, which involved close examination of more than 100 journal articles, books and reports, published between 1993 and 1999, will be used by many as a key source of references and as a stimulus for thinking about learning in environmental education. The article makes some key points about environmental education research. For example, he concludes the review by stating that 'the evidence base on learners and learning, while considerable in size, is less diverse in terms of methodological and theoretical approaches than the wider environmental education research field within which it is situated' (p. 207). This narrowness of approach is, I argue in this article, a weakness.

Rickinson identifies 'issues and challenges arising from the recent evidence on learners and learning for research users, researchers and future reviews of the field' (p. 208). I argue, in this article, that this statement is partially justified, but I would argue that it cannot claim to be a critical review for two fundamental reasons.

First, Rickinson chose not to critically examine the theories of learning on which the studies in his report are based. So, for example, in the 50,000-word review on learners and learning published in 2001, the word 'constructivism' does not occur, nor is there any mention of situated cognition. The review is not, therefore, a critical review in the accepted sense of the word, in that it fails to examine the key body of theory that underpins, or should underpin, research into learners and learning in environmental education. Making sense of the messages emerging from the literature that Rickinson reviews cannot successfully be achieved without understanding the theories of learning held by the researchers involved. Much early writing on environmental education emerging from the United States in the 1970s and 1980s, for example, can be appreciated more clearly when one understands the dominance, at that time, of behaviourist models of learning. It is important to note here that few of the papers that Rickinson reviewed make explicit reference to learning theories (Rickinson, personal communication). So, in effect, my criticism is more one for researchers to address rather than Rickinson, although, his review does fail to discuss this fundamentally important issue.

My second point, which follows on from and is inextricably linked to the first, derives from Rickinson's statement that his review 'makes a case for studies focused more explicitly on learning and the role learners play within this process' (p. 208). However, the case for that statement is weakened because the review ignores a major body of research into learning in what many refer to as 'informal' contexts. Now, Rickinson defines the scope of his review as 'learners and learning in primary or secondary school environmental education' but much of the learning in the literature that he reviews occurs in informal contexts (see, for example, Milton *et al.*, 1995; Hampel *et al.*, 1996; Emmons, 1997). In this article, I argue that the literature on learning in so-called informal contexts offers much to the readership of *EER* and, as such, cannot be excluded from any comprehensive review of learning about the environment. Not only is there an opportunity to compare and contrast learning in informal contexts with learning in formal contexts, but, I argue, the learning that takes place on a school field trip is similar, if not identical to, some of the learning that takes place in field centres and museums.

In the first section of the article, I begin by looking at theories of learning and show how they have implications for environmental educators. In the second section, I advance the claim that environmental educators have much to learn from the research done in museums, field centres, botanical gardens, etc.—the informal sector.

Theories of Learning

The first, and possibly the more critical, of my criticisms of Rickinson's review is the absence of any comment on theories of learning implicit or explicit in the 100 or so studies surveyed. This might seem particularly perverse in a study entitled 'Learners and Learning in Environmental Education'. It should be noted that there have been some papers that have concentrated on learning theories during the period of Rickinson's review, but as they were not empirical studies they were excluded from the review (see, for example, Robertson, 1994).

That researchers, particularly those engaged in empirical research, have failed to engage with learning theories to any depth is a concern and a weakness. Compare the situation in environmental education with that in science education, where learning theories are much more part of the discourse of significant numbers of researchers and the contrast is quite stark (see, for example, the *International Journal of Science Education*, the *Journal of Research in Science Teaching* and *Science Education*). This is not to say that all environmental education researchers *should* engage with learning theories but it might be of some advantage if the research community considered what is known about learning in the design of programmes, in researching teaching and learning in the environment, in epistemological considerations and in policy research. It would also be useful if more researchers focused on developing new models of learning in the context of environmental education, so that they were building theories and not just using them.

The informal learning community, mentioned below, has engaged more with learning theories, although it is only recently that researchers have devoted much time to researching learning or even to discussing the utility of various learning theories in researching museums, science centres and botanical gardens. Much writing about learning in the area is still vague and far too generalised to be of use to researchers. For example:

> Learning is a process of active engagement with experience. It is what people do to make sense of the world. It may involve an increase in skills, knowledge or understanding, a deepening of values or the capacity to reflect. Effective learning leads to change, development and a desire to learn more.
>
> (Campaign for Learning, 1998)

Some writers (and many educators) assume that knowledge is simply transmitted—passed on—in some way from expert to novice. This assumption, on which several of the papers that Rickinson reviews are based (see, for example, the review by Leeming *et al.*, 1993), is inadequate to explain the complex nature of learning. Other, more useful, views of learning claim that we build (construct) knowledge through social interactions—so that through dialogue, we become more knowledgeable. Such constructivist theories of learning have been around for some time—Piaget was, in effect, a constructivist. These theories are more generally accepted by a large number of educational researchers. There is, however, some debate about how much the context in which learning takes place actually matters and about the variety of ways in which we can make sense of the world, *based on what we already understand.* Some argue that context is all important—that learning is situated in specific contexts (hence, situated cognition)—i.e. it is difficult to transfer knowledge learned in one context to another; others think that this is not the case (for a good discussion of competing theories of learning, see Adey, 1997). Writing about learning in a way that I suspect many environmental educators would empathise with, Garrison warns that:

> Constructivism must be careful not to confine itself to the purely cognitive domain of human experience. Educators must strive to include the body, its actions, and its passions more prominently in the curriculum.
> (Garrison, 1998, p. 43)

More radical thinkers argue that we are effectively learning all the time during social interaction. By this they mean that you cannot separate out learning from social interaction itself. Lave and Wenger, key proponents of the idea of situated cognition, advocate this point of view:

> The notion of situated learning now appears to be a transitory concept, a bridge, between a view according to which cognitive processes (and thus learning) are primary and a view according to which social practice is the primary, generative phenomenon, and learning is one of its characteristics. There is a significant contrast between a theory of learning in which practice (in a narrow, replicative sense) is subsumed within processes of learning and one in which learning is taken to be an integral aspect of practice (in a historical, generative sense). In our view, learning is not merely situated in practice—as if it were some independently reifiable process that just happened to be located somewhere; learning is an integral part of generative social practice in the lived-in world. [. . .] Legitimate peripheral participation is proposed as a descriptor of engagement in social practice that entails learning as an integral constituent.
> (Lave & Wenger, 1999, p. 86)

This has significant consequences for anyone who thinks that they can have a major impact on learning (i.e. most teachers and education policy makers). Erickson expands the point forcibly:

> Another aspect of the social and cognitive ecology of interaction in the classroom—a system of relations of mutual influence among

participants—involves the combination of what I have called here the reciprocal and complementary aspects of the organisation of immediate social interaction in real time. Interaction it would seem, always involves the articulation of both successive and simultaneous actions, verbal and nonverbal, by those engaged in it.

(Erickson, 1996, pp. 33–34)

Educators who believe in the usefulness of this theory must pay careful attention to the many aspects of the social interaction that take place rather than focusing simply on the 'teaching' or the 'learning'. And so, too, it follows, must researchers—either in developing new theories or in using learning theories to plan interventions or evaluating programmes. The implications for researchers 'in' the environment are extensive. One of the arguments often put forward for environmental education is that, by locating itself in the 'field' it gains a relevance that is not achievable in the classroom. The implication of Erickson's point is that research that claims to look at learning in the environment that ignores 'both successive and simultaneous actions, verbal and nonverbal' might be missing the point completely. Put another way—assuming that because the body is in the field that the mind is there too might be naïve. As Lave and Wenger put it in arguing for discriminating between learning and intentional instruction:

> Such decoupling does not deny that learning can take place where there is teaching, but does not take intentional instruction to be in itself the source or cause of learning, and thus does not blunt the claim that what gets learned is problematic to what is taught [. . .]
>
> (Lave & Wenger, 1999, p. 89)

Even if Lave and Wenger's view is dismissed as extreme, we might be wise at least to take note of Garrison's views on the need for a pragmatic social constructivist perspective—recognising that knowledge is constructed socially and that both teachers and learners have a role in mediating the process.

> Teachers teach subject matter to students in some context. The minds and selves of teacher and student are distributed in a field of action only partially delineated by the subject matter. Pragmatic social constructivism urges educators to consider the entire context, the environmental ethos of schools and community within which the student as a creative individual must function in organic interconnection. Decentering minds and selves within fields of action has important implications for democratic education.
>
> (Garrison, 1998, p. 60)

Demonstrably successful innovation in science education, such as the Cognitive Acceleration through Science Education (CASE) project invariably begins with a model of learning, which seems, from Rickinson's review, not to be the case with innovation in environmental education. CASE, for example, is based on the work of both Piaget and Vygotsky. Vygotsky's theories provide the rationale for the use of discussion in CASE lessons aimed at encouraging children to develop their thinking within different Piagetian schema (Adey, 1997). Vygotsky's idea of a Zone of Proximal Development (ZPD) has gained in popularity with educationalists in recent years since the translation of the author's major work into English (Vygotsky, 1978). In essence, the theory posits that learners can be helped to move from a position of not knowing to a position of understanding with the aid of another person, usually the teacher. This might seem simple but, as Dewey points out, such teaching requires 'high organisation based upon ideas' (1966, pp. 28–29) and makes child-centred teaching more challenging than traditional pedagogies.

Rickinson fails to critically examine the literature that he found from theoretical perspectives such as that of Lave and Wenger and Erickson. A review of learners and learning has to acknowledge the power of theories of learning in critically examining researchers' assumptions, methodologies, methods, analyses and conclusions. Rickinson's categories of evidence (2001, p. 219) point to a recognition of the value of a better understanding of knowledge, attitudes, behaviours and students' experiences and perceptions. This is particularly true of those researchers who focus on what children do know and do not know in that the strategies advocated to 'remedy misconceptions' (as some might see it) are based on assumptions about teaching and learning. Rickinson's analysis examines the 'what' but, critically, not the 'how' of learning in environmental education. So, to return to the opening point made at the beginning of this article, Rickinson's review is only partially successful in delivering a thorough and grounded understanding of what studies have, and have not, been undertaken on students and learning.

Learning in Informal Situations

Erminia Pedretti (2002), herself reviewing informal (i.e. non-school) science education, draws attention to two earlier reviews of learning in informal situations. Rennie and McClafferty's (1996) study examined learning in interactive science centres whereas Hofstein and Rosenfeld's (1996) review examined 'school-based field trips; student projects; community-based science youth programs; museums; zoos; the press; and electronic media' (Pedretti, 2002, p. 1). Much has been written about children's learning about science and the environment in a range of contexts, using a range of phenomena to stimulate understanding and a range of pedagogic strategies aimed at improving instruction or facilitating learning, depending on your point of view. Researchers interested in what happens in classrooms or during fieldwork might gain much from studying this literature base, which is widely available in a range of journals (e.g. *Journal of Museum Education* and *Science Education*). It is worth making the point that there is *also* concern in the informal learning literature about a lack of awareness among academics and practitioners of the research evidence from the formal learning sector.

It is also worth pointing out that there is a long-standing debate in the field about the relevance and accuracy of the terms 'informal' and 'formal' and, in time, both might be replaced by better conceptualisations (e.g. mediated and unmediated). Further, some authors separate formal (school), informal (museums, environment centres, etc.) and non-formal (e.g. extra-mural courses) but this seems to beg even more questions than the formal/informal divide. The mediated/unmediated dichotomy is still problematic as it implies that the mediator plays a critical role whereas that might not always be the case— good learning happens in spite of poor teaching in many schools. As you read from, and, one hopes, learn from reading this article—who is doing the mediating—you or me? Or both of us?

Aside from the three reviews mentioned above, two key works stand out as sources of the research evidence in informal situations. In *Learning in the Museum,* George Hein (1998) discusses learning theories and the work of Dewey, Piaget and Vygotsky as they relate to the 'Constructivist Museum'. Hein provides practical advice on how museum designers and educators can develop exhibits and environments that might facilitate learning. As well as providing a carefully crafted summary of what is known about learning in informal contexts, Hein offers insights into the relationship between knowledge, learners and learning—which has relevance for many, if not all, environmental educators. An example of the kind of learning and experiences that Hein describes can be gauged from the following, randomly picked, paragraph:

> When the Boston Children's Museum developed 'Bridges', an exhibition based on the cultural differences found in the city and the ways in which various teenagers coped with them in their own lives, the museum capitalised on the common experience of its visitors to develop an exhibition that celebrated multi-culturalism. Science centres in India often start with building a science playground, where visitors can enjoy coupled swings and other modified playground equipment that illustrate physics principles while engaging in familiar playground activities.
>
> (Hein, 1998, p. 162)

A less well known book, *From Knowledge to Narrative: Educators and the Changing Museum,* is the work of Lisa Roberts (1997). Whereas Hein primarily reviews the research base on museum learning, Roberts focuses on an exhibition on the naturalist, Carolus Linnaeus, at the Chicago Botanic Garden, which she uses as a means to discuss the evolution of museum education from historical and philosophical perspectives concluding with a chapter on 'Education as a Narrative Endeavor' with sections on 'Rethinking "Learning" in Museums' and the '"Teaching" Function' of museums. Roberts offers environmental educators an opportunity to consider issues such as empowerment, ethics, entertainment and experience as they relate to learning. As an example of the insights into the value of diverse bodies of knowledge that Roberts' book has to offer environmental educators, particularly those concerned with the role of narrative in making sense of notions of identity in postmodern contexts, consider the following extract:

> Blythe Clinchy's work on the processes by which women acquire knowledge holds promising application to visitor experiences with its emphasis on connected, internalised methods of knowing. Even literary theorists like Umberto Eco and Jean Baudrillard, whose work is generally more descriptive than empirical, are relevant since the core of their work deals with the social construction of narrative.
>
> (Roberts, 1997, p. 141)

A caveat: although much has been written about learning in museums, the field is still new: deciding what is special about it as a field of enquiry and study is still a contested activity. For example, Anderson, reviewing the museum education scene in the UK, writes that 'Few museums evaluate the educational effectiveness of their galleries or other services, conduct learning research, or study the educational work of other museums, yet investment in these activities could significantly enhance the effectiveness of museums' (1999, p. 4). Anderson argues that:

> Museums have a responsibility to enable visitors to learn in ways that are different from and complementary to those offered by other educational providers such as schools and colleges. The unique characteristics of museum learning is that it is based on first-hand, concrete experience of real objects, specimens, works of art and other authentic resources in a social environment in galleries or at sites. Museum learning is more concentrated and deliberately structured than everyday life, and more diverse, informal and culturally rich than formal education.
>
> (Anderson, 1999, p. 31)

I am not convinced that such a clear delineation exists. Much education in the environment involves 'first-hand, concrete experience of real objects, specimens . . . in a social environment . . . at sites'. Much learning in schools happens in seemingly unstructured ways, judging from my observations of classrooms and fieldwork. Much museum learning involves school-like structured activities led by qualified education staff occupying roles as teachers (sometimes couched in terms such as 'explainers'). Having said that, much museum learning does not appear, from my experience, to be very concentrated, rather it is fleeting, often ephemeral and wonderfully random.

I would agree, however, with Anderson that 'A remarkable characteristic of museums is that so many people . . . have had exceptional and life-enhancing experiences, through encounters with beautiful, old, rare, spectacular, ingenious, well-realised or evocative objects in museum settings, which they can remember vividly many years later' (Anderson, 1999, p. 31). But many people also remember particular experiences from lessons or school trips. The claims of a special status for learning in museums seem to me to be rather more tenuous than some of its advocates assert. And as such, the research base of museum, botanical garden and national park studies offers much to the environmental education community than some of its advocates might imagine.

Take, for example, Hein's review of the museum education literature which, ostensibly, is an historical overview. Hein describes the evolution of 'visitor studies' pointing out that a lot is known about visitor behaviour, knowledge and attitudes. Museums and other centres have to facilitate visitor learning by proxy—through overt means, such as labels ('interpretation') or, more covertly, by juxtaposing artefacts. Hein draws some general conclusions from his reading of the 'sum total of the research carried out in [the 20th] century' (Hein, 1998, p. 153):

1   People 'learn' in museums. Whether learning is narrowly defined as absorbing specific pedagogic messages contained in exhibits or more broadly defined to include responding to the experience of a museum visit, there can be no doubt that visitors 'learn' in museums. People have enriching, stimulating, rewarding, or restorative experiences in museums. They learn about themselves, the world, and specific concepts; they have aesthetic, spiritual, and 'flow' experiences.

2   In order to maximize their potential to be educative, museums need first to attend to visitors' practical needs; degree of comfort influences the value of the museum experience. Comfort includes orientation, providing amenities, making the museum's agenda clear, and always maximizing the possibility that the intended interactions between the content of the museum and the visitor be as positive as possible.

3   People do attend to exhibits—they incorporate the content of museums into the agendas they bring with them, and their social interactions, attention, fantasies and feelings include, and often focus on, the content of museums.

4   People make unique, startling connections in museums.

5   Museums are not efficient places for traditional 'school education' learning specific facts and concepts, because people don't spend enough time and are not there primarily for that purpose.

6   Staff should never underestimate the value of wonder, exploration, expanding the mind, providing new, cognitively dissonant (intellectually shocking), and aesthetic experiences. Museums can do this well and these are an integral part of 'learning.'

7   For visitors to have a positive experience, their interactions with the contents of the museum must allow them to connect what they see, do and feel with what they already know, understand and acknowledge. The new must be incorporated into the old.

The evidence from studies of learning in informal contexts, as digested by Hein, would appear to have much in common with the literature quoted by Rickinson. Re-read the list above, substituting 'the environment' for 'museums' wherever possible.

Specifically, the informal context evidence base seems to offer environmental education researchers three types of potentially useful information:

Type 1   An understanding of how learners interact with everyday phenomena at a conceptual level;

Type 2   A description of children's ideas about a range of phenomena;

Type 3   An understanding of the design and efficacy of a range of pedagogic strategies, primarily ones which involve direct, hands-on experience with an object or a situation.

The type 2 evidence clearly overlaps with Rickinson's research category of 'Learners' perceptions of nature' (2001, p. 219). The other two categories seem to point to blind spots in the research base on learners and learning in environmental education. As Noel Gough (2002) points out, as well as filling in the blank spots through research, we can also benefit from investigating our blind spots.

As well as these overlaps and gaps, the informal education sector, as a whole, has more to offer the environmental education research community. First, it has been pointed out that a large proportion of teacher development and training takes place in museums (IRA, 1996). Studies of teacher development in museums and other informal centres offer something to those in environmental education concerned with in-service education and training (teachers as learners are another category not considered by Rickinson, for pragmatic reasons).

Secondly, Schauble *et al.* describe a theoretical framework for research that has been developed to organise the work of the members of the Museum Learning Collaboration (MLC) which might be of interest to environmental education researchers. The purpose of the MLC is to:

> develop then pursue together a research agenda that can support the development of a cumulative body of knowledge on museum learning, one that will build over time and become increasingly generative and generalisable. As it grows, such a knowledge base will transcend the concerns of museums to inspire new issues and questions about the nature of learning itself.

Schauble *et al.* comment that in order to pursue such an agenda, the MRC will need a 'broad but well-defined theoretical framework' because:

> Theory is essential to keep such an enterprise from spinning off into a mere collection of unrelated investigations, because theory highlights the

questions and issues worthy of exploration, points to what is central in the research findings, and provides the integrating frame that serves to define a coherent portrait from a series of independent investigations.

(Schauble *et al.*, 1999).

I would not argue that environmental education researchers *should* accept the need for such an integrating frame, or that 'a coherent portrait' should and could emerge. Rather, I would simply point out that here is a strategy, advocated by some members of another research community, which might be worthy of discussion. Reviews of environmental education research, such as Rickinson's, might benefit from juxtaposing the environmental education research community with that in allied educational spheres.

Conclusion

So what am I advocating? First, I am arguing that in reading and doing research in environmental education, there are benefits to be had in focusing more than has been the case in the past on learning theories. One might ask 'why is it the case that learning theories have been relatively ignored by environmental educators?' It might be that much of what has been and continues to be written about learning might well appear to be common sense (see, for example, Vosniadou, 2001). Few would argue that learning does not require 'the active, constructive involvement of the learner' or that 'people learn best when they participate in activities that are perceived to be useful in real life and are culturally relevant' (Vosniadou, 2001, unnumbered). However, the evidence provided by much of the environmental education literature is that this knowledge is neither common nor accepted yet as 'sense'. In concluding, I reiterate Rickinson's plea that shortcomings in environmental education research need to be addressed through 'a thorough and grounded understanding of what studies have, and have not, been undertaken on students and learning'.

Secondly, I am arguing that there are major advantages to environmental educators—whether researchers or otherwise—in looking at what research in informal contexts has to offer (part of this literature base addresses learning theories, but there is more to it than that). This point is exemplified by an article written by Michael Brody and Warren Tomkiewicz (2002) and published in the Special Issue of the *International Journal of Science Education* 24(11) devoted to 'Perspectives on environmental education–related research in science education'. In their article on park visitors' understandings, values and beliefs (using data collected in Yellowstone National Park) Brody and Tomkiewicz refer to various writers on learning including: Ausubel; diSessa and Sherin; Posner, Strike and Hewson; and, Minstrell. The research on learning in schools is linked by Brody and Tomkiewicz to the work on learning in informal settings and, specifically, in museums. The end result is an article that is conceptually rich, intellectually coherent and relatively commanding in its theoretical base. That the article has attained this quality is due not only to the authors but to the anonymous reviewers who were able to make a substantial contribution to the article's final appearance.

The two strands of my argument have implications not only for those who do research but for those who read research papers both after they have been published and, crucially, during the review process. Finally, there are implications here for graduate students too, many of whom would benefit from greater learning about learning.

Acknowledgements

Nancy Belfiore, formerly based at the Faculty of Environmental Management, Prince of Songkla University, Thailand and Mee Young Choi, a colleague at King's, have helped me with developing my own ideas about learning in environmental education. Kirsten Ellenbogen, another colleague, has, as ever, been an invaluable source of relevant material from the so-called informal sector.

References

Adey, P. (1997) It all depends on the context, doesn't it? Searching for general, educable, dragons, *Studies in Science Education,* 29, pp. 45–92.

Anderson, D. (1999) *A Common Wealth. Museums in the learning* age (London, The Stationery Office).

Brody, M. & Tomkiewicz, W. (2002) Park visitors' understandings, values and beliefs related to their experience at Midway Geyser Basin, Yellowstone National Park, USA, *International Journal of Science Education,* 24(11), pp. 1119–1141.

Campaign for Learning (1998) Quoted in D. Anderson (1999), p. 8. The quote is actually from a personal communication from the Director of CfL to Anderson.

Dewey, J. (1966) *Experience and Education* (London, Collier Books).

Emmons, K. M. (1997) Perceptions of the environment while exploring the outdoors: a case study in Belize, *Environmental Education Research,* 3(3), pp. 327–344.

Erickson, F. (1996) Going for the zone: the social and cognitive ecology of teacher–student interaction in classroom conversations, in: D. Hicks (Ed.) *Discourse, Learning, and Schooling,* pp. 29–62 (Cambridge, Cambridge University Press).

Garrison, J. (1998) Toward a pragmatic social constructivism, in: M. Larochelle, N. Bednarz & J. Garrison (Eds) *Constructivism and Education,* pp. 43–60 (Cambridge, Cambridge University Press).

Gough, N. (2002) Ignorance in environmental education research, *Australian Journal of Environmental Education,* 18, pp. 19–26.

Hampel, B., Holdsworth, R. & Boldero, J. (1996) The impact of parental work experience and education on environmental knowledge, concern and behaviour among adolescents, *Environmental Education Research,* 2(3), pp. 287–300.

Hart, P. & Nolan, K. (1999) A critical analysis of research in environmental education, *Studies in Science Education,* 34, pp. 1–69.

Hein, G. E. (1998) *Learning in the Museum* (London, Routledge).

Hofstein, R. & Rosenfeld, S. (1996) Bridging the gap between formal and informal science learning, *Studies in Science Education,* 28, pp. 87–112.

Inverness Research Associates (IRA) (1996) *An Invisible Infrastructure: institutions of informal science education* (Washington, DC, Association for Science–Technology Centers).

Lave, J. & Wenger, E. (1999) Legitimate peripheral participation, in: P. Murphy (Ed.) *Learners, Learning and Assessment,* pp. 83–89 (London, Paul Chapman).

Leeming, F. C., Dwyer, W. O., Porter, B. E. & Cobern, M. K. (1993) Outcome research in environmental education: a critical review, *Journal of Environmental Education,* 24(4), pp. 8–21.

Milton, B., Cleveland, E. & Bennett-Gates, D. (1995) Changing perceptions of nature, self, and others: a report on a Park/School Program, *Journal of Environmental Education,* 26(3), pp. 32–39.

Pedretti, E. (2002) T. Kuhn meets T. Rex: critical conversations and new directions in science centres and science museums, *Studies in Science Education,* 37, pp. 1–32.

Rennie, L. J. & McClafferty, T. (1996) Science centres and science learning, *Studies in Science Education,* 27, pp. 53–98.

Rickinson, M. (2001) Learners and learning in environmental education: a review of recent research evidence, *Environmental Education Research,* 7(3), pp. 207–317.

Roberts, L. C. (1997) *From Knowledge to Narrative: educators and the changing museum* (Washington, DC, Smithsonian Institution Press).

Robertson, A. (1994) Toward constructivist research in environmental education, *Journal of Environmental Education,* 25(2), pp. 21–31.

Schauble, L., Leinhardt, G. & Martin, L. (1999) A framework for organizing a cumulative research agenda in informal contexts, *Journal of Museum Education,* 22(2/3), pp. 3–7.

Vosniadou, S. (2001) *How Children Learn* (Brussels, International Academy of Education).

Vygotsky, L. (1978) *Thought and Language* (Cambridge, MA, MIT Press).

Section 4

Challenges and Opportunities—Science, the Environment and the Outdoors

This section draws on two synthesis papers, 'The value of outdoor learning: evidence from research in the UK and elsewhere' and 'On food, farming and land management—towards a research agenda to reconnect urban and rural lives' which show how new knowledge can be made from critiquing existing studies and why learning outdoors has so much to offer.

'Science, the environment and citizenship: teaching values at the Minstead Study Centre' is an empirical study that shows what can be done in terms of teaching students outdoors. Co-authored with Alan Reid, the chapter was published in a book that I co-edited with Debbie Corrigan and Dick Gunstone, both of Monash University, Australia. The book was the first in a series that reflects the work of a long-standing collaboration between King's, Monash and, latterly, Waikato University in New Zealand.

'Barriers and benefits to learning in natural environments: towards a reconceptualisation of the possibilities for change', provides a critique of reasons why schools say they cannot take classes outside the classroom. The paper was an invited contribution to *COSMOS*, the journal of the Singapore National Academy of Science. The ideas in the paper emerged from a very fruitful collaboration between King's and Natural England, an organisation that provides advice to the UK government on environmental issues. This collaboration still continues and provides an opportunity to engage with policy and practice.

11   The Value of Outdoor Learning

Evidence from Research in the UK and Elsewhere

Justin Dillon, Mark Rickinson, Kelly Teamey,
Marian Morris, Mee Young Choi, Dawn
Sanders and Pauline Benefield

Dillon, J., Rickinson, M., Teamey, K., Morris, M., Choi, M. Y., Sanders, D., & Benefield, P. (2006). The value of outdoor learning: evidence from research in the UK and elsewhere. *School Science Review*, 87(320), 107–111.

This article summarises the key findings of a review that critically examined 150 pieces of research on outdoor learning published between 1993 and 2003 (Rickinson *et al.*, 2004). The Field Studies Council and partner organisations commissioned the review in response to the growing concern that opportunities for outdoor learning by school students in England have decreased substantially in recent years (Harris, 1999; Barker, Slingsby and Tilling, 2002).

We found substantial evidence to indicate that fieldwork, properly conceived, adequately planned, well taught and effectively followed up, offers learners opportunities to develop their knowledge and skills in ways that add value to their everyday experiences in the classroom. In this article we distil some of the review's findings of particular relevance to secondary school teachers. We look first at the impacts of fieldwork and outdoor educational visits, and then discuss what is known about effective practice before concluding with a look at barriers to fieldwork.

The Impacts of Fieldwork and Outdoor Educational Visits

Not surprisingly, research suggests that students remember fieldwork and outdoor visits for many years. Dierking and Falk (1997) found that 96 per cent of a group (128 children and adults) could recall field trips taken during their early years at school. However, simply recalling a visit does not mean that it was an effective learning experience or that the time could not be more usefully spent in the classroom.

Evidence for the relative efficacy of fieldwork comes from a study of secondary students from 11 Californian schools that used an environmentally focused curriculum. The students scored higher in 72 per cent of the academic assessments (reading, science, maths, attendance rates and grade point averages) than students from traditional schools (SEER, 2000). Eaton (2000) found that outdoor learning experiences were more effective for developing cognitive skills than classroom-based learning. Such comparative studies, though important, are rare and very difficult to carry out.

In terms of the impact on students' attitudes, Mittelstaedt, Sanker and Vanderveer (1999) looked at the impact of a week-long experiential programme on 46 US children. The children (31 male, 15 female) attended a five-day summer-school programme of biodiversity activities. The authors found that '*even though the children arrived with a positive attitude toward the environment, they left with an even stronger positive attitude*' (p. 147). Uzzell and colleagues, however, sound a note of caution about making too many assumptions about the relative permanency of attitudinal changes (Uzzell, Rutland and Whistance, 1995). The researchers point out that environmental attitudes are fairly well entrenched: '*What they learn . . . both in the classroom and in the field, only serves to strengthen their views and perhaps heighten their sense of action paralysis*' (p. 1 77).

In terms of changing students' behaviours, Bogner (1998) tested one-day and five-day versions of a long-established outdoor ecological programme with 700 students aged 11–13 in a German national park. Bogner reported that '*the 5-day program explicitly provoked favorable shifts in individual behavior, both actual and intended*' (p. 17).

What Counts as Effective Practice?

There is considerable evidence indicating that longer programmes are more effective than shorter ones. A study by Emmons (1997) of an outdoor environmental education programme in Belize argued that:

> *the length of time that students spent at Cockscomb (five days for most) appeared to be important in the reduction of negative perceptions of the environment, including fears . . . A shorter environmental education programme may not have had the same effect.*

> (p. 342)

Bogner's (1998) evaluation, mentioned above, found that '*only the residential five-day programme had any effect on behavioural levels*' (p. 26).

The value of preparatory work prior to outdoor learning is another factor well-evidenced in the literature. For example, in their study of nature-based excursions in Queensland, Ballantyne and Packer (2002) found significant differences between students who had done pre-visit activities and those who had not. The former both looked forward to, and enjoyed, their visit more than the latter. Work by Orion and Hofstein (1994) in Israel provides a strong rationale for preparatory work that introduces students to the cognitive (field trip concepts and skills), geographic (field trip setting), and psychological (field trip processes) aspects of fieldwork. The benefit of preparatory meetings, discussions, explanations and materials for creating accessible and inclusive field courses is stressed by Healey *et al.* (2001).

Several studies highlight the importance of carefully designed learning activities and assessment of students' outdoor learning. Ballantyne and Packer (2002:228) warn against over-structuring learning activities. They found that *'the use of worksheets, note-taking and reports were all unpopular with students, and did not appear to contribute greatly to [their] environmental learning'*. They suggest that touching and interacting with wildlife is a more effective strategy. Emmons' (1997) study of a five-day field course in Belize found that students' learning was facilitated by their shared and direct experience of the surroundings, as well as their teachers' role-modelling of their interests and likes about the forest environment.

The ability to choose between different kinds of learning activity appears to be an important requirement for students. Openshaw and Whittle (1993) comment upon the need for teachers and outdoor educators to balance *'the students' desire for a structure within which they can feel comfortable and not threatened and the added excitement caused by the unexpected'* (pp. 63–64).

The need for effective follow-up work after outdoor experiences is stressed by several authors (for example, Orion and Hofstein, 1994). Uzzell *et al.* (1995) emphasise the need for clear links to be made between outdoor activities (*'the world of our physical surroundings'*) and indoor activities (*'the world of the school'*).

Factors Influencing Outdoor Learning and its Provision

So far we have looked at the impacts of fieldwork and at what constitutes effective practice. It is the case, though, that there is substantial variation between students and schools in terms of opportunities to experience the outdoors and in the subsequent learning that takes place. So what are the factors that affect how much learning takes place outdoors and the amount and quality of provision of experiences for students? Notable barriers include:

- fear and concern about health and safety;
- teachers' lack of confidence in teaching outdoors;
- school curriculum requirements;
- shortages of time, resources and support;
- wider changes within and beyond the education sector.

As well as these external factors, a range of personal influences on learning have been identified as indicated below.

Age

An Australian study of school students' perceptions of learning in natural environments found significant differences between the primary and secondary school age group (Ballantyne and Packer, 2002). Primary school students were found to be significantly more enthusiastic than their secondary counterparts, both before and after the experience. The two groups were also looking forward to different aspects of the experience:

> *Primary school students tended to focus on specific features of the programme . . .*
> *Secondary school students gave a more varied range of responses, including get-*
> *ting out of school, experiencing nature . . . and experiencing something new or*
> *different.*

(p. 221)

Prior Knowledge and Experience

Students' learning can be strongly influenced by their previous field and classroom-based experiences (Orion and Hofstein, 1994; Lai, 1999). Openshaw and Whittle note that '*if students have been accustomed to a diet of "experiments" based on well tried recipes that "work", then real experimental practical ecology is likely to prove a difficult experience for them*' (1993: 64).

Fears and Phobias

Several studies suggest that outdoor settings can be the source of genuine fear and concern for young people. Simmons (1994a, b) found children in Chicago expressed concerns about a variety of nature scenes: possible natural hazards; threats from other people; and inconveniences for their physical comfort. Similar worries about getting lost and encountering snakes or poisonous plants are reported by others (Bixler *et al.*, 1994; Wals, 1994). The important point is that such fears '*pose barriers to enjoying and learning [in and] about wildlands*' (Bixler *et al.*, 1994: 31). This phenomenon is seen in students with a high 'disgust sensitivity' who are found to prefer activities that do not involve handling of organic matter, and fieldwork sites with clear water, no algae and easy lakeshore access (Bixler and Floyd, 1999).

Learning Styles and Preferences

There is growing appreciation of the importance of students' learning styles and preferences in outdoor learning, especially fieldwork. Lai's (1999) in-depth study of Hong Kong secondary school students on a geography field trip found marked differences in individuals' responses to the two parts of the day. While some preferred the teacher-guided tour of local physical features in the morning, others were much happier with the student-led field investigation in the afternoon when they could '*work on their own and hence have more freedom*' (p. 248).

Physical Disabilities and Special Educational Needs

Recent work in the UK has highlighted the many barriers that disabled students can face to participating fully in fieldwork, and the ways in which institutions, departments and tutors can help to reduce them (Healey *et al.*, 2001). This challenge is also pertinent to organisations conducting horticultural and gardening activities with school students (Marsden, 2003).

Ethnic and Cultural Identity

Recent research in Australia suggests that young people's ethnic and cultural identities can be important factors in their outdoor learning. Purdie, Neill and Richards (2002) found that learning outcomes varied significantly with individuals' cultural identities: '*Most of the gains were made by students who rated themselves as totally Australian, and not by students who expressed somewhat of a lesser affiliation with an Australian identity*' (p. 38). They recommend that outdoor educators '*need to devise strategies to counter the psychological discounting and disengagement processes that are typical of how individuals attempt to cope with stereotype threat*' (p. 39).

The Setting

The importance of the setting is not a new theme in outdoor education research, especially on fieldwork (see, for example, Martin, Falk and Balling, 1981). A number of more recent studies have emphasised the importance of the location as a factor affecting students' outdoor learning. A recurring idea is that outdoor environments can place on students learning demands and emotional challenges, the impacts of which are not always sufficiently recognised by teachers and outdoor educators. Australian researchers reporting on a study of high school science students during visits to a marine theme park argued that '*teachers need to ensure that students are not distracted by the novelty of the location*' (Burnett, Lucas and Dooley, 1996: 63).

There is, however, clearly a balance to be struck between novelty and familiarity. In their study of students' perceptions of nature-based excursions, Ballantyne and Packer (2002) found that '*students who had not visited the particular site before were looking forward to their visit more than those who had*' (p. 221). Emmons (1997) saw significance in the fact that the programme that she evaluated '*did not completely remove students from all that was familiar to them, as might a nature experience for inner-city children in the USA, for example*' (p. 342). Instead, in her view, it was an environment that '*although certainly novel*' was also one that the students could link with, due to '*their own experiences in rural Belize*' and this contributed to its ability to challenge participants' environmental perceptions (p. 342).

In Conclusion

There is a concern that the amount of fieldwork in secondary schools is under threat. However, the evidence from research carried out around the world is that fieldwork can have a range of beneficial impacts on participants. To be effective, fieldwork needs to be carefully planned, thoughtfully implemented and followed up back at school. In planning activities, teachers and outdoor educators need to take account of factors such as students' fears and phobias, prior experience and preferred learning styles.

References

Ballantyne, R. and Packer, J. (2002) Nature-based excursions: school students' perceptions of learning in natural environments. *International Research in Geographical and Environmental Education,* **11**(3), 218–236.

Barker, S., Slingsby, D. and Tilling, S. (2002) *Teaching biology outside the classroom: is it heading for extinction? A report on biology fieldwork in the 14–19 curriculum.* FSC Occasional Publication 72. Preston Montford, Shropshire: Field Studies Council.

Bixler, R. D., Carlisle, C. L., Hammitt, W. E. and Floyd, M. F. (1994) Observed fears and discomforts among urban students on field trips to wildland areas. *Journal of Environmental Education,* **26**(1), 24–33.

Bixler, R. D. and Floyd, M. F. (1999) Hands on or hands off? Disgust sensitivity and preference for environmental education activities. *Journal of Environmental Education,* **30**(3), 4–11.

Bogner, F. X. (1998) The influence of short-term outdoor ecology education on long-term variables of environmental perspective. *Journal of Environmental Education,* **29** (4), 17–29.

Burnett, J., Lucas, K. B. and Dooley, J. H. (1996) Small group behaviour in a novel field environment: senior science students visit a marine theme park. *Australian Science Teachers' Journal,* **42**(4), 59–64.

Dierking, L. D. and Falk, J. H. (1997) School field trips: assessing their long-term impact. *Curator,* **40**(3), 211–218.

Eaton, D. (2000) Cognitive and affective learning in outdoor education. *Dissertation Abstracts International—Section A: Humanities and Social Sciences,* 60, 10-A, 3595.

Emmons, K. M. (1997) Perceptions of the environment while exploring the outdoors: a case study in Belize. *Environmental Education Research,* **3**(3), 327–344.

Harris, I. (1999) Outdoor education in secondary schools: what future? *Horizons,* 4, 5–8.

Healey, M., Jenkins, A., Leach, J. and Roberts, C. (2001) *Issues in providing learning support for disabled students undertaking fieldwork and related activities.* Available: http://www.glos.ac.uk/gdn/disabil/overview/index.htm (accessed 13 January 2004).

Lai, K. C. (1999) Freedom to learn: a study of the experiences of secondary school teachers and students in a geography field trip. *International Research in Geographical and Environmental Education,* **8**(3), 239–255.

Marsden, D. (2003) *Observations on the use of horticulture, gardening and environmental work in the learning and caring establishments that work with children and young people with special educational needs (SEN).* Reading: Thrive.

Martin, W. W., Falk, J. H. and Balling, J. D. (1981) Environmental effects on learning: the outdoor field trip. *Science Education,* **65**(3), 301–309.

Mittelstaedt, R., Sanker, L. and Vanderveer, B. (1999) Impact of a week-long experiential education program on environmental attitude and awareness. *Journal of Experiential Education,* **22**(3), 138–148.

Openshaw, P. H. and Whittle, S. J. (1993) Ecological field teaching: how can it be made more effective? *Journal of Biological Education,* **27**(1), 58–66.

Orion, N. and Hofstein, A. (1994) Factors that influence learning during a scientific field trip in a natural environment. *Journal of Research in Science Teaching*, **31**(10), 1097–1119.

Purdie, N., Neill, J. T. and Richards, G. E. (2002) Australian identity and the effect of an outdoor education program. *Australian Journal of Psychology*, **54**(1), 32–39.

Rickinson, M., Dillon, J., Teamey, K., Morris, M., Choi, M. Y., Sanders, D. and Benefield, P. (2004) *A review of research on outdoor learning*. Preston Montford, Shropshire: Field Studies Council.

SEER (State Education and Environment Roundtable) (2000) *The effects of environment-based education on student achievement*. Available: http://www.seer.org/pages/csap.pdf (accessed 23 January, 2004).

Simmons, D. A. (1994a) A comparison of urban children's and adults' preferences and comfort levels for natural areas. *International Journal of Environmental Education and Information*, **13**(4), 399–413.

Simmons, D. A. (1994b) Urban children's preferences for nature: lessons for environmental education. *Children's Environments*, **11**(3), 194–203.

Uzzell, D. L., Rutland, A. and Whistance, D. (1995) Questioning values in environmental education. In *Values and the environment: a social science perspective*, ed. Guerrier, Y., Alexander, N., Chase, J. and O'Brien, M. Chichester: John Wiley.

Wals, A. E. J. (1994) Nobody planted it, it just grew! Young adolescents' perceptions and experiences of nature in the context of urban environmental education. *Children's Environments*, **11**(3), 177–193.

12 On Food, Farming and Land Management

Towards a Research Agenda to Reconnect Urban and Rural Lives

Justin Dillon, Mark Rickinson, Dawn Sanders and Kelly Teamey

Dillon, J., Rickinson, M., Sanders, D., & Teamey, K. (2005). On food, farming and land management: towards a research agenda to reconnect urban and rural lives. *International Journal of Science Education*, 27(11), 1359–1374.

Introduction

By 2015, it is estimated that 53% of the world's population will live in urban areas (Chawla, 2002, p. 33). They will depend on those working in the countryside to provide food and other items considered to be essential or desirable. There is, however, evidence that the urban public's knowledge of food production and associated land management issues (such as the use of pesticides) is woefully inadequate (see, for example, Kuhlemeier et al., 1999; Trexler, 2000). This appears to be an issue in many countries around the world. Desmond et al. (1990, p. 151) describing 'new approaches for a better understanding of agriculture', point out that 'paradoxically, the United States has one of the world's most plentiful food supplies and possibly the least agriculturally-informed public'. In Europe, a study of 686 Greek primary schoolchildren reported that 'the pupils were ignorant about the significant impact of farmers on the food chain' (Paraskevopoulos et al., 1998, p. 58).

In some countries, such as England, farmers and other people who live and work in the countryside have expressed their dissatisfaction with the way that they are perceived by those who live in towns and cities. In some cases, this dissatisfaction has led to country-dwellers taking direct political action such as marches and disruption of motorway traffic. Given the potential impact of consumer knowledge and attitudes on farmers and those involved in the food chain, efforts are being made by a range of governmental and non-governmental bodies to address the issue (see, for example, Groundwork, 2000).

Policy Responses and the Need for Research Evidence

It has been known for some time that young people's knowledge of how their food is produced and how it gets to their plate seems limited. The

UK Policy Commission on the Future of Farming and Food argues that 'the key objective of public policy, in this area, should be to reconnect consumers with what they eat and how it is produced' (Policy Commission on the Future of Farming and Food 2002: 6). The Commission suggested a number of strategies including demonstration farms and the need for schools to develop stronger links with farms. As part of its response, the UK Government said that it recognized the importance of young people experiencing the 'outdoor classroom' and noted that 'children benefit from hands-on experiences of plants and animals, within school grounds, through visits to farms, woodlands or field study centres' (House of Commons, 2002, p. 47).

Assuming that reconnecting consumers with food and how it is produced is a reasonable aim, then in order to make wise decisions and to utilize resources appropriately, research evidence about the public's knowledge and understanding is desirable if not essential. Concerned, *inter alia,* about the apparent deficit in public knowledge and its impact on the rural community, the Countryside Agency, the Department for Education and Skills (DfES) and Farming and Countryside Education (FACE) commissioned King's College London and the National Foundation for Educational Research (NFER) to review the literature on what is known about young people's (3–19) views towards, and learning about, food, farming and land management. The study was carried out between November 2002 and April 2003, and the subsequent report drew together the evidence from almost 200 pieces of research published internationally in English between 1960 and 2002 (Dillon et al., 2003).

Although both the review and a summary of its findings are available online, we begin the paper with the research questions and their source. The way in which the critical analysis of the literature was carried out and some examples of the research reviewed are provided, together with a commentary on the review process. We discuss in some detail what we mean by the expression 'critical analysis' in this context. The second part of the paper draws on the findings of the review to posit a research agenda that might address some of the weaknesses in the evidence base. We are optimistic that there is the desire among funding bodies and policy-makers to facilitate at least some of the research that we propose and, subsequently, to engage with any relevant findings. We explain the reasons for our optimism in our concluding comments.

Because of the nature of the topics considered, we are addressing this agenda to the science education community. Although the focus of the discussion in this paper is on food and its production, we are aware that the level of urban public knowledge about rural affairs also has implications in the tourism and recreation sectors. A final caveat: although we use the terms 'land management' and 'countryside' throughout the paper, neither term adequately encapsulates all the issues that we think are important to be studied. There are, for example, many opportunities to teach students about ethical issues inherent in the management of the countryside that have yet to be researched in any depth.

A Review of the Literature on Food, Farming and Land Management

The aims of the review, which reflects the interests of the funders and a more general concern in the field, were to identify and appraise the extent and quality of research and statistical evidence relating to:

- School-age (3–19) children's knowledge of, and attitudes towards, food, farming and the countryside.
- School-age children's learning about food, farming and land management in a range of contexts.
- The impact of such learning on pupils' achievement, progression and other educational/behavioural outcomes (for all pupils and particular groups).
- How such learning experiences can be delivered most economically and effectively.
- The factors that can impede or facilitate pupil learning about food, farming and land management in a range of contexts.

The review team undertook a critical analysis of empirical research and statistical evidence based on criteria negotiated with the funding agencies. The criteria included work published internationally (in English) between 1960 and 2002 in articles, books and monographs, research theses, statistical evidence, and government/international publications covering early years, primary and secondary schools.

Methodologically, the focus was on establishing not only what was known, but also what was not known. Partly because of our own backgrounds (in science, geography and environmental education) and interests, we were interested in evidence relating to learning about the production and origins of food and about the links between producers and consumers through the food chain.

Comprehensive details of how the review was carried out can be found in the report that is available online (Dillon et al., 2003). Relevant research literature was identified using a number of complementary search methods, including: electronic bibliographic database searches; hand searches of journals and other documents; and e-mail requests for information to students and researchers working in the field. In essence, 5000 potential studies identified by various means, mainly electronic database searching, were whittled down to 190 relevant pieces of work. These were read and summarized under various headings (e.g. 'Knowledge and concerns about agriculture') that related directly to the research questions. In the following section we discuss aspects of the evidence base illustrating our points with examples from the literature.

A Framework for Analysis

An initial inspection of the 5000 or so studies identified through electronic searches allowed us to discard studies that were not empirical or that did not seem relevant to the review. Of 270 potentially relevant studies identified by the searches, 190 studies were eventually identified as being directly relevant to the review's aims. A common framework was used by the team to analyse the final selection of papers, chapters and reports. The framework included space for cataloguing, reporting, and evaluating the evidence base.

In terms of evaluation, the framework captured information about the depth of detail provided about the different aspects of each study (conceptual/theoretical framework, sample, methodology, validity measures, methods, main findings, key conclusions, and author's view of implications). Any particular strengths and potential weaknesses that were apparent to the reviewer within the work as reported were noted.

In terms of the evidence base, the purpose of the framework was to enable the generation of ideas about two issues. First, contributions that individual papers made to the evidence base (i.e. the main findings, key conclusions, author's view of implications and the researcher's view of the implications), and, second, cases of agreement and disagreement between the evidence generated by different papers (i.e. links with other studies).

The Evidence Base

We do not intend to summarize the 190 studies examined in details in the review. Rather, in the following section we aim to provide a flavour of the methodologies, the findings and the issues that we encountered. Subsequently, we will discus what we considered to be two types of gaps in the literature with a view to framing a realistic research agenda.

What Young People Know and How They Feel About Issues

In terms of school students' agricultural knowledge and concerns, there is a large body of empirical evidence. One area that has been examined is the awareness of agricultural production and its environmental impacts. Trexler (2000), for example, found that most of his small sample of elementary school pupils had little or no understanding of pests and pest protection. In a much larger study, a lack of knowledge about agriculture was also found among 9000 Dutch 15 year olds (Kuhlemeier et al., 1999). Only one-half of the sample, for example, knew that chemical pesticides and fertilizers are used more in 'regular' agriculture than in 'alternative' agriculture. Mabie and Baker, in a study of 147 Los Angeles pupils, found that 'very few children could give a basic definition of the word agriculture itself [and] most could not name crops grown by farmers in their state' (Mabie & Baker, 1994, p. 77). Few of the studies examined what or how students had been taught, they simply reported what they appeared to know.

As might be expected, where people live seems to play a part in their knowledge and their attitudes. For example, Matthews' and Falvey (1999) surveyed 550 year 10 students in Victoria, Australia and found that significantly more non-metropolitan than metropolitan students felt that agriculture has a negative impact on the environment. Age (Harbstreit & Welton, 2002) and socio-economic status (Kotrlik et al., 1986) also seem to be relevant factors.

Young people's awareness of and views about the use of biotechnology and genetic engineering in food production have been a focus for research. Hill et al. (2000), in an exploratory study with 270 16–19 year olds in north-west England, found that students see these two phenomena differently, and tend to associate genetic engineering (rather than biotechnology) with agriculture. Students' concerns varied, too, with biotechnology being seen as 'risky' whereas genetic engineering was more likely to be seen as 'ethically wrong'. Having said that, the study also found that 'many students had positive attitudes to biotechnology and genetic engineering' (Hill et al., 2000, p. 82).

Gunter et al. (1998) found that 16-year-old to 19-year-old students ($n = 138$) in north-west and south England had limited awareness about biotechnology and wanted more information on the topic. They were most mistrusting of information from government, food retailers and manufacturers. The authors concluded that 'as far as today's teenager's are concerned, reassurance is needed from food producers and retailers that food is safe to consume and that it has not been manufactured with environmentally unsafe or ethically dubious procedures' (Gunter et al., 1998, p. 111).

In terms of young people's relative feelings about issues generally, a comparative study of English and German 14 year olds carried out by Prelle and Solomon (1996) found that food and farming was seen as considerably less important than 'the hole in the ozone layer' and 'cutting down the rain forest'. Further evidence comes from a survey of over 1000 young people in Birmingham that included some items relating to food (Birmingham City Council Education Service 2000). When asked to select their four most important issues from a list of 10, almost one-third (32.9 per cent) selected 'too much fats and sugars in what I eat'. Overall, however, this was ranked seventh out of the 10, below other issues such as litter, air pollution, crime, and so on. More research is needed to understand the implications of these findings for policy-makers and educators.

A small number of studies of young people's understandings of ecological concepts (for example) food chains, food webs and ecosystems are available in the science education literature. A recurring theme is the difficulty young people have in understanding the inter-connected nature of a food web, as opposed to the more linear concept of a food chain. Barman et al.'s (1995) study of 96 US, Australian, and Canadian high school students found that their descriptions/representations of food chains were based on predator/prey ideas, rather than the transfer of energy. These findings echo those of Adeniyi

(1985) in Nigeria, Webb and Boltt (1990) in South Africa and Griffiths and Grant (1985) in Newfoundland. Again, however, many of these studies are surveys rather than explanatory research studies. They raise important questions, some of which we address in our research agenda in the second half of the paper.

Other research has focused on students' knowledge and views of particular kinds of food products, such as genetically modified foodstuffs, and locally produced/organic foods. Hill et al. (1998, 1999) found that students in England reported the advantages of genetically engineered food as improved storage and increased productivity, but not improved taste, reduced price or health benefits. Such foodstuffs were seen as 'unnatural', but not necessarily unsafe for the environment or for consumers. There are a lack of high-quality ethnographic studies of the role that food plays in young people's lives and the influence that advertising and school education play in helping young people to decide what to eat. We suggest, later in the paper, that ethnographic studies might advance the field significantly.

Some evidence about students' views is provided by 'Teaching and Learning about Food and Nutrition in Schools', a report by Burgess and Morrison (1995). The study found that boys and girls voiced differing attitudes towards food and drink, and that vegetarianism was seen as a 'female thing' to do. Worryingly for educators, factors such as 'home, family and the media' were seen by students as more important influences on their food consumption than school, particularly by secondary-age students.

A study of 647 New York City high school students (Bissonnette and Contento 2001) found that students were generally positive about organic foods, believing that they were healthier, tastier and more environmentally friendly. They were, however, less knowledgeable about the issue of locally grown foods and in favour of organic and local food as an abstract ideal, but did not necessarily see it as important to purchase such foods themselves. They were, as many parents of teenagers would also be aware, more concerned with taste and the ability to eat their favourite foods all year round, than with where or how food had been grown. The study's key finding was that:

> while 'these adolescents were not knowledgeable about [. . .] the environmental impacts of food production practices and were uncertain in their attitudes, [. . .] the beliefs and attitudes that they did have about these issues were correlated with their food choices, at least to a moderate degree'.
>
> (Bissonnette & Contento, 2001, p. 81)

This finding, Bissonnette and Contento argue, highlights the 'need to make salient to adolescents the environmental impact of food production practices through both cognitive and experiential approaches' (2001, p. 72). In other words:

adolescents should be provided with field trips to farmers' markets [. . .] and visits to both organic and conventional farms to observe how foods are grown and to talk with farmers.

Furthermore, in view of the important influence of parents, such strategies:

> should be accompanied by the involvement of parents, for example through take-home materials or through participation of parents in field trips.
>
> (Bissonnette & Contento, 2001, p. 81)

However, no follow-up studies seem to have been carried out to see whether the word 'should' in the recommendations is actually justified.

Factors Impeding or Facilitating Learning

As well as distilling what is known about students' knowledge of and attitudes towards the topics under consideration, the team also examined what the research said about factors that might impede or facilitate the quality of young people's learning in a range of contexts. We were able to classify our findings under four themes; emotions, fears and phobias; teacher planning and training; learners' motivational and identity factors; and the effects of settings.

In terms of emotions, fears and phobias, impediments to learning often reside within individual teachers and learners rather than being intrinsic to the topic itself. Fears and phobias, for example, were mentioned by several studies of both teachers (Simmons, 1998), and pupils (Bixler et al., 1994). Teachers and pupils, not surprisingly, had a strong fear of getting lost in unfamiliar locations. Teachers reported poisonous plants as primary fears whereas pupils reported fearing snakes and insects. Dirt and mud and touching phobias also rated quite highly.

What are termed disgust factors appear to have implications for participation in activities that involve handling organisms and/or organic matter. Bixler and Floyd (1999), in a study of 450 middle-school students in Texas, noted a correlation between feelings for handling organisms and preferences for fieldwork settings. Students with high disgust sensitivity were significantly more likely to prefer activities that did not involve handling organic matter, and fieldwork sites with clear water, no algae and easy lakeshore access. Bixler and Floyd (1999) reported that aversions to slugs, snails and crawling insects are primarily due to disgust sensitivity rather than fear factors. They suggest that further work should be done to test whether disgust sensitivity is predictive of actual behaviour of students participating in 'hands-on' fieldwork.

With respect to planning and training, teachers' lack of familiarity with the field-trip setting is a common theme explored by several studies. In our experience, teachers sometimes value the novelty of a situation (e.g. residential experience or a novel geographical location) but may be unaware that this might inhibit student learning for some time. Martin et al. (1981) concluded that novel environments make learning demands that have not been adequately considered by educators. This concern could necessitate a perceptual shift for teachers, but also suggests the need for a stronger interface between educators and researchers working in this field—few teachers are up-to-date with the research literature.

The impact of previous fieldwork experiences on students can be a factor in assisting student engagement with study topic. Openshaw and Whittle, in their study of fieldwork in the UK, comment upon the role that experiences in previous settings has in facilitating learning on field trips:

> If students have been accustomed to a diet of 'experiments', based on well tried recipes that 'work'; then real practical ecology is likely to prove a difficult experience for them.
>
> (Openshaw & Whittle, 1993, p. 64)

Turning now to learners' motivation and identity, the impact of a young person's cultural identity on their learning needs to be considered in planning out-of-school experiences. This is particularly so if learning about food, farming and land management is to provide appropriate and meaningful experiences for the multicultural urban communities present in many inner-city schools. Several papers have examined motivational factors and/or questions of identity as elements that can influence the learning process within a range of relevant settings. For example, in a US study of high school agriculture students, Turner and Herren (1997) concluded that educators must provide agricultural learning opportunities that are informed by different motivation factors and the culture of motivation. For example, female students are motivated to participate in different ways than male students. Crucially, they also acknowledge differences in motivation between urban and rural students.

Finally, with respect to the effects of settings, there is some evidence that young people do not see aspects of the environment in the way that adults might assume. There is also evidence that urban settings may limit young people's opportunities to forage, and in doing so inhibit their ability to develop empathy for flora and fauna. In a Canadian study (Chipeniuk, 1995) on childhood foraging, two groups of teenagers in two distinct landscapes were given the opportunity to forage for artifacts. The study highlighted foraging as a process that enabled children and young people to come into contact with local biodiversity. However, the author was concerned at the possible gaps in research on 'megalopolitan' children and their relationships with nature and the process of foraging. He questioned whether 'some environments are too artificial to develop children's sense of biodiversity properly' (Chipeniuk, 1995, p. 507).

Respecting Research Traditions—What Makes a Review 'Critical'?

We have used the term 'critical analysis' several times. Rather than assume that readers understand what we mean by this term, we would like to discuss how we interpreted it and how we strove to ensure that the review respected research traditions while maintaining an independent, critical nature. To us, being critical means that the importance of different research paradigms should be recognized and that evidence that is more conclusive needs to be distinguished from evidence that is less conclusive. We hope that this discussion will be clear in the light of the examples of the studies reviewed that we have provided earlier.

We attempted to review work *from within* the research tradition (or paradigm) that it had been conceived and undertaken. Specifically, quantitative (pre-test/post-test) programme evaluations were considered in terms of positivistic research traditions, while qualitative case studies were examined from the perspective of interpretivistic inquiry. The concern was to examine how well the researchers had carried out what they had intended to according to the paradigm in which they were operating.

Distinguishing between evidence that is more conclusive and less conclusive can be achieved by identifying methodological strengths and weaknesses of individual studies. It was thus possible to distinguish between evidence that was more reliable and conclusive, and evidence that was more questionable or preliminary. Specifically, this process involved distinguishing between findings based on empirical evidence and those based on anecdotal reflection or unjustified prior assumptions. It also involves distinguishing between claims based on empirical findings and those based on speculation about empirical findings; statistically significant results and those based on description of trends; and survey findings based on very small samples and those based on larger representative samples.

Another important aspect of being critical, however, is the idea that strengths and weaknesses in evidence can be identified not only at the level of individual studies but also at the level of the evidence base as a whole. Each of these levels involves quite different issues and concerns. At the level of individual studies, strengths and weaknesses were in terms of methodological coherence, research design, analytical procedures, validity of claims, and so on. At the level of the evidence base, issues such as the diversity of evidence types, the nature and variety of substantive foci, and the extent of interconnections between types of evidence and types of foci were more important. We pointed out, earlier, that there is a lack of ethnographic studies in the area, which we suggest is a weakness.

An Analysis of the Evidence Base—'Blank Spots' and 'Blind Spots'

Having looked at 190 individual studies in a critical manner, we identified what we considered a range of strengths and weaknesses in the evidence base.

Although many surveys have been carried out into knowledge and attitudes, there was a general lack of recognition of theories of learning or of links to broader conceptual frameworks. This was partly a result of examining studies dating back to the 1960s and 1970s when learning theories were less well developed than now but many more recent studies suffered from the same lack of theory. Equally disturbing, we noted a general lack of concern about validity or reliability, a lack of critical reflection and a lack of convincing evidence substantiating claims for better learning or improved attitudes resulting from teaching strategies. We address these weaknesses in our proposed agenda.

There is a danger, in looking at studies carried out many years ago, that the judgements of today might seem unfair to researchers of the past. Richly contextual case studies might end up being discarded because of their lack of generalizability or, more probably, there lack of evidence to support their assertions. We acknowledge this possibility and have tried to ensure that important messages did not get lost. However, in the final analysis, policy-makers and practitioners often need convincing evidence rather than richly woven rhetoric, and we have erred on the side of caution when reporting on studies that we decided others might reasonably find unconvincing.

In the light of the aforementioned caveat, we identified what we described as key messages emerging from the literature review. These messages were as follows:

- School-age students' knowledge and understanding about various aspects of food and farming appear to be poor.
- While young people are concerned about food issues such as genetic engineering, organic/local products, there is also evidence of ambivalence and confusion in students' views, and inconsistencies between attitudes and behaviours.
- For food and farming issues in general, some studies suggest that young people see these as less serious than other environmental issues such as ozone depletion and tropical deforestation.
- In several studies, levels of concern were found to differ between boys and girls, with girls attaching greater seriousness to issues such as food additives and pesticides, genetic engineering of farm animals, and the importance of organic/local foods.
- Young people's perceptions and experiences of the countryside are complex and varied: while some children can have a very positive attitude towards the countryside, others focus on the possibility of boredom and isolation in rural areas.
- The research on young people's knowledge and attitudes suggests that there is a strong case for improving teaching and learning about food, farming and land management. (Dillon et al., 2003, p. iii–iv)

In taking a broad look at the evidence base and by looking at these messages, we identified what Wagner (1993) refers to as 'blank spots'; that is, gaps in the knowledge base (e.g. few studies address the UK context whereas there are many from the US). Sometimes these blank spots are 'obvious': sometimes they emerge in comparison with other areas of science or environmental education. We were also interested in identifying 'blind spots'; that is, topics that have been neglected by researchers for whatever reason. The 'blind spots' that we identified were:

- Studies of young people's understanding of the connections between food, farming and land management in terms of the food chain.
- Students' perceptions of the countryside as a context for food production and land management.
- Changes in children's thinking about food, farming and the countryside over a period of several years resulting from one or more learning activities in a range of contexts such as farms and botanical gardens.
- The sources of, and the factors that can influence and shape, young people's knowledge, attitudes and concerns about food, farming and the countryside.
- Teachers' aims for work in and visits to farms and botanic gardens.
- Students' learning experiences resulting from farm visits and other food and farming-related activities. (Adapted from Dillon et al., 2003, p. vii)

Again, the blind spots emerge from a comparison of the literature base with that in other fields in the social sciences. So, for example, although we reviewed studies in which successful teaching was described, researchers were rarely able to identify what aspects of a particular programme helped to yield positive impacts. While there is strong evidence about the impacts of some programmes, there is little evidence that can help curriculum planners and developers not to mention teachers and others who engage with students at farms, botanic gardens, and so on.

A Research Agenda

In order to address the weaknesses and gaps in the literature base, and thus to provide evidence for informed policy initiatives, we have identified the following research agenda. For each of the three broad topics on the agenda we have identified several research questions. We think that answers to those questions would be useful to policy-makers. We have identified why we think such research is needed in terms of policy change. We have also suggested a methodology and methods for some of the studies we propose.

We regard it of paramount importance that any study aiming to address the questions outlined in the following should discuss issues of validity and reliability and should provide convincing evidence to support any findings supporting particular teaching strategies or resources. We would hope that this stipulation would be redundant but our experience of reviewing almost 200 studies confirms that it is not. We would also expect that studies reported in refereed journals would contain some element of critical reflection. Again, we have not found this to be equally true of all studies.

Research Area 1: Identifying Learning about Food, Farming and Land Management—the Content and the Process

Research Questions

- Where do children learn about food, farming and land management?
- What does learning about these topics involve cognitively and affectively?
- What are teachers' aims for visits to farms and school gardens?
- Where and what do children learn about the connections between food, farming and land management in terms of the food chain and the processes by which food gets from the farm gate to the dining table?
- What are students' perceptions of the countryside as a context for food production and land management?
- What does progression look like in terms of understanding food, farming and land management issues?
- How does children's thinking about food, farming and the countryside change over a period of several years after involvement in learning activities in a range of contexts such as farms and botanic gardens?
- What factors (including social and historical) can influence and shape young people's knowledge, attitudes and concerns about food, farming and the countryside?

Methodology and Methods

This set of questions would require studies that aimed to address the lack of recognition of theories of learning or of links to broader conceptual frameworks in the literature reviewed. A range of learning theories from behaviourist, constructivist, socio-cultural and situative perspectives might be used to inform the design of the studies as well as the interpretation of findings. Mixed methods of research, using the strengths of qualitative and quantitative methodologies, would, we believe, maximize the chances of yielding rich and useful findings. Studies using ethnographic approaches into, for example, how young people's thinking about food, farming and the countryside change over a period of several years might provide evidence of the complex overlap between the learning that takes place in school and the influence of children's wider patterns of socialization.

In terms of identifying what happens already in schools, desk-based research looking at the curriculum, textbooks, websites and inspection reports would need to be supplemented by ethnographic studies utilizing school visits and observations of learning activities out-of-school. We envisage in-depth, mixed-method studies of students' learning in terms of processes and outcomes from particular learning activities such as farm visits, ICT-based programmes and fieldwork.

In terms of progression, we envisage small-scale, classroom-focused studies of classroom and out-of-classroom learning of related topics at different ages might develop our knowledge base about what progression, in this context, looks like. Longitudinal studies of small numbers of students might be possible although few funders might support such a strategy.

We believe that there would be some benefits of quantitative and qualitative comparative studies of curriculum organization and implementation in countries outside the UK. In the US, for example, there is a long tradition of agricultural education that might yield interesting comparisons with that found in other countries.

Rationale

Any curriculum change aimed at developing better understanding of food, farming and land management issues needs to be based on sound knowledge of what is appropriate for children at different ages and stages of development, and based on what we know about learning. A better understanding of learning in formal and informal contexts might lead to better learning out-of-doors.

Research Area 2: Removing the Barriers to Learning

Research Questions

- What are the barriers to learning about food, farming and land management in terms of provision (e.g. the status of the topics; numbers of suitably qualified staff; financial resources for going on trips; requirements of curriculum and/or assessment schemes)?
- What are the barriers to learning about food, farming and land management in terms of learning (e.g. teachers' attitudes towards different sites for fieldwork; children's fears and phobias)?
- How do such barriers vary between different kinds of students?
- How can the provision of learning opportunities related to these topics be increased?
- How can the non-provision barriers related to learning about these topics be reduced or overcome for different kinds of students?
- What lessons can be transferred from experience in the informal learning environments to formal learning environments, and vice-versa?

Methodology and Methods

We mean here physical barriers such as cost and safety considerations as well as psychological barriers (fears, phobias, etc.). We envisage that ethnographic and more quantitative comparisons of schools in similar situations (location, size, funding, etc.) might identify strategies for coping with the physical barriers. Such studies might be carried out at different levels of the education system: local education authorities, schools, departments and in non-school contexts.

In terms of psychological barriers, we would envisage that knowledge from disciplines other than education might offer avenues to reduce the fears and phobias possibly through better preparation of children for what they might reasonably expect to see. Intervention studies might provide an effective strategy for testing out new approaches. An end product might be learning theories imbued with insights from a range of disciplines: cognitive psychology, neuro-psychology and social psychology, for example,

Rationale

Teachers and others are quick to identify 'practical' reasons why they cannot take children on visits (safety, cost, etc.). However, some schools appear to find ways to minimize the barriers while others treat them as immutable. In terms of *psychological* barriers, better understanding of what stops people from learning effectively might help students to get more out of the rare opportunities that they get to experience the outdoors.

Research Area 3: Measuring Cost-Effectiveness

Research Questions

- Can the cost-effectiveness of particular teaching strategies be measured?
- What strategies already exist to identify the costs and benefits of particular learning strategies (visits, etc.)?
- What perceptions of 'value for money' do headteachers, parents and teachers hold?
- Do some strategies or sites for learning offer better value for money than others?

Methodology and Methods

In some ways this is the most challenging of the questions. Identifying the long-term benefits of learning is not something that is easily done, and disaggregating the impacts of particular episodes of learning is certainly almost impossible. We believe that some work identifying 'best value' in terms of some aspects of education provision has been carried out by some local education authorities in England although it is not in the public domain. Such studies might well inform research into the costs and benefits of educating people about food, farming and land management. Such studies might be of interest to the DfES funded Centre for the Economics of Education.

In terms of methodology, research studies carried out within a more positivist, economics-based framework, while potentially controversial, might shed some interesting light on subjects that often seem taboo in contemporary educational discussions.

Rationale

Despite the inherent difficulties in addressing such research questions, schools have budgets and need to know how best to spend them. If researchers ignore these issues they face charges of failing to engage with the real world of educational finance. Some notion of relative costs and benefits of particular strategies (and an awareness of the limitations of such estimates) would help policy-makers at both national and school levels.

Concluding Comments

We have set out what we feel is a realistic agenda for research into the teaching and learning in the context of food, farming and land management. Subsequent to the publication of our report, the Countryside Agency, the DfES and FACE invited tenders for research into 'The Outdoor Classroom in a Rural Context'. A team drawn from King's College London, NFER and the University of Bath carried out the research. This project explored the processes and impacts of outdoor learning and the planning and evaluation of teaching activities. The project finished in April 2005 and is due to report later in the year. We are convinced that other sources of funding, such as the Economic and Social Science Council small awards scheme (in England) are available to researchers interested in other aspects of the agenda.

If science education is to contribute towards a more sustainable future then researching the teaching and learning of food, farming and land management would seem to be a prerequisite step. The same is true for citizenship, too: in England and Wales, pupils in the final phase of compulsory schooling should be taught about 'the wider issues and challenges of global interdependence and responsibility, including sustainable development and Local Agenda 21' (Qualifications and Curriculum Authority, 2004). We believe that this global dimension of science that requires an understanding of complex socio-scientific issues (see, for example, Ratcliffe & Grace, 2003) is something that policy-makers and researchers need to address urgently if the growing gap between urban and rural populations is to be reduced.

Acknowledgements

The authors would like to acknowledge the significant input made to the research review by Pauline Benefield of NFER. They would also wish to acknowledge the financial support received from the Countryside Agency, the DfES and Farming and Countryside Education.

References

Adeniyi, O. E. (1985). Misconceptions of selected ecological concepts held by some Nigerian students. *Journal of Biological Education, 19*(4), 311–316.

Barman, C. R., Griffiths, A. K., & Okebukola, P. A. O. (1995). High school students' concepts regarding food chains and food webs: a multinational study. *International Journal of Science Education, 17*(6), 775–782.

Birmingham City Council Education Service. (2000). *Year of the environment: Survey of the views and concerns of young people.* Birmingham: City Council Education Service.

Bissonnette, M. M., & Contento, I. R. (2001). Adolescents' perspectives and food choice behaviors in terms of the environmental impacts of food production practices: application of a psychosocial model. *Journal of Nutrition Education, 33*(2), 72–82.

Bixler, R. D., Carlisle, C. L., Hammitt, W. E., & Floyd, M. F. (1994). Observed fears and discomforts among urban students on field trips to wildland areas. *Journal of Environmental Education, 26*(1), 24–33.

Bixler, R. D., & Floyd, M. F. (1999). Hands on or hands off? Disgust sensitivity and preference for environmental education activities. *Journal of Environmental Education, 30*(3), 4–11.

Burgess, R. G., & Morrison, M. (1995). *Teaching and learning about food and nutrition in educational settings (Final Report).* Swindon: Economic and Social Research Council.

Chawla, L. (Ed.) (2002). *Growing up in an urbanising world.* London: UNESCO and Earthscan Publications.

Chipeniuk, R. (1995). Childhood foraging as a means of acquiring competent human cognition about biodiversity. *Environment and Behavior, 27*(4), 490–512.

Desmond, D. J., Leising, J. G., King, N. J., Rilla, E. J., & Coppock, R. (1990). New approaches for a better understanding of agriculture. In H. O. Carter, & C. F. Nuckton (Eds.) *Agriculture in California: On the brink of a new millennium* (pp. 151–158). Davis, CA: University of California, Davis, Agricultural Issues Center.

Dillon, J., Rickinson, M., Sanders, D., Teamey, K., & Benefield, P. (2003). *Improving the understanding of food, farming and land management amongst school-age children: A literature review. Research Report 422* Department for Education and Skills. Available online at: http://www.dfes.gov.uk/research/data/uploadfiles/RR422b.pdf (accessed 26 March 2005).

Griffiths, A. K., & Grant, B. A. C. (1985). High school students' understanding of food webs: identification of a learning hierarchy and related misconceptions. *Journal of Research in Science Teaching, 22*(5), 421–436.

Groundwork (2002). *Farmlink: Connecting children with the countryside.* Available online at: www.groundwork.org.uk/what/doc/farmlink.doc (accessed 26 March 2005).

Gunter, B., Kinderlerer, J., & Beyleveld, D. (1998). Teenagers and biotechnology: a survey of understanding and opinion in Britain. *Studies in Science Education, 32*, 81–112.

Harbstreit, S. R., & Welton, R. F. (1992). Secondary agriculture student awareness of international agriculture and factors influencing student awareness. *Journal of Agricultural Education, 33*(1), 10–16.

Hill, R., Stanisstreet, M., Boyes, E., & O'Sullivan, H. (1998). Reactions to the new technology: students' ideas about genetically engineered foodstuffs. *Research in Science & Technological Education, 16*(2), 203–216.

Hill, R., Stanisstreet, M., & Boyes, E. (1999). Genetically engineered foodstuffs: school students' views. *International Journal of Environmental Studies, 56*, 785–799.

Hill, R., Stanisstreet, M., & Boyes, E. (2000). What ideas do students associate with 'biotechnology' and 'genetic engineering'? *School Science Review, 81*(297), 77–83.

House of Commons (2002). Response to the Report of the Policy Commission on the Future of Farming and Food by HM Government (Cm. 5709). London: The Stationery Office.

Kotrlik, J. W., Parton, G., & Lelle, M. (1986). Factors associated with knowledge level attained by vocational agriculture 11 students. *American Association of Teacher Educators in Agriculture Journal, 27*(2), 34–43.

Kuhlemeier, K., Van Den Bergh, H., & Lagerweij, N. (1999). Environmental knowledge, attitudes and behavior in Dutch secondary education. *Journal of Environmental Education, 30,* 4–14.

Mabie, R., & Baker, M. (1994). *Strategies for improving agricultural literacy and science process skills of urban fifth and sixth graders in the Los Angeles unified school district.* Paper presented at the Annual Western Region Agricultural Education Research Meeting, Honolulu, HI.

Martin, W. W., Falk, J. H., & Balling, J. D. (1981). Environmental effects on learning: the outdoor field trip. *Science Education, 65*(3), 301–309.

Matthews, B., & Falvey, L. (1999). Year 10 students' perceptions of agricultural careers: Victoria (Australia). *Journal of International Agricultural and Extension Education, 6*(1), 55–67.

Openshaw, P. H., & Whittle, S. J. (1993). Ecological field teaching: how can it be made more effective? *Journal of Biological Education, 27*(1), 58–66.

Paraskevopoulos, S., Padeliadu, S., & Zafiropoulos, K. (1998). Environmental knowledge of elementary school students in Greece. *Journal of Environmental Education, 29,* 55–60.

Policy Commission on the Future of Farming and Food (2002). Farming & food: a sustainable future. Available online at: http://archive.cabinetoffice.gov.uk/farming/index/commissionreport.htm (accessed 26 March 2005).

Prelle, S., & Solomon, J. (1996). Young people's 'general approach' to environmental issues in England and Germany. *Compare, 26*(1), 91–101.

Qualifications and Curriculum Authority (QCA) (2004). *The Programmes of Study for Citizenship at Key Stage 4.* Available online at: www.nc.uk.net/webdav/servlet/XRM?Page/@id=6001&Session/@id=D_5tUMAqvYc22VM76mHQks&POS[@stateId_eq_main]/@id=4188 (accessed 26 March 2005).

Ratcliffe, M., & Grace, M. (2003). *Science education for citizenship.* Maidenhead: Open University Press.

Simmons, D. (1998). Using natural settings for environmental education: perceived benefits and barriers. *Journal of Environmental Education, 29*(3), 23–31.

Trexler, C. J. (2000). A qualitative study of urban and suburban elementary student understandings of pest-related science and agricultural education benchmarks. *Journal of Agricultural Education, 41*(3), 89–102.

Turner, J., & Herren, R. V. (1997). Motivational needs of students enrolled in agricultural education programs in Georgia. *Journal of Agricultural Education, 38*(4), 30–41.

Wagner, J. (1993). Ignorance in educational research: or, how can you not know that? *Educational Researcher, 22,* 15–23.

Webb, P., & Boltt, G. (1990). Food chain to food web: a natural progression? *Journal of Biological Education, 24*(3), 187–190.

13 Science, the Environment and Citizenship

Teaching Values at the Minstead Study Centre

Justin Dillon and Alan Reid

Dillon, J., & Reid, A. (2007). Science, the environment and citizenship: teaching values at the Minstead Study Centre. In, D. Corrigan, J. Dillon, & R. Gunstone (eds), *The Re-emergence of Values in Science Education*. Rotterdam: Sense Publishers, pp. 77–88.

> . . . using science and the natural world as a vehicle for something else. And that something else is citizenship, being a good human being and the world could be a better place.
>
> (Chris, Teacher, Minstead Study Centre)

Introduction

Whereas values are usually implicit rather than explicit in much school science education, some environmental study centres deliberately promote a range of personal, communal and environmental values. In this chapter we examine the views of two educators who have worked with thousands of young people at an unusual although not unique residential study centre in England.

In describing the strategies used at the centre and in discussing a series of issues, we use a conceptual framework that has emerged from our shared reflections on the interview data. We regard this study, using Stake's (2000) terms, as both an *intrinsic* and an *instrumental* case study. An intrinsic case study:

> is undertaken because one wants better understanding of [a] particular case. It is not undertaken primarily because the case represents other cases or because it illustrates a particular trait or problem, but because, in all its particularity and ordinariness, this case itself is of interest.
>
> (Stake, 2000, p. 237)

An instrumental case study provides 'insight into an issue or refinement of theory' (Stake, 2000, p. 237). As we have written elsewhere:

> This is a familiar function: a case study of the particular is drawn upon to address and/or elucidate more concisely some essential, underlying principle, issue or point the author seeks to highlight, even to the extent

that general features or lessons may be drawn. It is about deepening understanding and knowledge of the issue at focus.

(Dillon & Reid, 2004, p. 26)

Because of the unusual nature of the centre, the chapter does not focus on what should or should not be taught based on theoretical positions. Rather, our approach is to give substantial voice to the thoughts of the two educators, Jane and Chris, because of their broad experience actively promoting values to young people of many ages and backgrounds through their teaching. Their views, values and pedagogy are of interest in themselves and provide an opportunity to explore some general features concerned with the relationship between the curriculum of the centre, science and the environment. We are particularly interested, in contributing to this book, to reflect on a range of issues including: the process of choosing teaching activities to promote values; the use of metaphors to promote learning, and, the relationship between environmental values and science education. Thus, in effect, with Jane and Chris we have developed what we term a 'Janus' case study, simultaneously looking inwards and outwards by integrating the intrinsic and instrumental functions that a case study methodology can afford.

The Minstead Study Centre

> Minstead Study Centre is a magical place—a place where you can discover all sorts about yourself and the plants, animals and people that you share the Planet Earth with.
>
> (Minstead Study Centre website: http://www.wildwoodweb.co.uk/)

Minstead Study Centre occupies the grounds and buildings of a Victorian village school and offers accommodation for up to 30 primary age (5–11 years-old) children. The centre is owned by the local education authority (LEA) which funds 50% of the cost of school visits. Minstead is situated in a national park in the south of England and occupies a prime site for the study of a wide range of habitats in an area of outstanding natural beauty. The centre grounds have been developed to provide a number of unique features including an extensive organic vegetable and herb garden, a miz-maze (a pattern cut in the turf), willow tunnel, ponds, an astronomical circle, an armillary (a skeletal celestial sphere) and a Celtic roundhouse. On arrival, each child is given a badge with the Latin name of an English animal or plant, and children learn to reflect on the world by empathising with this, their 'creature teacher'.

The activities that children carry out are curriculum-related because the centre is owned by the LEA and because schools tend to choose residential experiences that support the English national curriculum (QCA, 2006). The centre's curriculum is based on science and geography though the activities involve other school subjects such as art, drama, music, religious education and history. Two qualified teachers plan and deliver diverse learning experiences ranging from lunar modelling, animating, egg collecting and spinning wool to organic gardening and building compost heaps. Even if activities might not be novel, the approach that the staff take certainly is, as we will discuss later.

The centre aims to provide both environmental education and education for sustainable development. The ethos at the Centre is to promote respect for all living things and to encourage children to work together:

> Developing personal responsibility and nurturing positive attitudes towards each other ranks high amongst the aims of the Centre. Such diversity allows us to reach a wider audience. We feel it is through such experiences that children are able to establish and understand their connection, influence and responsibilities towards the people, plants and animals of Planet Earth.
>
> (Minstead Study Centre website: http://www.wildwoodweb.co.uk/)

As well as establishing and understanding their connections with the wider world, students are expected to value other people with whom they come into contact during the (usually) five days that courses last.

In this section, we provide an analysis of the Minstead experience as presented to us by Jane, the director, and Chris, who undertakes the majority of the teaching. Our relationship with the staff at Minstead is grounded in our earlier work with them on an action research project looking at learning in the outdoor classroom (Dillon et al. 2005). An interview involving one of the authors (JD) and Jane and Chris forms the primary source of data for this chapter. The interview focused on the relationship between Jane's and Chris's values and on how they used activities to promote particular ideas and behaviours in the primary school aged pupils who visit the centre. We have also included text from the centre's website and have referred back to our notes from previous visits and action research work. We have additional data from interviews with children that were recorded during previous research visits but, for the purposes of this chapter, and with Jane and Chris's agreement, we focus on the values of the two key staff at the centre. We emphasise here that the interview involved both Jane and Chris responding to questions and discussing their answers with each other, something that needs to be borne in mind when reading the chapter.

Our framework for analysing the Minstead experience involves several categories that emerged from our reading of the transcript of the interviews with Jane and Chris (Table 13.1), interpretations of which were cross checked and discussed with the two of them. In analysing the data, we did not use the start list approach (Miles & Huberman, 1994, p. 58), rather, we adapted a more grounded approach based on that of Glaser and Strauss (1987). As Bliss et al. (1983) point out, the meanings of words are not contained in them but depend on their significance in given contexts, so, our coding

Table 13.1 Conceptual framework emerging from the interview analysis

Code	Explanation
Values	Underpinning assumptions and beliefs
Metaphors	Comparison between ideas seemingly unrelated
Issues	Environmental or scientific concerns
Activities	Tasks undertaken by students
Strategies	Generic pedagogical approaches
Outcomes	The end product of the activities

departs from a strictly inductive grounded theory methodology in the sense that we openly acknowledge that our selections and interpretations are informed by, and reflect, our views of the meanings of the words, in the light of their use in the context of the interview, our foreknowledge of the work and staff at Minstead, and our wider engagement with the literature and field in these areas.

In this chapter we use the framework to illustrate a range of themes that we think are worthy of further discussion. We draw extensively from the transcript in order to illustrate the work at Minstead in detail, and to prevent this chapter being too focused on our interpretations of the interview data.

To illustrate the framework, we start with an annotated version of a transcription of an answer given by Jane to the question about what she and Chris had done with the group of pupils that had left about thirty minutes before the interview took place. Shortly after arriving at the centre, the children are taken into the Wild Woods (see http://www.wildwoodweb.co.uk/) which gives them the opportunity to immerse themselves in the immediate natural environment, and to appreciate the tranquility and beauty of their surroundings. Stillness and reflection are used to engage the children. Chris sets the tone of the week: the respectful way that staff and visitors work together, the search for the 'secrets of the earth', and the values the staff attach to 'our work'. Much of the programme is based around circles: a circle of equality where all have equal place and equal voice. Once back in the centre the staff focus on how the children can make a difference in helping the planet regain a balance. They speak of how the planet has 'a problem, a headache, a mountain of landfill to deal with'. The children are encouraged to think of themselves as 'waste watchers' reducing, reusing, recycling towards a zero waste outcome:

> Jane: I think the simple fact of going into the woods for their first immersion activity, 'Into The Wild Woods', you are allowing them that opportunity to immerse [STRATEGY], and appreciate tranquillity and the natural environment and just the beauty of surroundings—to allow this opportunity and then putting it forward as this is something to value

[VALUE]. If you are going to be quiet in it you will appreciate more [STRATEGY]. If your behaviour is such that you allow yourself to be receptive, then the children, I think, can absorb more of that [STRATEGY]. And the style of the way that Chris would work with them in the woods is kind of setting the tone of how we would expect the children to listen to us and respect us and listen to each other in the circle—the 'circle of equality' [STRATEGY]. And most of our activities are based on this kind of circle of equality [STRATEGY]. So the teacher has a say but the children have a say too [VALUE], and they have got a place to contribute to that [STRATEGY].

And I think, when they are coming back to the building, we take on board more the idea that the planet is threatened [ISSUE], and talk through various scenarios of how the planet has got a headache or back ache, eczema or whatever [METAPHOR]. And the children could be, should be, valuing the planet [OUTCOME], and they have a part to play in its recovery or remediation, whatever [VALUE]. And so when they are first back here I would be talking about reducing, reusing and recycling as ways to solving some of the burgeoning problems that are ahead [STRATEGY]. And I put the problems down to mankind [ISSUE]. I say it is mankind that is mucking up this delicate equation [VALUE].

Two of the key strategies that characterise the Minstead experience are the implicit and explicit use of metaphors and stories. In effect, we see the values expressed by Jane and Chris as underpinning their overall pedagogic strategies (such as using hands–on activities). We also assume that the activities that they plan and deliver are guided, implicitly and explicitly, by the centre's (predominantly Jane's and Chris's) underlying values and strategies. Their activities are aimed at promoting specific outcomes such as reflection and subsequent behaviour change (sometimes the behaviour change precedes the reflection). Issues, such as the influence and impact of humans on the environment, provide the content warp to complement the weft of values. In the next part of the chapter, we discuss the Minstead experience in more detail focusing particularly on how, in theoretical and practical terms, values, pedagogies and outcomes can become linked.

The Minstead Experience

The centre's ethos is made explicit as soon as the students arrive: helping each other out and working as a team are stressed as being of great value. The visitors are encouraged to realise that their week will be enriched if they co-operate and if they look out for each other.

Jane: The first thing we talk to them about is when they are unpacking their belongings and you are trying to establish the idea that they might work better if they co-operate as a team. So you highlight the issue of teamwork and co-operation is going to make a more successful week. Looking after each other, caring for their friends and keeping a happy attitude. I think probably it is then that we tend to give the first input of intent. Just thinking how we move through from the moment of arrival and they are getting out their things—personal responsibility as well as caring for the group. We try and establish that and then we establish the links with places far away and people far away.

Chris: Within ten minutes of being here they are aware that they are expected to work as a team. And value opinion, and respect opinion, and listen to opinion. Then fairly quickly we are dropping into . . . they are looking to something broader than the group themselves, there is a group connection.

Although the core of the curriculum is a mix of subject work in science and geography, the key outcomes that the centre promotes are better team-work, responsibility for self and others, respect for opinions and connections with the wider context. Fundamental to both Jane's and Chris's shared view of education for sustainable development and environmental education is that although it is the local environment that is studied, the *activities* are designed so that the students make connections with wider contexts. Metaphors are used to encourage children to see the world in different ways, as Chris explains:

There are elements of getting perspective from the small and personal to the enormous and global. And beyond that, the bigness of it all, the vastness of it all, we come to see how small this vast planet is, compared to that scale. So when you compare problems and pleasures of being an individual on Planet Earth, and the delights that has and the difficulties that has, you can scale that up to a planetary level quite quickly.

You can put a child in role as 'planet': "Pretend you are a planet: What were you proud of?" "Pretend you are a planet: what upsets you, what cheeses you off, what makes you angry?" And that works very effectively. And the idea that you have personal problems that can be overcome—and most of us have experienced that, where something has cheesed you off and you do something to make it better—and you can apply that to the planet as well . . . [JD: How?] . . . with small actions. It is continual messages throughout and that would almost be the final message as well. If you do nothing it is not going to make a difference or it could change it for the worse. If you do something it can make a positive difference.

The intended outcome here is that students realise that their actions can make a difference to the planet. For young people used to seeing the moon and photographs of the Earth, it might be easier to grasp the idea of a planet rather than something more abstract such as 'the environment'. Implicit in this approach is that there is an underpinning *value* that people should do something positive for the planet in order for it to be sustained.

Jane and Chris have analysed all their activities so as to emphasise education for sustainable development. As Jane says: "So whilst the children are enjoying themselves, having a great time with immediate hands on learning, we can link in the broader global message of how their actions can make a difference." The overall educational approach, and how it came about, are described by Jane: "It's trying to scale down issues to a point where children feel they can handle the responsibility for change" To Chris: "It's not something that you should be panicking about and crying over, but if you take this small, tiny step in this direction and another small step in that direction it can have a positive effect". Jane adds:

> Children can be completely daunted by the enormity of what they could, should, maybe, want to do, and don't feel empowered to do. So it is giving them little routes that all of them, as human beings, can take, at school, in their family, in their personal lives that doesn't just swamp them and drown them.

An issue here is the degree of assent that has been given to Jane and Chris by the parents of the children, by the schools and by the LEA. The centre is inspected both by local and national inspectors but to what extent are the underlying values of the centre examined? Who really knows what values are being promoted, in what ways and why? It is worth noting here that the children's teachers are in attendance throughout the children's time at the centre. We are *not* questioning the values or the approaches used, rather we are raising the question of to what extent parents should be informed of the values implicit and explicit in any educational experience, particularly those that are designed to change children's attitudes and values. It would be difficult to imagine that parents would not want their children to value the planet, resources, or other people, but there may well be values that are implicit in activities carried out in the name of science in schools or study centres that parents might wish to be informed of in advance. We recognise that this is an issue that is much more controversial in some countries, such as the USA, than in others.

We referred earlier to the children being encouraged to value and respect opinion—a key *strategy* used during the week. One way that this is done is through a practical *activity,* a walk, barefoot, through an area of bog land. Chris describes how the *activity* is used to prompt reflection on a particular issue: human impact on the environment:

> So, for example, if you walk onto a bog land, a very fragile habitat and take children there and they squash plants, do we or don't we go onto the bog land? Is it worth saving? Is it not worth saving? Or allowing them an opportunity to see that there are two sides—several sides to a problem.

The bog land walk is an example of another *strategy* that is used during the week—experiential learning. Jane explains why this strategy is particularly appropriate for the particular issue of habitats and their degradation and conservation [NB We illustrate our conceptual framework (see Table 13.1), again, throughout this explanation from Jane]:

> I think to experience a new habitat [ISSUE], for the first time, is a very powerful moving moment [VALUE]. To simply stand on that Bogland and feel the water squidge in between your toes creates [ACTIVITY] some sort of personal link with the place, for some, an affection for that place [OUTCOME]. Then we hope that children would want to protect such a place and consider it an absolute horror to think of damaging it [OUTCOME/VALUE]. This is where you can suggest how they can help out and give them actions that they can positively do [OUTCOME].
>
> So it is important that we take the groups to these amazing, wonderful places, including the centre grounds [STRATEGY], so that they can make a personal connection with their environment, and start to place value on it [VALUE]. To sit in the classroom and talk about these amazing bog lands [ACTIVITY] and how endangered they are [ISSUE], wouldn't inspire children in the same way, it would be less immediate, less relevant, more forgettable. It is very powerful, just being there [VALUE].

It is interesting that Jane uses the expression "so they can make a personal connection with the environment". We interpret this as meaning that Jane sees that the notion of valuing, in this case, the plants in the bog, can be understood by the students to be applicable to other species in other habitats. The 'powerful' nature of the experience promotes a better grasp of the value of a habitat than might be achieved simply by talking in the classroom about bogs. The residential nature of the experience, the story-telling abilities of Jane and Chris and the lack of conventional school environment pressures mean that children here can reflect on their experiences and can relate what they have seen and felt to other situations or realise how unique they are. It is this aspect, we believe, that makes such experiences special and, therefore, more likely to be memorable than the day-to-day experience of conventional schooling.

We indicated earlier that at the core of the centre's curriculum is science and geography. The centre is not a field studies centre *per se* as the focus is on broader issues. We discussed with Jane and Chris the role of ecology in the centre's activities, and how it linked with the rest of the curriculum.

> JD: How much of the experience is about learning some science, learning some ecology, learning about what the names of things are, how things are connected, in a scientific way?
>
> Jane: I was thinking about that too, that again, quite deliberately we are trying to introduce the groups to the systems and processes of life through the overarching theme of the Earth Secret beads [children are

given coloured beads during the week which represent key concepts]—the *Circles, Cycles, Interconnectedness, Energy Flows*—these key concepts should be valued as a point of view. So we are saying we value these principles and we are trying to find these principles in everything we do through the weeks. There is a way, if we keep on saying—this is important, let's just think about this, where can you see this? You are modelling the fact that maybe there are things that the children should be taking on and finding important too. There is a huge amount of science, going on, through the week, finding things out through your 'creature teacher', through first-hand investigations and through your understanding of the earth secrets.

Chris: There is [a huge amount of science, going on], and it is not always specified. Sometimes it seems a little bit intangible, because they are not following a science activity as laid down in a textbook . . . but simulating a living food chain with children and their interactions are parallels to ecological community interactions. You play games, the food chains are running around and eating each other or finding food supplies. I think the gist of it is almost using science and the natural world as a vehicle for something else. And that something else is citizenship, being a good human being and the world could be a better place.

There are three key pedagogic points here that show how the philosophical positions underpinning Jane and Chris' work are represented through their pedagogic actions and choices. The first is that beads and badges are used as reminders and signifiers (in a Saussurian sense) of important ecological principles. The second issue is that by repeating the importance of the principles, 'modelling' behaviour, as Jane puts it, children develop an awareness of the *value* of the ideas. Thirdly, as Chris points out, science and the natural world are used as a vehicle for the promotion of citizenship—a subject in the English National Curriculum.

The children are, as we pointed out earlier, from primary schools, often aged between five and nine-years-old. It might be argued that the *issues* and *metaphors* used by Jane and Chris are too complex for the pupils. Over the years, though, Jane and Chris have developed a range of strategies that are aimed at facilitating learning even when the subject matter is difficult. One particular strategy is the use of stories to promote awareness of *values*, as Jane explains:

I think, as well, the use of stories is probably significant, in that you are picking stories from many different cultures, and you are presenting different value systems simply through the stories. Not explicitly always, you do reflect sometimes on the story you have told, or what they have understood of it. But I think by giving them this real wide scope of different tales, different places, different people, different expectations, I think they are taking on board the oneness of the world.

Story-telling is a way of sharing knowledge and perspectives from other cultures about scientific and other ideas. It is also a way of showing children the different ways in which knowledge can be construed or passed on. A key issue, for us, is the value which is ascribed directly or otherwise to knowledge produced by Western scientists:

> JD: Again, there might be a danger of the view that you are stereotyping, if you are saying these groups around the world, they believe that the 'big bear' constellation came about this way. And maybe look at 'The Sky at Night' [a well-known, long-running television programme] and see there are scientific explanations to the world. Is there a danger that you might be promoting an image of a culture that sees things in a wrong way? Or are you trying to argue that people see things in lots of different but equally valuable ways?
>
> Jane: My conclusion to a story like that would be that you can see the world in many different ways and each should be valued as a point of view. Our scientific take on the world reflects a transient moment. It will be a different view in two weeks' time, because something else has been discovered, changed, evolved. Likewise it is the same all over the world.

It would seem to all four of us however, particularly when we were all discussing earlier drafts of this chapter, that there are right and wrong ways to look at the world. Indeed, Jane and Chris clearly have ideas about what should be taught to children and what views they should hold if the environment is to be sustained. Jane's point about science is illustrated rather aptly by the fact that scientists announced that a tenth planet in the solar system had been discovered almost as these words were being processed. One might hypothesise that if it were astrologers that had made the announcement that the news would have been received rather more sceptically by the majority of people in the UK and Australia at least if not in the USA.

Chris explains how stories can be used to encourage reflection on the way that knowledge is created:

> Chris: In the stories you use some reference, for example, to an Aborigine who believes that a rainbow serpent came down and created the earth. This particular Aborigine is also a lawyer working in the city. Hang on, some twist on the tail there. It is not the outback, it is not deep in the rainforest, it is not deep in a Ugandan village, it is people like you and me.
>
> There is a Minister of Transport in Iceland that will not allow a road to be built past a place where fairies have been seen. It was on telly, I believed it, don't think it was a hoax. So you can twist the tail, that there are groups of people, diverse in themselves, who have different beliefs. There is a core or a stem or a history, this is where the story has

come from, and these people have been dispersed or lost things or gained things, they have maintained things, and they are just people like you and me with tales to tell, with beliefs.

Where stories become religion, it's much the same really. You put stories into a historical context because you might have been beheaded if you said otherwise. If you didn't believe the story you were an outcast, if you didn't believe in Odin, you couldn't live in the village. So what is story, and what is fake? They are very, very, similar. And that throws things at children to contemplate at a later time—in a contemplative moment, years later, perhaps, when they suddenly realise some of these [everyday] words [or place-names] are old English words or old Viking words. Remember the story and who is to say what the response is.

There are two issues, here, that we wish to address. The first is that people need tools to be able to decide on truth and falsehood—otherwise they will be duped, hoodwinked and bamboozled in the same way that people traditionally have always been. To what extent this need is implicit in Jane and Chris's ethos is hard to ascertain. The second issue is that Chris sees the *outcomes* of the activities as being very long-term. The overall strategy of sowing seeds, to use a metaphor of our own, through giving students powerful stories, which are often easy to remember, is a critical aspect of the Minstead experience. Chris explained how stories might lead to children making better decisions (again, there are right answers!):

So you are looking for powerful moments that are likely to stay, or likely to enthral, and likely to allow them to make better decisions, a story at the bog lands or whatever, when they are in the garden centre with their parents or themselves in twenty odd years, to buy a bag of peat, will they just buy the bag of peat or will they stop and say—hang on. Something is happening. I remember a reason why I shouldn't buy this bag of peat. How do you measure that?

Chris is referring here to, generally middle-class, people buying peat for use in their gardens as a fertiliser—a frequently unsustainable activity generally frowned upon by most environmentalists. Chris continued by explaining the links between science, the environment, stories and values:

Going back to where science pervades and permeates: if you asked a child about the structure of the bog land they would be able to tell you, because the build up has been talking about a layer of sphagnum moss and plants and a deep dank layer, and you could push a stick down through, and the bog bodies and the potential for bog spirits. And this kind of blend of science, non-science, storytelling, religion and faith is all there. What you make of it will vary.

A question remains as to why Jane and Chris have evolved their philosophy. Here Jane provides part of the explanation:

> I think I've become increasingly saddened at how disconnected the children are to the planet, and how discordant some visiting groups are. And how much you have to do, sometimes, even to become a working group before you can start to entertain any of the wider environmental thoughts that we have been talking about today. It seems a sad situation that children are so out of sorts with themselves and their environment that our week needs to spend such an amount of time trying to socialize a group before moving on any further.

Both Jane and Chris have taken part as leaders and participants in a range of earth education courses (Van Matre, 1990) which promote a distinctive set of ecocentric, ecological and human values similar to their own. In terms of how this form of education might be spread, Jane hypothesised about taking on a different role:

> It makes me think whether we should be more actively political. Whether your best tool of change is to work with individuals and hopefully make some small personal change or whether we should be lobbying more at parliamentary level and having a voice in that way. It is a constant frustration to me how little focus has been given to education for sustainable development at a national level.

Summary

In some ways we hope that the words of Jane and Chris speak for themselves and for the ethos that characterises the Minstead Study Centre. However, we recognise that our choice of questions and our selection of quotes tell a story that is partial and open to challenge. We hope that we have provided enough by way of description and illustration for readers to identify and engage with the issues. Because Jane and Chris are practically engaged in promoting values, their experiences provide a rich source of ideas and experience that might help to inform the debate about school-based science education that this book is meant to encourage. In that light, we commend to readers the relatively simple sensitising framework that emerged from our reading of the interview transcript, conditioned as it is by our own experience as environmental educators and researchers with backgrounds in science and geography. Values, metaphors, issues, activities, strategies and outcomes seem, to us at least, to be powerful ways of opening up spaces for discussing and debating new and exciting trends in science, citizenship and environmental education in formal and informal contexts.

Acknowledgement

Justin Dillon's contribution to the research described in this chapter was funded by the US National Science Foundation grant number 0119787 to the Centre for Informal Learning and Schools, a collaboration of King's College London, the San Francisco Exploratorium and University of California Santa Cruz.

References

Baldwin, R. G. (1996). Faculty career stages and implications for professional development. In D. Finnegan, D. Webster, & Z. F. Gamson (Eds.), *Faculty and faculty issues in colleges and universities* (2nd ed.). Boston, MA: Pearson Custom Publishing.

Bliss, J., Monk, M., & Ogborn, J. (1983). *Qualitative data analysis for educational research: a guide to uses of systemic networks.* London: Croom Helm.

Dillon, J., Morris, M., O'Donnell, L., Reid, A., Rickinson, M., & Scott, W. (2005). *Engaging and learning with the outdoors—The final report of the Outdoor Classroom in a Rural Context Action Research Project.* Slough: National Foundation for Educational Research.

Dillon, J. & Reid, A. (2004). Issues in case study methodology in investigating environmental and sustainability issues in higher education: Towards a problem-based approach? *Environmental Education Research, 10*(1), 23–37.

Glaser, B. G. & Strauss, A. L. (1967). *The discovery of grounded theory.* Chicago, Illinois: Aldine.

Miles, M. B. & Huberman, A. M. (1994). *Qualitative data analysis* (2nd ed.). Thousand Oaks, California: Sage.

Qualifications and Curriculum Authority (QCA) (2006). *National Curriculum online.* Retrieved January 30, 2006, from http://www.nc.uk.net/webdav/servlet/XRM?Page/@id=6016)

Stake, R. E. (2000). Case studies. In N. K. Denzin & Y. S. Lincoln (Eds.), *Handbook of Qualitative Research* (2nd ed., pp. 237–247). Thousand Oaks, California: Sage.

Van Matre, S. (1990). *Earth education: A new beginning.* Warrenville, Illinois: Institute for Earth Education.

14 Barriers and Benefits to Learning in Natural Environments

Towards a Reconceptualisation of the Possibilities for Change

Justin Dillon

Dillon, J. (2013). Barriers and benefits to learning in natural environments: towards a reconceptualisation of the possibilities for change. *COSMOS*, 8(2), 1–14.

1. Introduction—The Need to Strengthen the Connections Between People and Nature

Our ambition is to strengthen the connections between people and nature. We want more people to enjoy the benefits of nature by giving them freedom to connect with it.

> Everyone should have fair access to a good-quality natural environment. We want to see every child in England given the opportunity to experience and learn about the natural environment. We want to help people take more responsibility for their environment, putting local communities in control and making it easier for people to take positive action.[29]

In June 2011, the Secretary of State for Environment, Food and Rural Affairs presented to the UK Parliament the first white paper on the natural environment for 20 years. "The Natural Choice: securing the value of nature," otherwise known as the Natural Environment White Paper (NEWP), outlined an ambitious plan to tackle a number of inter-linked issues including biodiversity loss, the threat of climate change and a decline in young people's engagement with the natural environment.

Concerns about biodiversity loss extend far beyond England. Earlier in 2011 the United Nations General Assembly declared the period 2011–2020 to be "the United Nations Decade on Biodiversity". As a follow-up to NEWP, the UK's Department for Environment, Food and Rural Affairs (DEFRA) published "Biodiversity 2020: A strategy for England's wildlife and ecosystem services". The strategy, which set out a decade-long agenda for action, identified four key outcomes by which its success might be judged in time. Outcome 4, the most relevant to this paper, states that "By 2020, significantly more people will be engaged in biodiversity issues, aware of its value and taking positive action".[17]

The rationale for identifying and prioritising this outcome included the following statement:

> The level of direct contact with nature is a factor in influencing attitudes towards it suggesting that the more we can stimulate interest in and access to nature, the more people will be willing to contribute to its protection and enhancement.
>
> (ibid., p. 15)

This desire to give more people greater contact with the natural environment can be addressed in a number of ways from increasing the number and quality of urban parks to funding school visits to farms. Many of these initiatives have been taking place for a long time and a number of case studies of their effectiveness can be found in the literature.[29]

Despite these initiatives, increasingly robust evidence suggests that children are losing their connection with nature. Worse still, children in urban environments are particularly disadvantaged.[68] For example, nowadays 10% of children play in the natural environment compared to 40% of adults when they were young.[20] This "extinction of experience"[58] has a detrimental long-term impact on environmental attitudes and behaviours. In order to achieve its aim of helping people "take more responsibility for their environment" the government needs to address this disconnection as a matter of urgency.

The scale of the ambition in NEWP, which spelt out government policy with a view to bringing in appropriate legislation to effect change, is suggested by the Secretary of State's commitment to a strategy that 'places the value of nature at the centre of the choices our nation must make: to enhance our environment, economic growth and personal well-being' (ibid. p. 2). The value of research in providing evidence of the need for action and informing future policy was recognised in the following succinct statement:

> Science, economics and social research have broken new ground, demonstrating that, year by year, the erosion of our natural environment is losing us benefits and generating costs. This knowledge must be the spur for a new policy direction, nationally and internationally.
>
> (ibid. p. 7)

This paper outlines some of the thinking behind these new policies, particularly in terms of the public's engagement with biodiversity and, in particular, it discusses a need to reconceptualise some of the perceived barriers to learning in the natural environment (LINE), which is seen as a prerequisite for taking action. The key issue that this paper addresses is directly linked to the extract from the white paper above and, specifically, these two sentences:

> Everyone should have fair access to a good-quality natural environment. We want to see every child in England given the opportunity to experience and learn about the natural environment.

There is no doubt that, currently, access to a good quality natural environment is not equitable. Similarly, children's access to LINE varies from school to school and from class to class. The degree of variation is, as I have stated before, a national disgrace.[37] In this paper, which draws heavily on a report prepared for Natural England, a nondepartmental UK government body, I first describe what we know about the benefits of LINE: I believe that we have frequently underestimated the potential impact on students. I then take a critical look at the established canon of "barriers" to taking children outdoors which, I believe, do not really explain the substantial variance that exists in terms of access to LINE.

2. Part I. The Benefits of Learning in the Natural Environment

Learning in the natural environment affords *direct* benefits as diverse as educational, health and psychological and *indirect* benefits ranging from social to financial. For too long, though, research into the value of LINE has failed to address the full range of benefits. Instead, there has been a narrow focus on easily measurable outcomes and a desire to seek answers to simplistic questions such as "does LINE raise standards more than learning in the classroom?" One consequence of this narrow-minded thinking is that too many children have been denied the rich educational experiences that have been available to others. In the current financial situation, and at a time when education systems around the world are under review, it is opportune to set out the full range of benefits which are available to all students in schools across the country.

At this point it is worth clarifying some key terms that are used in this paper. The term *learning in the natural environment (LINE)* encompasses a range of provision, including:

- activities within a school's or college's own buildings, grounds or immediate area;
- educational visits organised within the school day; and
- residential visits that take place during the school week, weekends or holidays.[52]

Natural environments are those which, in contrast to the built environment, contain living and nonliving material. They include school grounds, local open spaces, parks, rivers, lakes, forests, coastlines, caves, mountains and the atmosphere. *Fieldwork,* for the purposes of this paper, refers to all teaching and learning activities that are carried out in natural environments.

The most authoritative survey of research into learning outside the classroom was carried out by Rickinson *et al.* in Ref. 60. The review concluded that: "Substantial evidence exists to indicate that fieldwork, properly conceived, adequately planned, well taught and effectively followed up, offers learners opportunities to develop their knowledge and skills in ways that add value to their everyday experiences in the classroom".[60] The Rickinson *et al.* review identified four areas of impact on students: cognitive, affective; social/inter-personal; and physical behavioural.

Cognitive benefits include developing an understanding of the biological, chemical and physical processes that are taking place in the environment all the time. It is one thing to talk about photosynthesis, oxidation and erosion in the classroom—it is much more meaningful to witness the results of these processes in urban and rural locations. Water testing in ponds, streams and rivers becomes more memorable when carried out *in situ* instead of in the laboratory or even virtually through software simulations. This might particularly be the case if the environment concerned is polluted as in the case of a river that provides water for the local community.

Other benefits, beyond the individual, have been examined although not in great depth. A study to begin to assess the economic value of LINE in England identified benefits arising from educational attainment, attitudes to other children, awareness of environment and natural science skills, behavioural outcomes and social cohesion, health benefits, school staff morale, and a more attractive school (aesthetically and to prospective parents).[19] Furthermore, complementarity between these benefits means that the overall value of LINE to society is probably greater than the sum of these parts. The qualitative evidence linking LINE to such benefits is compelling; however, quantitative evidence linking LINE and changes in these benefits is lacking.

Even in the absence of such quantitative links, it is possible to use monetary value evidence to illustrate that LINE's contribution is significant. For example, the costs to society of the problems that are encountered in the absence of health, community cohesion, higher educational attainment and so on range from tens of millions to billions of pounds. Even if LINE has only a very small impact on these costs (e.g. reducing the relevant impacts by 0.1%), its value in reducing costs would be very large—of the order of £10 m to £20 m per year in England. Greater percentage reductions in impact would give proportionately greater reductions of costs.

The benefits accruing from LINE can be reduced remarkably easily by a lack of adequate preparation, weak pedagogy and inadequate follow-up back in school. Fredericks and Childers note that "Effective field trips require planning, preparation, and follow-through upon returning to school as well as coordination between the host site, school, and chaperones".[24]

Many of the outcomes are inter-related and mutually reinforcing. In a seminal study of the impact of residential fieldwork on upper primary school students, Nundy[49] identified a positive impact on long-term memory due to the memorable nature of the fieldwork setting as well as affective benefits of the residential experience (e.g. individual growth and improvements in social skills). Perhaps more importantly, Nundy also reported reinforcement between the affective and the cognitive outcomes which resulted in students being able to access higher levels of learning.

> Residential fieldwork is capable not only of generating positive cognitive and affective learning amongst students, but this may be enhanced significantly compared to that achievable within a classroom environment.[48]

Nundy's findings are supported by report from the English schools' inspectorate (Ofsted) which stated that 'learning outside the classroom contributed significantly to raising standards and improving pupils' personal, social and emotional development'.[52] So, while the benefits outlined below are organised into categories, it must be borne in mind that many of them do not occur in isolation and, indeed, a class of 30 students exploring their local surroundings may well have 30 different individual experiences resulting in a complex and hard-to-measure set of personal outcomes.

The outcomes listed below are organised as follows: Benefits to individual participants (knowledge and understanding; skills; attitudes and behaviours; health and well-being; self-efficacy and self-worth); benefits to teachers, schools and the wider community, and benefits to the natural environment sector.

2.1. Increasing Knowledge and Understanding

By far the greatest proportion of research findings focus on the impact of LINE on participants' knowledge and understanding. Specifically, students who have experienced LINE have been shown to perform better in reading, mathematics, science and social studies and show greater motivation for studying science.[46] For example, in a comparative study in the USA, Randler et al.[59] found that students aged 9–11 who had taken part in conservation action "performed significantly better on achievement tests" and that pupils "expressed high interest and well-being and low anger, anxiety, and boredom" compared with students who had been taught using more traditional methods (p. 43).

A number of activities have been designed specifically with urban children in mind such as the "Schools in a Park" project[26] that involved identifying local open spaces and the "Thinking Beyond the Classroom" project.[4]

The impact of visits to the Eden Project in Cornwall has been reported by Bowker who examined pre- and post-visit drawings of tropical rainforests made by 9–11 year-old children. Bowker reported that the "post-visit drawings [. . .] demonstrated far greater depth, scale and perspective than the pre-visit drawings" (2007, p. 75). In an earlier paper, Bowker[7] interviewed children (n = 72) from eight primary schools about one month after they had been on a one-day school visit to the Eden Project. He noted that the children's "opinion of plants changed, they understood the link between plants to their own daily lives and took delight in finding out where chocolate came from" (ibid., p. 241). In another study, Hamilton-Ekeke[27] compared three groups of Nigerian school students. Students who were taught ecology by taking them to the school farm, pond, and nearby stream performed better than a matched group who were taught only in the classroom.

2.2. Developing Skills

A broad range of skills ranging from the technical to the social have been identified as outcomes of LINE, particularly when it is integrated with the everyday school curriculum. In a major report on the work of outdoor education centres, Ofsted found that participating students "develop their physical skills in new and challenging situations as well as exercising important social skills such as teamwork and leadership".[51] Peacock's evaluation of the National Trust Guardianship scheme, which involved students making multiple trips to sites, was that participating students developed social skills such as tolerance, caring, group awareness and self-discipline as well as research skills involving understanding and management of the natural environment. Specific skills were developed which ranged from gardening and cooking to using digital cameras and microscopes.[55]

Cowell and Watkins[15] describe the outcomes of a museum outreach programme, "Spring Bulbs for Schools", which was established in Wales in 2006. The scheme involved setting up 160 monitoring sites across the principality. The authors, one of whom was a project officer and the other a schoolteacher, evaluated the project and found that the students became "aware of the world around them and the idea that human activity can have noticeable effects, even on a local scale in the school garden" adding that "the project enabled them to undertake pattern-seeking and observational activities—aspects of scientific enquiry that are often underdeveloped throughout the science curriculum" (p. 27).

Relatively few studies have looked at the experience of early-year education. However, Jones[31] reported on the development of children aged 3–5 on a school programme in Minnesota, USA. Jones noted that the "children learn to work collaboratively, socially construct knowledge, and develop social skills while cooperating, helping, negotiating, and talking with others" (2005, p. 86). Possick[56] reported on a small-scale study involving her kindergarten class and another first-grade class. A month-long project culminated in turning their school hall into a "forest". The project 'was based on observing, questioning, taking field trips, conducting library research (including the internet) and asking experts' (p. 30). Possick reports that the children in the two primary classrooms 'developed skills in forming questions about what they thought they knew, wanted to know, and had learned' (ibid., p. 30).

2.3. Changing Attitudes and Behaviours

There is abundant evidence of the positive impact of LINE on a range of attitudinal and behavioural dimensions. Childhood experience has been linked to adults' attitudes by Wells and Lekies:

> . . . childhood participation in "wild" nature [. . .] as well as participation with "domesticated" nature such as picking flowers or produce, planting trees or seeds, and caring for plants in childhood have a positive relationship to adult environmental attitudes. "Wild nature" participation is also positively associated with environmental behaviors.
>
> (2006, p. 1)

Environmental-based education makes other school subjects rich and relevant and gets apathetic students excited about learning.[46] Research has identified such impacts resulting from a range of experiences including school gardening and environmental improvement; visits to local parks; farm visits and residential visits.[44] Coskie et al.,[14] for example, describe the impact of a five-week intervention in which students aged 8–10 were taught how to write a field-guide to identify plants in a small area of woodland near to the school. The authors found that students "came to understand and care for the natural world in their immediate environment" (p. 26).

Few researchers have looked at the long-term impact of LINE. An exception is a US study by Pace and Tesi[54] that involved interviewing four men and four women between the ages of 25 and 31 about their "field trip" experiences while attending school from K-12 (that is, kindergarten through to twelfth grade (age 17–18)). Most of the participants revealed that they experienced 'enhanced camaraderie with fellow students, teachers, and chaperones [accompanying adults]' (p. 30) as a result of their experiences. In another long-term impact study, Farmer et al.[23] evaluated *Parks as Classrooms,* an environmental education programme in the Great Smoky Mountains National Park, USA. The programme focused on the impact of nonnative species and humans on local biodiversity. The primary school participants were aged 9–10. 15 of the 30 students agreed to be interviewed a year after their visit. The authors reported that 'many students remembered what they had seen and heard and had developed a perceived pro-environmental attitude' (p. 33).

Evaluation of a woodland-survival skills course that Warwickshire Children and Voluntary Youth Services ran with Groundwork for young people who were "neither employed nor in training", found that they gained more than just measureable skills.[13] As well as developing their confidence, leadership skills, and perseverance, they became more motivated and tolerant of their environment, staff and each other, as well as learning to live away from their families and to create their own entertainment.

In terms of changing attitudes to studying, Thompson[68] argues that teachers and principals 'should not overlook the role educational travel can play in motivating students to achieve'. Using a case study of a middle school in Michigan, USA, Thompson describes benefits to both the students and the school 'that come from linking trips to the science and social studies curricula' (ibid., p. 2). Finally, Chawla's[11] review found that adults who had significant and positive exposure to nature as children were more likely to be environmentally sensitive, concerned, and active.

2.4. Heath and Well-Being Benefits

Links between contact with the environment and personal health are well-established. Studies have shown that exposure to the natural environment can lower the effects of various mental health issues that can make it difficult for students to pay attention in the classroom. In particular, Kaplan[34] proposes the Attention Restoration Theory—the theory that exposure to nature reduces directed attention fatigue, restoring the ability to concentrate at will. The symptoms of Attention Deficit/Hyperactivity Disorder are less severe when individuals (both children and adults) are regularly exposed to natural outdoor environments.[38,67]

The publication in 2005 of *Last child in the woods*, by Richard Louv, appeared to touch a nerve in the public consciousness in the US and elsewhere. Louv described a "Nature Deficit Disorder" which was meant to be a way of thinking about a society-wide problem of disconnectedness with the natural environment. The book stimulated the formation of a "No Child Left Inside" movement which has had substantial success in influencing policy makers. Environmental literacy appeared in the US Department of Education budget for the first time in 2010.

Children are more likely to have hands-on contact with the natural environment during their time at primary schools than while they are attending secondary schools. A study in Australia found that hands-on contact with nature in primary school "can play a significant role in a cultivating positive mental health and wellbeing".[43] The study involved a postal survey of 500 urban Melbourne primary schools, a more in-depth study of 12 schools and interviews with seven "key industry informants". Reporting only on the interviews, Maller found that 'hands-on contact with nature in primary school, regardless of the type, is an important means of connecting children with nature and can play a significant role in cultivating positive mental health and wellbeing' (pp. 21–22). Maller concluded that such contact was not only 'essential for protecting the environment' but that it also appeared to be "a means of cultivating community and enhancing the mental health and wellbeing of children and adults alike" (ibid.). Maller found that her respondents identified what she describes as structured and unstructured hands-on activities, and that while structured activities "result in greater benefits to children's mental health and wellbeing" it was the case that "unstructured activities were thought to be important for connecting children with nature and fostering an interest in the environment that may emerge later in adult life" (ibid., p. 21). Maller also claims that structured activities, "such as those commonly occurring in sustainability education", were seen as being 'powerful catalysts for creating a stronger sense of community—both within and beyond school boundaries' (ibid., p. 21).

Bird[6] highlights the links between mental health and the natural environment. He found over 100 studies supporting the role of the natural environment in "attention restoration" (when indirect attention allows concentration to be held with little or no effort, allowing the brain to recover for more direct attention usage),[33] as it provided the most effective location for promoting indirect attention.

In 2009, following a study of sustainability education in schools, Ofsted recommended that schools should 'ensure that all pupils have access to out-of-classroom learning to support their understanding of the need to care for their environment and to promote their physical and mental well-being'.[53]

2.5. Self-Efficacy and Self-Worth

The mental and physical health benefits are closely linked to other impacts such as improvements in feelings of self-worth and self-efficacy. Swarbrick et al.[65] report on a forest school initiative in Oxfordshire. Although acknowledging that research into the project is in its "infancy", the authors do report that a questionnaire sent to schools, early years settings and individuals using the forest school approach 'revealed that the project was viewed very favourably by participant adults', adding that they mentioned the 'increased ability of quiet children to express themselves, an increase in confidence, and positive participation from disruptive children' (p. 145). There was also evidence of increased speaking and listening skills during the one-year involvement in the forest school programme.

A child who had severe language difficulties (i.e. needed to attend a speech unit for four sessions a week) was extremely quiet in the nursery environment and seldom initiated conversations with other children or adults. However in the forest environment her speech was clearer and much louder. She also displayed more self-confidence and interacted with a wider circle of peers. In the nursery environment her interactions tended to be on a one-to-one basis.

Amos and Reiss's evaluation of the 2004 London Challenge Residential Initiative, which involved 51 schools from five relatively deprived London boroughs sending groups of 11–14 year-olds to field centres, found that pupils 'surpassed their own expectations of achievement during the courses, and both pupils and teachers felt that the general levels of trust in others and the self-confidence shown by the pupils on the courses were higher than in school subjects' (2006, p. 37).

An unusual and very thorough approach to evaluating the impact of an outdoor experience was reported by Whittington.[73] The participants in this doctoral study were a group of adolescent girls who took part in a 23-day canoe expedition as part of an all-female wilderness programme in Maine, USA. Whittington interviewed the girls twice following the expedition, once 4–5 months afterwards and the second time after 15–18 months had elapsed. Whittington reported that the experience enabled the participating girls to challenge 'conventional notions of femininity in diverse ways' (p. 287) including: (1) perseverance, strength, and determination; (2) challenging assumptions of girls' abilities; (3) feelings of accomplishment and pride; (4) questioning ideal images of beauty; (5) increased ability to speak out and leadership skills; and (6) building significant relationships with other girls.

In a study of a 10-week expedition by 14 young people to Ghana organised by Raleigh International, Beames found that 'Interpersonally, young people developed an increased facility for working and living with people they did not know before' (2005, p. 14). It was also noted, perhaps unsurprisingly, that participants gained a greater appreciation of the modern conveniences they were accustomed to and learned about the economic and democratic differences between the UK and Ghana. Beames noted that the participants 'developed a certain mental resilience, became more willing to undertake challenges, and gained a greater understanding of themselves' (p. 14).

Larson[40] examined the effects of an adventure camp programme on the self-concept of 61 adolescents with behavioral problems aged between 9 and 17. Using an experimental/control group design, Larson found that the 31 participants who voluntarily attended an adventure camp demonstrated a statistically significant and positive difference in terms of their self-concept compared to the control group. Similarly, Lan *et al.*[39] reported significant long-term effects of participation in a wilderness programme including greater participant self-actualisation and decreased hopelessness. Lan *et al.* reported that: 'Police recidivist data indicated that 42 of 56 youth who had prior convictions did not re-offend in the two years following the wilderness intervention' (p. 37).

2.6. Benefits to Schools, Teachers and the Wider Community

Teachers benefit from LINE, becoming more enthusiastic about teaching and bringing innovative teaching strategies to the classroom.[46] Schools also benefit from teachers taking more ownership and leadership in school change. Several of the studies mentioned above have already highlighted possible benefits of LINE beyond those felt by the individual. These inter-related benefits include social, economic, health and crime reduction.[13]

Maller, whose study was mentioned above, identifies a number of aims for engaging children in hands-on contact with nature noting its increasing popularity:

Many schools, both in Australia and internationally, are including hands-on contact with nature in their curricula, usually to meet sustainability education, environmental education or science learning objectives. However, other reasons cited for the recent growth in these types of activities include beautification of school grounds, habitat restoration, and to foster qualities of stewardship and nurturing in children.

(2005, p. 16)

Another Australian study, this time by Davidson (2005), described the experiences of schools that took part in the Sustainable Schools Initiative. The initiative, which is similar to many other environmental initiatives in the UK and elsewhere, focuses on waste, water, biodiversity/school grounds and energy management.

Stepath[63] reported on the impact of a marine education research project carried out on in 2002/3 on the Great Barrier Reef, Australia. Noting the lack of impact of knowledge on behavior (that is, there is little evidence of a simple cause-effect relationship), Stepath advocates community-based environmental monitoring in conjunction with experiential environmental education which 'can work to improve responsible behavior when used in coordination with a comprehensive education strategy and media campaign' (p. 1).

One of the most well-known examples of cross-community education aimed at intergenerational mentoring is the Garden Mosaics project. Kennedy and Krasny[36] describe the mission of the project which is 'connecting youth and elders to explore the mosaics of plants, people, and cultures in gardens, to learn about science, and to act together to enhance their community' (2005, p. 44).

The National Trust's Guardianship scheme involved school-age students paying multiple visits to sites. An evaluation of the long-term benefits of the scheme, which involved over 100 schools, found that they saw great benefits from having a "classroom in the park".[55] Headteachers reported a development of "community spirit" and valuing what was "in their own back yard" as a result of the scheme. A rarely reported finding was that the scheme resulted in an increased willingness of parents to come into school for events and meetings.

2.7. Benefits to the Natural Environment Community

The evidence suggests that the more young people engage with the natural environment, the more they appreciate and care for it.[14] Schaaf (2005) describes how four classes of primary-aged children engaged with a water quality project. By the end of the yearlong project the students had not only learned how to monitor water quality but they had 'raised salmon in the classroom for release into the river' (p. 5). Few attempts have been made to quantify the impact of LINE on the natural environment or the benefits, financial or otherwise of being providers of education and training in LINE.

The economic or environmental benefits of educational providers have not been adequately studied. An exception is the review commissioned from eftec[19] which found that LINE makes a significant contribution to environmental education in the current UK National Curriculum. Its value is estimated in the National Ecosystem Assessment by Mourato *et al.*[45] through its contribution to greater lifetime earnings associated with educational qualifications in relevant subjects. The estimated annual value of environmental knowledge in 2010 was £2.1 billion (£1.6 billion for GCSE (exams taken at age 16) subjects and £0.5 billion for A-Level (exams taken at age 18)), to which LINE makes a vital and necessary contribution.

3. Summary

Substantial evidence exists to indicate that LINE, properly conceived, adequately planned, well taught and effectively followed up, offers learners opportunities to develop their knowledge and skills in ways that add value to their everyday experiences in the classroom. Specifically, several studies indicate that students perform better in reading, mathematics, science and social studies and show greater motivation for studying science. A broad range of skills ranging from the technical to the social have been identified as outcomes of LINE, particularly when it is integrated with the everyday school curriculum. Environmental-based education can make other school subjects rich and relevant and get apathetic students excited about learning.

Links between contact with the environment and personal health are well-established. Studies have shown that exposure to the natural environment can lower the effects of various mental health issues that can make it difficult for students to pay attention in the classroom. Hands-on contact with nature is not only essential for protecting the environment but appears to be a means of cultivating community and enhancing the mental health and well-being of children and adults alike. Structured activities, such as those commonly occurring in environmental monitoring and sustainability education, are powerful catalysts for creating a stronger sense of community—both within and beyond school boundaries.

Teachers benefit from LINE, becoming more enthusiastic about teaching and bringing innovative teaching strategies to the classroom.[46] Schools also benefit from teachers taking more ownership and leadership in school change. Increasingly, providers of LINE are marketing what they have to offer more effectively so that schools realise that there is more to learning outside the classroom than going to a museum or a science centre.

Such is the strength of the evidence base that the Teaching and Learning Research Programme (TLRP)[9] concluded as one of its ten principles for effective teaching and learning that learning in informal contexts 'such as learning out of school, should be recognised as at least as significant as formal learning and should therefore be valued and appropriately utilised in formal processes'.

The research has had an impact on those individuals and institutions working in the natural environment sector. The English Outdoor Council, for example, summarising a number of research studies claims that 'learning outside the classroom raises educational standards' and that 'it offers for many their first real contact with the natural environment' (2010, p. 2). A survey by the Countryside Alliance Foundation reported 'huge enthusiasm for outdoor education among children and teachers' with 85% of children and young people wanting to take part in countryside activities with their school.[22]

The Learning Outside the Classroom Manifesto[18] and the Quality Badge scheme have both raised the profile of LINE. However, it has been clear for some time that children's access to still depends far too much on where they go to school and who teaches them. In Part II we examine the barriers to LINE which are frequently identified by researchers and practitioners.

4. Part II. A Critical Examination of the Barriers to LINE

> There is a lot written about the problem of declining opportunities for outdoor education in this country . . . There is, however, considerably less published research into the factors (both real and perceived) that might help to explain such trends.[60]

Two groups of barriers to LINE in natural environments can be identified. One set of barriers challenge the sector and the other set challenges schools and teachers.

5. Barriers and Challenges to the Sector

5.1. A Common Vision of LINE in Natural Environments

The natural environment sector contains a substantial number of groups and organisations providing a diverse range of materials, training, resources and experiences. Although the diversity of the sector is a strength, in that schools can choose providers, resources and the level of support that they need, a lack of a common vision of the value of LINE and a tendency to work in isolation means that the diversity may also be a weakness.

5.2. Continuing Professional Development (CPD)

Tabbush and O'Brien note that 'schools and teachers cannot be expected to take total responsibility for environmental and outdoor education' (p. 22) and the role of providers in providing coherent CPD must not be neglected. Developing teachers' confidence and competence as well as their self-efficacy and awareness of LINE requires high quality CPD which will probably be school-based and mainly organised during the statutory inset days, after-school, on weekends and during school holidays.

Reviews of research into teacher CPD have established that it takes about 30 hrs to make a substantial difference in pedagogy.[1] To be effective, CPD must be focused on strategies for teaching inside and outside the classroom and involve coaching and feedback.[32]

Teachers are more receptive to changing their pedagogy if they are dissatisfied with some aspect of their teaching.[16] A recent survey found that although 97% of teachers believed that schools needed to use their outside spaces effectively to enhance their pupils' development, 82% did not agree that their own school was making as much use as it can of this valuable resource'.[41] The survey also found that only 12% of respondents saw lack of support for LINE from senior management as a major issue in their schools. Training for LINE needs to focus on developing the confidence and competence of all teachers, not just those who are already committed.

6. Challenges to Schools

The House of Commons Education and Skills Committee's report 'Education outside the classroom (Second report)'[61] identified five groups of barriers to Learning Outside the Classroom (LOtC): risk and bureaucracy; teacher training; schools; cost; centres and operators.

6.1. Risk

The risks of LINE have been exaggerated over many years.[25] They form part of what has been called 'a prevailing social trend, not only towards making things safer, but also towards seeking compensation for acts or omissions that result in personal injury'.[28] Schools and providers need to ensure that they inform parents about outdoor activities and reassure them that adequate safety procedures are in place.

> Many of the organisations and individuals who submitted evidence to our inquiry cited the fear of accidents and the possibility of litigation as one of the main reasons for the apparent decline in school trips. It is the view of this Committee that this fear is entirely out of proportion to the real risks.[61]

6.2. Teacher Training

> While in-service training has been very effective in recent years, we are not convinced that initial teacher training does a good enough job in terms of giving trainee teachers the confidence they need to take their pupils out of the classroom.[61]

The evidence supporting the English Outdoor Council's submission to the Select Committee which said that 'in-service training has been very effective' has to be put into context: teachers continue to report that their access to professional development is very limited.[71] A wide-ranging survey of initial teacher training (ITT) institutions published in 2006 found 'substantial variation' in the amount of training for LOtC across courses and institutions.[35] The three main factors that respondents felt had hindered training were funding, curriculum changes/pressures and the demands/expectations of the ITT course. However, the variation between the best and the worst providers cannot easily be explained by those factors.

6.3. Schools

The Select Committee concluded that LINE was most effective 'where it is well integrated into school structures, in relation to both curriculum and logistics (for example, the organisation of timetables and supply cover where necessary)' (ibid., para. 48). The question, though, is why is it that the most effective schools are able to integrate LINE into school structures? The Select Committee commented that 'Positive and reliable evidence of the benefits of outdoor activities would help schools determine the priority to afford to such work' (ibid., para. 13). However, that evidence exists but what is not clear is why some schools prioritise LINE while others do not. Part of the problem might be that no reliable mechanism for measuring the full impact of LINE activities exists as yet. Work needs to be done to establish the full value of LINE to learners, schools and the broader community.

6.4. Costs

Though frequently mentioned as a barrier to LINE, the Select Committee noted that 'we do not believe that cost alone is responsible for the decline of education outside the classroom, or that simply throwing money at the problem would provide a solution' (ibid., para. 56). There are many examples of schools with relatively restricted budgets providing exemplary LINE and relatively well-funded schools doing very little.

> This conclusion is supported by evidence from the DfES London Challenge programme. As part of this initiative, the Field Studies Council offered full funding to schools to support an off-site educational visit. One third of schools did not take up this offer despite it being effectively free of charge. It seems therefore that an increase in funding alone would not be enough to persuade schools to change their behaviour . . .
>
> (ibid., para. 56)

6.5. Centres and Operators

Provision for LINE varies for a range of historical, geographical and other reasons. Some local authorities have outstanding levels of provision of service while others offer very little support. In the latter cases, private sector and voluntary sector organisations provide access to LINE. A small number of local authorities have increased their support over the years and have found that demand often exceeds supply. Again, children's access to LINE depends far too much on where they live and often those children in the poorest parts of the country have the least access to LINE.[57,68] A recent survey (CAF, 2010) reported that over 60% of children polled felt they did not learn enough about the countryside at school.

> This disparity of opportunity is . . . particularly tragic in that most disadvantaged pupils have potentially most to gain from the transformative impact that outdoor education has for many young people.
>
> (p. 5)

While the Select Committee noted that 'any attempt to raise the quantity and quality of outdoor education depends crucially on the skills and motivation of the teachers involved' (ibid., para. 44) it neither addressed the issue of what constitutes effective CPD nor the issue of teacher motivation to take part. It is evident, particularly within the emerging picture of school funding, that if LINE is to be more accessible to more students then the focus of efforts needs to be on teachers' needs, motivations and pedagogies.

7. Towards a Reconceptualisation of the Possibilities for Change

The variation between teachers and schools in terms of commitment to LINE is partly explained by perceptions of risk, cost of activities and curriculum pressures. There is no doubt, for example, that much of the difference in provision found in primary and secondary schools can be explained by systemic factors. However, another set of barriers must exist to explain the differences between individual teachers and schools. These barriers are likely to be centred around the following factors:

- Teachers' views of the nature of their subject[2]
- Teachers' views of the role of education[64]
- Teachers' views of effective pedagogy[42]
- Teachers' self-efficacy[10]
- Teachers' working practices (planning, teaching and evaluation)[70]
- Teachers' and school leaders commitment to school-community links[62]
- The relationship between schools and providers[47]

Teachers who see their subject as primarily laboratory-based may be less likely to exploit LINE in their teaching compared to those who see it as involving fieldwork. Teachers who see the role of education as being to engage students with the outside world are more likely to value LINE and to see fieldwork as effective pedagogy compared to those who see the purpose of education somewhat more narrowly. Teachers' self-efficacy may well be higher when they use familiar methods of teaching than when they are faced with novel situations, for example, in unfamiliar environments. Teachers who plan lessons collaboratively and who watch each other teach may be more likely to try out new pedagogies than other teachers. Schools that know and value their local communities may be more likely to value LINE than other schools. Finally, those providers who build relationships with schools and teachers and who share common purposes are more likely to find that they are valued and that the relationship grows.

For LINE to become mainstream for all pupils, there must be a greater awareness that without teacher commitment and adequate CPD, there will be no progress. Given the current funding arrangements and the levels of resources available to schools, the onus for prioritizing CPD for LINE will fall on schools and, specifically, on their senior management teams. Consequently, the natural environment sector will need to work more closely together to provide a coherent message to school leaders, and services more likely to meet their needs. Schools should be able to see how their provision compares with the leading schools in terms of LINE and they need to see a clear framework of provision matched to learning and other outcomes.

> It was apparent that some schools and subgroups/departments within schools had developed quite sophisticated and effective professional development learning communities, others just as clearly had not.[30]

Despite a range of initiatives over a long period of time, the use of school grounds and local parks for LINE remains very variable. Schools with seemingly poor provision have made the most of their limited space while other schools have done very little. The focus for future developments including CPD will probably start with the immediate environment.

> . . . well-designed school grounds could make outdoor learning a daily possibility. However, the continued rarity of such use in the secondary sector, partly due to the inadequate design of grounds as well as the classroom-biased philosophy prevalent in most schools, means that there is no evidence into the effect of sustained use of the school grounds for learning.[12]

7.1. Sources of Information

There is no shortage of advice for teachers about using the outdoors. Sources of information include websites, practitioner journals and external providers. Much of the advice on offer would tally with research findings, for example,

'Effective field trips require planning, preparation, and follow-through upon returning to school as well as coordination between the host site, school, and chaperones'[24] What teachers do not have is a lot of time to keep up-to-date with new and existing resources. A mechanism needs to be found to make access to such resources quick and easy.

8. Summary

Several barriers exist to the effective delivery of learning in natural environments. These barriers can be grouped into those that challenge the Natural Environment sector and those that challenge schools. The challenges facing the sector include a lack of a coordinated effective approach to working with schools at a local level. The challenges facing schools include those frequently mentioned such as the risk of accidents, cost and curriculum pressures. However, another set of challenges exists, at local, institutional and personal levels. These challenges include teachers' confidence, self-efficacy and their access to training in using natural environments close to the school and further afield.

In terms of a way forward, the natural environment sector should take action to provide schools with a compelling rationale for LINE that sets out the evidence for impact and shows how barriers, both institutional and individual, can be overcome. There would also appear to be a need to support staff in schools locally to develop their capacity to use activities and resources that promote LINE within their vision of effective education. Finally, there would appear to be a need to develop working practices to provide schools with coherent, effective services for LINE that overcome barriers and facilitate collaboration between providers that reflect local needs and opportunities.

References

1 Adey P, Hewitt G, Hewitt J, Landau N, *The Professional Development of Teachers: Practice and Theory,* Dordrecht: Kluwer, 2004.

2 Akerson VL, Cullen TA, Hanson DL, Fostering a community of practice through a professional development program to improve elementary teachers' views of nature of science and teaching practice, *Journal of Research in Science Teaching,* **46**(10): 1090–1113, 2009.

3 Amos R, Reiss M, What contribution can residential field courses make to the education of 11–14 year-olds? *School Science Review,* **88**(322):37–44, 2006.

4 Astra Zeneca Science Teaching Trust (AZSTT), (2013) Thinking Beyond the Classroom. Available at http://www.azteachscience.co.uk/resources/continuing-profes sional-development/thinking-beyond-the-classroom.aspx.

5 Beames S, Expeditions and the social construction of the self, *Australian Journal of Outdoor Education,* **9**(1):14–22, 2005.

6 Bird W, *Natural Thinking—Investigating the links between the Natural Environment, Biodiversity and Mental Health,* Sandy: RSPB, 2007.

7 Bowker R, Children's perceptions of plants following their visit to the Eden project, *Research in Science & Technological Education,* **22**(2):227–243, 2004.

8 Bowker R, Children's perceptions and learning about tropical rainforests: An analysis of their drawings, *Environmental Education Research,* **13**(1):75–96, 2007.

9 Cambridge Primary Review, *Learning and Teaching in Primary Schools: Insights from TLRP.* Cambridge: Cambridge University, 2008.

10 Carrier SJ, The effects of outdoor science lessons with elementary school students on preservice teachers' self-efficacy, *Journal of Elementary Science Education,* **21**(2):35–48, 2009.

11 Chawla L, Significant life experiences revisited: A review of research on sources of environmental sensitivity, *The Journal of Environmental Education,* **29**(3):11–21, 1998.

12 Chillman B, *Do School Grounds Have a Value as An Educational Resource in the Secondary Sector?* Winchester: Learning through Landscapes, 2003.

13 Connexions Coventry and Warwickshire, Connexions NEET's Bushcraft Project in partnership with Groundwork—Evaluation Report, 2009.

14 Coskie T, Hornof M, Trudel H, A natural integration, 2007.

15 Cowell D, Watkins R, Get out of the classroom to study climate change—the Spring Bulbs for Schools project, *Primary Science Review,* **97:** 25–28, 2007.

16 Davis NT, Looking in the mirror: Teachers' use of autobiography and action research to improve practice, *Research in Science Education,* **26**(1): 23–32, 1996.

17 Department for Environment, Food and Rural Affairs (DEFRA), *Biodiversity 2020: A Strategy for England's Wildlife and Ecosystem Services,* 2011.

18 Department for Education and Skills (DfES), Learning Outside the Classroom Manifesto. London: DfES, 2006.

19 Eftec (2011). *Assessing the Benefits of Learning Outside the Classroom in Natural Environments,* Final Report for King's College London.

20 England Marketing (2009). Report to Natural England on childhood and nature: A survey on changing relationships with nature across generations.

21 English Outdoor Council (EOC), *Time for Change in Outdoor Education,* Nottingham: EOC, 2010.

22 Environment, Food and Rural Affairs Committee (2010), Memorandum submitted by the Countryside Alliance. Available at: http://www.publications.parliament.uk/pa/cm200910/cmselect/cmchilsch/418/10030302.htm.

23 Farmer J, Knapp D, Benton GM, An elementary school environmental education field trip: Long-term effects on ecological and environmental knowledge and attitude development, 2007.

24 Fredericks AD, Childers J, A day at the beach, anyone? 2004.

25 Gill T, *Nothing Ventured . . . Balancing Risks and Benefits in the Outdoors,* Lifton: English Outdoor Council, 2010.

26 Glackin M, Jones B, Park and learn: improving opportunities for learning in local open spaces. *School Science Review* **93**(344):105–113, 2012.

27 Hamilton-Ekeke J-T, Relative effectiveness of expository and field trip methods of teaching on students' achievement in ecology, *International Journal of Science Education* **29**(15):1869–1889, 2007.

28 Harris I, Outdoor education in secondary schools: What future? *Horizons* 4:5–8, 1999.

29 HM Government, *The Natural Choice: Securing the Value of Nature,* London: The Stationery Office, 2011.

30 Hustler D, McNamara O, Jarvis J, Londra M, Campbell A, Howson J, *Teachers' Perceptions of Continuing Professional Development* (DfES research report no 429). London: Her Majesty's Stationery Office, 2003.

31 Jones NP, Big jobs: Planning for competence, *Young Children,* 60(2):86–93, 2005.

32 Joyce B, Showers B, *Student Achievement through Staff Development* (2nd ed.). New York: Longman, 1995.

33 Kaplan R, Kaplan S, *The Experience of Nature,* New York: Cambridge University Press, 1989.

34 Kaplan S, The restorative benefits of nature: Toward an integrative framework, *Journal of Environmental Psychology,* 15(3):169–182, 1995.

35 Kendall S, Murfield J, Dillon J, Wilkin A, *Education Outside the Classroom: Research to Identify What Training is Offered by Initial Teacher Training Institutions,* Research Report 802. London: DfES, 2006.

36 Kennedy AM, Krasny ME, Garden Mosaics, *The Science Teacher,* March, 44–48, 2005.

37 King's College London (2011), More outdoor education needed. Available at: http://www.kcl.ac.uk/newsevents/news/newsrecords/2011/01Jan/Moreoutdooreducationneeded.aspx.

38 Kuo FE, Taylor AF, A potential natural treatment for attention-deficit/hyperactivity disorder: Evidence from a national study, *American Journal of Public Health* 94: 1580–1586, 2004.

39 Lan P, Sveen P, Davidson J, A Project Hahn empirical replication study, *Australian Journal of Outdoor Education* 8(1):37–43, 2004.

40 Larson BA, Adventure camp programs, self-concept, and their effects on behavioral problem adolescents, *Journal of Experiential Education,* 29(3):313–330, 2007.

41 Learning through Landscapes (LTL) (2010). Research shows benefit of outdoor play.

42 Lotter C, The influence of core teaching conceptions on teachers' use of inquiry teaching practices, *Journal of Research in Science Teaching,* 44(9):1318–1347, 2007.

43 Maller C, Hands-on contact with nature in primary schools as a catalyst for developing a sense of community and cultivating mental health and wellbeing. *Eingana* 28:16–21, 2005.

44 Malone K, *Every Experience Matters: An Evidence Based Research Report on the Role of Learning Outside the Classroom for Children's Whole Development from Birth to Eighteen Years,* Report commissioned by Farming and Countryside Education for UK Department Children, School and Families, Wollongong, Australia, 2008.

45 Mourato S, Atkinson G, Collins M, Gibbons S, MacKerron G, Resende G, *Economic Assessment of Ecosystem Related UK Cultural Services, Report to the Economics Team of the UK National Ecosystem Assessment,* London: London School of Economics, 2011.

46 National Environmental Education & Training Foundation (NEETF), *Environment-Based Education—Creating High Performance Schools and Students.* Washington: NEETF, 2000.

47 Nicol R, Higgins P, Ross H, Mannion G, *Outdoor Education in Scotland: A Summary of Recent Research,* Perth: Scottish Natural Heritage, 2007.

48 Nundy S, The fieldwork effect: The role and impact of fieldwork in the upper primary school, *International Research in Geographical and Environmental Education,* 8(2):190–198, 1999.

49 Nundy S, *Raising Achievement Through the Environment: A Case for Fieldwork and Field Centres* Peterborough, UK: National Association of Field Studies Officers, 2001.

50 O'Brien L, Murray R, A marvellous opportunity for children to learn: A participatory evaluation of Forest School in England and Wales, Farnham: Forest Research, 2006.

51 Office for Standards in Education (Ofsted), *Outdoor Education: Aspects of Good Practice*, London: Ofsted, 2004.

52 Ofsted, *Learning Outside the Classroom: How Far Should You Go?* London: Ofsted, 2008.

53 Ofsted, *Education for Sustainable Development. Improving Schools—Improving Lives.* London: Ofsted, 2009.

54 Pace S, Tesi R, Adult's perception of field trips taken within Grades K-12: Eight case studies in the New York Metropolitan Area, *Education,* **125**(1): 30–40, 2004.

55 Peacock A, *Changing Minds: The Lasting Impact of School Trips*, Exeter: University of Exeter, 2006.

56 Possick J, An artful forest. *Science and Children,* **44**(6):30–32, 2007.

57 Power S, Taylor C, Rees G, Jones K, Out of school learning: Variations in provision and participation in secondary schools, *Research Papers in Education,* **24**(4):439–460, 2009.

58 Pyle RM, The extinction of experience, *Horticulture* **56**:64–67, 1978.

59 Randler C, Ilg A, Kern J, Cognitive and emotional evaluation of an amphibian conservation program for elementary school students, *Journal of Environmental Education* **37**:43–52, 2005.

60 Rickinson M, Dillon J, Teamey K, Morris M, Choi MY, Sanders D, Benefield P, *A Review of Research on Outdoor Learning,* Preston Montford, Shropshire: Field Studies Council, 2004.

61 Select Committee (2005). House of Commons Education and Skills Committee. Second Report. Available online: http://www.publications.parliament.uk/pa/cm 200405/cmselect/cmeduski/120/12002.htm.

62 Sosu EM, McWilliam A, Gray DS, The complexities of teachers' commitment to environmental education, A mixed methods approach, *Journal of Mixed Methods Research,* **2**(2):169–189, 2008.

63 Stepath C, (2004). *Awareness and Monitoring in Outdoor Marine Education,* Paper presented at the Tropical Environment Studies and Geography Conference (Cairns, Australia, June 7).

64 Stevenson RB, Schooling and environmental education: Contradictions in purpose and practice, *Environmental Education Research,* **13**(2):139–153, 2007.

65 Swarbrick N, Eastwood G, Tutton K, Self-esteem and successful interaction as part of the forest school project, *Support for Learning* **19**(3):142–146, 2004.

66 Tabbush P, O'Brien L, Health and Well-being: Trees, woodlands and natural spaces, Edinburgh: Forestry Commission, 2003.

67 Taylor AF, Kuo FE, Sullivan WC, Coping with ADD—the surprising connection to green play settings, *Environment and Behavior* **33**(1):54–77, 2001.

68 Thomas G, Thompson G, (2004). *A Child's Place: Why Environment Matters to Children.* http://www.green-alliance.org.uk/publications/ PubAChildsPlace_page195. aspx.

69 Thompson D, Including travel in your academic plans, *Middle Ground* **8**(2):31–32, 2004.

70 Vescio V *et al.*, A review of research on the impact of professional learning communities on teaching practice and student learning, *Teaching and Teacher Education* **24**(1):80–91, 2008.

71 Wellcome Trustm, *Believers, Seekers and Sceptics, What Teachers Think About Professional Development*, London: Wellcome Trust, 2006.

72 Wells NM, Lekies KS, Nature and the life course: Pathways from childhood nature experiences to adult environmentalism, *Children, Youth and Environments,* **16**(1):1–24, 2006. Available online at: www.colorado.edu/journals/cye.

73 Wittington A, Challenging girls' constructions of femininity in the outdoors, *The Journal of Experiential Education* **28**:205–221, 2006.

Section 5

Classroom Issues—the Emergence of Science | Environment | Health

Many years ago, I wrote a relatively short paper with a King's colleague, Peter Gill, a science educator with experience of teaching physics and mathematics in school. 'Risk, environment and health: aspects of policy and practice' was published in the *School Science Review (SSR)*, a journal aimed at school teachers which is published by the Association for Science Education. For me, as for many researchers with a teaching background, professional journals such as *SSR* provide an opportunity to disseminate research findings and engage in debate with classroom practitioners.

'Approaching "soft disasters" in the classroom: teaching about controversial issues in science, technology, society, and environment education' is based around the controversial issues research that I was involved with in the 1990s and continues the theme of showing how research can inform classroom practice. This work led to a developing interest in the notion of scientific literacy which is reflected in 'On scientific literacy and curriculum reform', an invited paper, published in the *International Journal of Environmental and Science Education*. The final contribution in this section, 'Science, environment and health education: towards a reconceptualisation of their mutual interdependences' argues for the value of Science | Environment | Health as a conceptual and pedagogic framework. The chapter developed from ideas that I had developed for an invited conference in Switzerland organised by Albert Zeyer and Regula Kyburz-Graber.

15 Risk, Environment and Health

Aspects of Policy and Practice

Justin Dillon and Peter Gill

Dillon, J., & Gill, P. (2001). Risk, environment and health: aspects of policy and practice. *School Science Review*, 83(303), 65–73.

Recent newspaper headlines reflect a high level of public concern about a range of scientific and environmental issues, such as BSE, GM crops, HIV-AIDS and global climate change. Controversy and uncertainty over such issues can contribute to what are known in the discourse of risk as 'soft disasters', that is, *'environmental and political crises that emerge Only slowly but at high cost to society'* (ESRC Global Environmental Change Programme, 2000: 3). Soft disasters can include the loss of public trust and confidence in industry, scientists and decision-makers.

Media coverage of environmental issues, particularly those that directly affect human health, often simplifies or misrepresents the science behind the headlines. The scientific knowledge necessary to understand contemporary issues might be complex, contested or both. Much of the content may well be addressed, albeit at a simplified level, in school science. However, making sense of many of the 'headline' issues often requires an understanding of risk and probability. In this article we consider the issue of risk and relate it to the curriculum and to the classroom.

Crisis and Confusion

The BSE crisis is a good example of a soft disaster that continues to be felt around the world—with Japan banning European beef imports as recently as December 2000. Gerrard and Petts (1998) point out that when the crisis was at its height, the UK Environment Agency issued a press release explaining the results of trial burns it had conducted at two coal-fired power stations to destroy animal carcasses. The Environment Agency (1997) reported that:

> *The risk of human infection resulting from burning cattle cull wastes in power stations would be negligible. A detailed risk assessment, carried out by the Agency, based on test rig trial burning of meat and bonemeal and tallow from cattle slaughtered under the Government's Over Thirty Month Scheme, shows that the risk of an individual contracting CJD (Creutzfeldt Jacob Disease) would be as low as 1 in 30,000 million. This is 3000 times less than the risk of death by lightning.*

The press release continued: '*Every application [for cattle incineration] will be rigorously assessed on its own merits and there will also be widespread public consultation before a decision is made*'. Gerrard and Petts point out that it is indicative of the times that a regulatory agency, which has carried out a quantitative risk assessment that has judged the risk to be negligible, is still concerned to offer public consultation. Some may ask what is the point of the consultation if the risk is so small; whereas others might ask, if the consultation is necessary, what is wrong with the risk assessment? The situation was mirrored months later when people in some parts of the country were exposed to carcinogens, emitted from the pyres of roasting livestock, as part of the strategy to control the foot and mouth outbreak by destroying predominantly uninfected animals. The absurdity of the situation—the lack of a consensus about the most appropriate solution and the impact of media images of a single calf on the overall strategy—graphically illustrate the curious relationship between science, public understanding and government policy.

Risk and Health

The Environment Agency's press release quoted above takes for granted the notion that the risk of being struck by lightning is perceived as being remote. In the circumstances this seems reasonable. But how good are we at using and interpreting such information? Here is a list of some causes of death that might kill a man within one year of his fortieth birthday. Take a look and mentally arrange the risks in order of magnitude; then give them a numerical value—a probability that they would kill our 40-year-old man sometime during the year.

- an accident on the railway
- murder
- smoking 10 cigarettes *per* day
- hit by lightning
- road accident
- all natural causes

What you have been doing might be called risk assessment. The first activity, ranking the risks in order of their relative severity, is qualitative risk assessment, whereas giving numerical values to the probability of adverse effects resulting from exposure to specific activities or substances is quantitative risk assessment. Table 15.1 gives the actual figures which show that for every 200 males, one will die in his fortieth year from smoking-related illness. This figure is over four times greater than the figure for deaths from all natural causes. On average, only one in ten million males will die in the year after their fortieth birthday as a result of being struck by lightning.

Table 15.1 Probability of death from different causes within one year for a 40-year-old male in the UK (from Harrison, 1998: 70)

Descriptor	Risk estimate	Example	
Moderate	1:100 to 1:1000	Smoking 10 cigarettes/day	1:200
		All natural causes	1:850
Low	$1:10^3$ to $1:10^4$	Influenza	1:5000
		Road accidents	1:8000
Very low	$1:10^4$ to $1:10^5$	Leukaemia	1:12 000
		Playing soccer	1:25 000
		Accident at home	1:26 000
		Accident at work	1:43 000
		Homicide	1:100 000
Minimal	$1:10^5$ to $1:10^6$	Accident on railway	1:500 000
Negligible	$<1:10^6$	Lightning strike	$1:10^7$
		Release of radiation from nuclear power station	$1:10^7$

Assessing Risks

'*Risks are defined as the probabilities of physical harm, due to given technological or other processes*' (Lash and Wynne, 1992: 4). Many of us are familiar with the concept of risk in the context of substances hazardous to health. Traditionally, risk assessment for substances has consisted of the following steps:

- Hazard identification (Does a substance pose a threat?)
- Dose-response assessment (What is the link between the dose absorbed and the effect on health?)
- Exposure assessment (What is the likelihood of exposure to the substance?)
- Risk characterisation (What is the level of risk posed to an individual or a group?)

Whereas some people might argue that *any* risk can be accurately quantified, others might disagree. Your own position might well depend on your view of the nature of science and of scientific inquiry. One extreme position, which results from a logical positivist philosophy, is that risk assessment can be carried out accurately and independently using the appropriate scientific methods. A second extreme, resulting from a cultural relativist position, is that no estimate of risk can be carried out independently: there is no such thing as objective science or independent results. A third view, which is gaining ground in the field, is the 'scientific proceduralism' perspective, which accepts that science is value-laden and that scientific experiments cannot be separated from the social context, but that nevertheless useful and reliable information can be obtained about specific risks.

| Scientific risk assessment ——— Risk assessment policy ——— Risk management |
| (*facts*)   (*values*) |

Figure 15.1 A model of the relationship between risk assessment, risk management and risk policy.

Risk assessment is related to another concept: risk management. Assessing the risk of death from smoking is one thing; deciding on policies for taxing cigarettes or on smoking in public places is another. In the early 1980s, concern grew among scientists, particularly in the US, about the incursion of extraneous policy issues into scientific risk assessment. The scientists thought that there was too much interference in their work on risk assessment. One high-level US study even advocated the separation of facts and values in decision-making (see Figure 15.1). The scientists' argument was that they could deliver objective data that could be used to formulate risk assessment policy (Gerrard and Petts, 1998).

Making such assessments is notoriously difficult for non-specialists and is heavily influenced by experience and a number of sociological factors. It has been shown that there is a tendency to overestimate the risk of statistically low-risk activities and to underestimate the danger from statistically high-risk causes (Fischhoff *et al.*, 1981). However, the evidence from a range of studies is that this is not the case. Different teams of scientists faced with the same data and the same contexts produced risk assessments that varied by up to an order of magnitude, depending on the scientists' prior assumptions. The US National Academy of Sciences listed up to 50 opportunities where scientists may have to make discretionary judgements in coming up with risk assessments—such as which hazards to study and which exposure pathways to take into account.

Similar concerns to those expressed in the US emerged in the UK scientific community. Some UK scientists were frustrated by the incursion of social and political considerations into risk assessment procedures. At the same time, scientists and others have been genuinely surprised by the level of criticism and mistrust of a range of technologies, ranging from nuclear power to genetic engineering. This is partly explained by issues to do with perception and communication. It seems to be the case that risks mean more to people than measurable fatalities or injuries, for example. We have moved from a situation whereby scientists told the public about risks, to one where there was a dialogue between scientists and the public, and on to a further situation where the stakeholders are involved in the risk assessment procedures in some way.

Perception of Risk

So far we have written in a way that would allow the equating of risk with likelihood. However this raises problems in the non-scientific (and indeed much of the scientific) community. Those of us who have tried to teach probability (a significant element of the National Curriculum for mathematics) will know just how difficult it is for pupils to estimate likelihoods. The popularity of the National Lottery reinforces this view. Surely, the argument goes, no one in their right mind would put money on a one in fourteen million chance? Yet they do—in their millions. So why do they? The answer is one that illuminates the understanding of risk perception: 'consequence'. The likelihood may be small but the consequences are enormous for the winner.

The same is true with risk. It is not just the likelihood that is important, it is the consequences. Quite clearly, then, there will be events of small likelihood that are perceived as high risk because of the catastrophic consequences. In some cases numerical estimates of the consequences can be made. It is not difficult to forecast the outcome of an airliner losing a wing during flight. On the other hand, there will be other events for which the outcome is very difficult to predict and this is often the case in the 'soft disasters' referred to earlier. In such events the temptation is increasingly to apply the so-called *precautionary principle*. This principle, although often mentioned by name, is not well defined. It is best illustrated by the recent activity by eco-groups in destroying GM crops because no one really understands what they could lead to. Although initially an attractive idea in risk management, it actually reduces to 'If in doubt don't' and taken to its logical conclusion could lead to the abandonment of much scientific research.

Another difficulty generated by the incorporation of consequence into our definition of risk is the personal reaction to possible consequences, which while it may not be 'logical' in the scientific sense, it is no less real. It is well known that an event which kills ten people in one accident is perceived as worse than ten single events killing one person each, even though the likelihoods may be the same. A train crash, such as the Hatfield 'disaster', makes headlines even though more people are likely to have been killed in road traffic accidents on the same day.

Dimensions of Risk Perception

Research carried out on the public perception of risk (Slovic, Fischhoff and Lichtenstein, 1980) resulted in a two-factor model generally referred to as the 'unknown risk factor' and the 'dread risk factor' (see Figure 15.2). At the low end of the known/unknown axis are events that are observable, known to those exposed, of immediate effect and scientifically understood, for example smoking or car travel. At the other end of the axis are events that are unobservable, unknown to those exposed, of delayed effect, possibly new and unknown to science, for example food irradiation and DNA research.

The dread factor has, at its low-risk end, activities that are perceived as easily controllable, not globally catastrophic, without fatal consequences,

Hazard locations in relation to unknown and dread risk

Not observable; unknown to
those exposed; effects delayed;
new risk; unknown to science

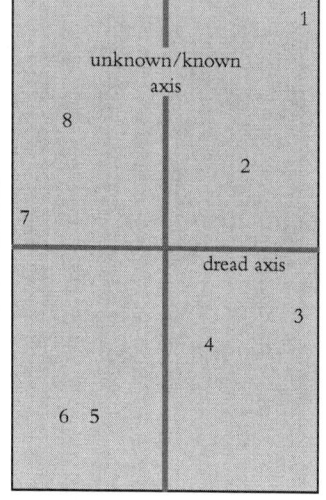

Easily controllable, not dread;
not global catastrophic;
consequences not fatal;
equitable; low risk to future
generations; risk easily
reduced; voluntary

Not easily controllable, dread;
global catastrophic;
consequences fatal; not
equitable; high risk to future
generations; risk not easily
reduced; involuntary

Observable; known to those
exposed; effect immediate;
old risk; risk known to science

Figure 15.2 Two dimensions of risk (after Slovic *et al.*, 1980).

Key to hazards: 1—nuclear power, 2—asbestos, 3—terrorism, 4—smoking, 5—motor vehicles, 6—alcoholic beverages, 7—sunbathing, 8—water fluoridation

equitable, of low risk to future generations, taken on voluntarily and easy to reduce, for example using home appliances or sunbathing. At the other end of this axis are events that are uncontrollable, possibly fatal or leading to global catastrophes, inequitable, involuntary and possibly of high risk to future generations, for example the use of nerve gas or nuclear power (nuclear power features quite highly on the unknown risk factor as well). The dread factor seems to have the greater influence on attitudes, and, when respondents were asked which risks should be the subject of statutory control, they chose those high on the dread factor (on the right-hand side of Figure 15.2).

Figure 15.3 shows a theoretical model of the relationship between risk assessment and risk management. In this model, the role of the public is less to do with examining risk assessments *per se* and more to do with the process as quality assurance—looking at the data inputs, the underlying assumptions and the uncertainty of specific techniques and strategies.

But what of the role of government in assessing and managing risk? In the UK, the Health and Safety Executive (HSE) has statutory powers that affect

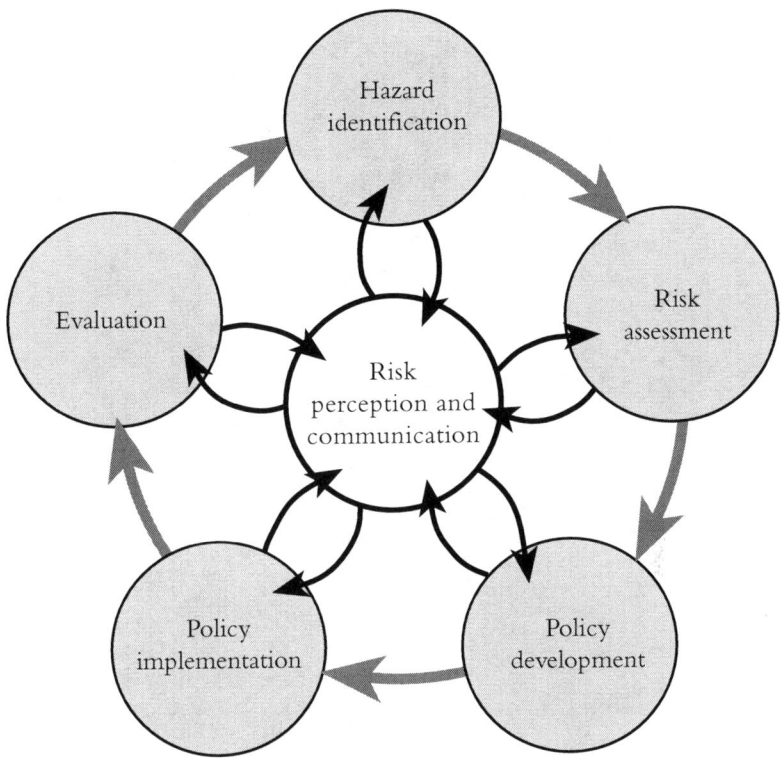

Figure 15.3 A model of the cycle of risk management and risk assessment revolving around consultation (Gerrard and Petts, 1998: 6).

the living and working conditions of all of us. Ironically, in an era when the public seems to want less interference from government and less regulation, it expects government agencies to set and maintain strict safety limits for all manner of substances and practices. Figure 15.4 shows the HSE framework for decisions on the tolerability of risk.

The uppermost region of the triangle indicates risks that are unacceptable except in extraordinary circumstances. In the middle area, the 'tolerability region', where the risk is as low as reasonably practicable (ALARP), a risk is taken *only* if there is a clearly identifiable benefit. At the upper end of this region, a risk is tolerated if its reduction is impracticable or the cost of its reduction in relation to the improvement gained is perceived as being hugely expensive. At the lower end of the ALARP region, a risk is tolerable if the cost of reduction would exceed the improvement gained by taking the risk. In the lower, third, section the risk is broadly acceptable and no detailed working is necessary to demonstrate ALARP.

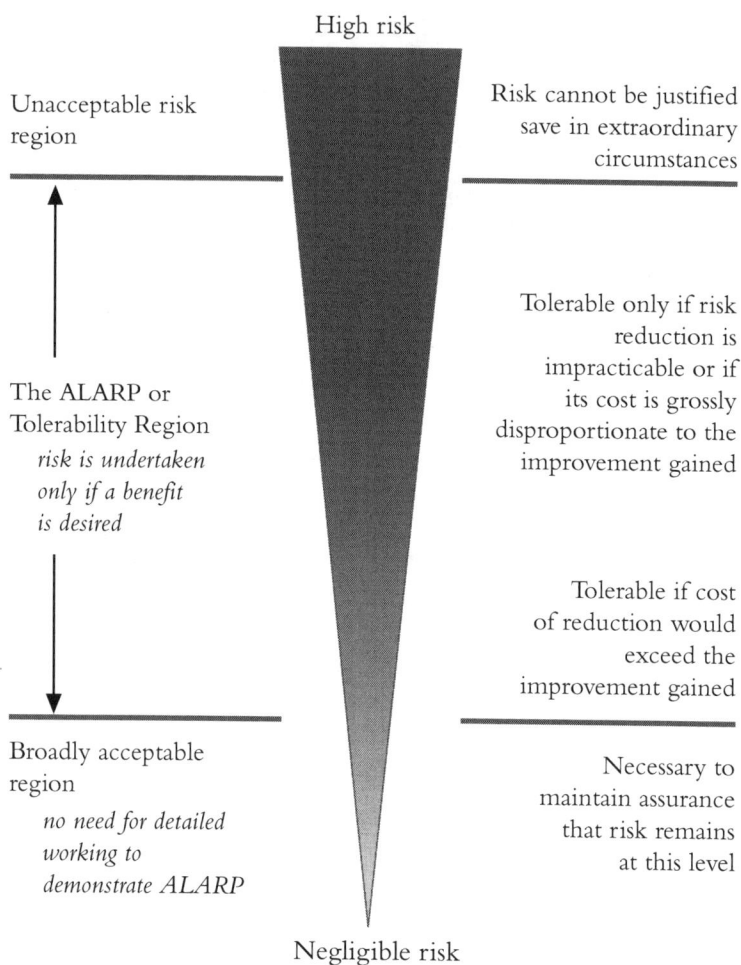

High risk

Unacceptable risk
region

Risk cannot be justified
save in extraordinary
circumstances

The ALARP or
Tolerability Region
*risk is undertaken
only if a benefit
is desired*

Tolerable only if risk
reduction is
impracticable or if
its cost is grossly
disproportionate to the
improvement gained

Tolerable if cost
of reduction would
exceed the
improvement gained

Broadly acceptable
region

*no need for detailed
working to
demonstrate ALARP*

Necessary to
maintain assurance
that risk remains
at this level

Negligible risk

Figure 15.4 The Health and Safety Executive framework for decisions on the tolerability
of risk (from McQuaid and le Guen, 1998: 28).

Environment, Risk and Uncertainty

Harrison (1998) answers the question '*What is a tolerable level of risk ?*' by
pointing out that '*there is no universally agreed answer to this question . . . society
is notably more tolerant of self-imposed risks (e.g. cigarette smoking) than of risks
which are perceived as externally imposed such as outdoor air pollution*' (p. 69).
Traditionally, environmental policy decisions have involved some element
of cost–benefit analysis such as that identified in the preceding section. In
environmental decision making, major difficulties have arisen owing to the
problem of allocating values to natural resources—land, rivers, etc. Recently,

a process of environmental valuation has been developed in an attempt to produce better decisions.

To illustrate the complexity of the risk assessment and management procedures, consider the case of the disposal of carcasses resulting from the foot and mouth culling. The options for disposal included burning and burying. Burning offers a relatively quick solution but the impact of airborne pollutants on people downwind are obvious, albeit hard to quantify. Burying, on the other hand, may lead to the contamination of water supplies that may not take place for many years after the event. What is clear from the example is the importance of using scientific knowledge in order to understand the complexity of the environment. As well as knowing about concepts such as transport mechanisms in the specific contexts mentioned above, it is necessary to understand the connectivity of the systems.

Researchers are generally well aware of the methodological problems inherent in the techniques that they use in risk assessments. Policy makers, in contrast, rarely seem to appreciate the uncertainty inherent in science and particularly in something as complex as risk assessment. For example, in recent years, there has been an over-reliance on the accuracy of multi-level modelling (such as is used in global climate change research and policy making) even though the limitations of the technique are reasonably well known.

One criticism that can be levelled at researchers is that the calculation of risk has too often involved treating substances in isolation—'single-substance science'—whereas what we find in the real world is a vast complexity of interactions with often unknown consequences. Another criticism has been that risk assessment has often focused on the risks to healthy, white European males (or caged white rats), which do not reflect the make-up the world's population. A particular example that illustrates the complexity of the real world as opposed to the laboratory is environmentally-mediated intellectual decline (EMID), which is caused by the effects of toxic substances on the brain. Recent research seems to indicate that EMID has been neglected as a health issue because of its complexity, whereas, in general, health priorities appear to be driven by less complex science and by visible illnesses, such as cancer or heart disease (ESRC Global Environmental Change Programme, 2000: 12).

Cancer provides a good example of the complexities of risk assessment and management. Although a family of several diseases, cancer is often treated as though it was a single disease, and it is often perceived as such by the public. To give an idea of its significance to the public in developed countries, 39% of Canadians will contract cancer during their lifetime and 25% of the Canadian public will die of cancer (National Cancer Institute of Canada, 1997: 45). But what do we know about the causes of cancer? The proportion of cases of cancer attributed to extrinsic (i.e. non-genetic) factors has been set, by different studies, at between 75% and 90%. Doll and Peto (1981) calculated that pollution might be responsible for <1–5% of all cancer deaths in the US, with their best estimate being 2%. Tobacco they calculated as causing 25–40% of cancer deaths and diet as causing 10–70%. Environmental pollution, it seems, is not a major cause of cancer. More fibre and less tobacco would do more for prolonging life than almost any other initiatives.

A range of issues emerges from a study of risk assessment in the cancer context (Hrudey, 1998). The most important issue is the assumption that has been made that there is no threshold in the dose–response relationship for a carcinogen. That is, it has frequently been asserted that there is no safe level for a substance that is known to produce cancers. Anyone with a knowledge of statistics should be able to work out that it would be necessary to find no tumours in 460 000 000 laboratory animals to conclude with a 99% certainty that the risk of a tumour occurring was less than 1 in 100 000 000. It is very difficult, if not impossible, to 'prove' that something is completely 'safe'— instead, we can only judge if something is relatively risk-free.

Risk and the Curriculum

There is little doubt that issues pertaining to risk are high in the public mind and we hope that what we have written so far has cast a little light on them. The question now is where, if at all, does it fit in the school curriculum? A simplistic answer would be 'either in science or PSHE/citizenship'. However we think that the real answer is more complex.

Recent work by science education researchers (for example, Millar and Osborne, 1998; Osborne and Collins, 2000; Levinson, 2001) has raised questions about the aims of the science curriculum. There seems to be general agreement that the National Curriculum, and the views of many science teachers, are rooted in a philosophical approach that we referred to earlier as logical positivism. From that perspective, the science classroom is not the place for issues surrounding risk and 'soft disasters'. The question then is, where *are* the issues to be addressed? Levinson (2001) constructs a case for teaching such things through the humanities, by which he clearly means English. But is that really the best place? Teachers without a science background are wary of teaching science and mathematics even when they are dealing with essentially factual material. The whole problem with risk is that it is a mix of hard science and mathematics, sociology, psychology and politics. Do we wish to leave that in the hands of teachers trained to discuss Shakespeare and Jane Austen? Well some of it yes, but all of it?

What about teaching about risk in PSHE and citizenship? The arguments are similar, but at least English has the respect of both teachers and pupils. The same cannot be said of PSHE which, despite efforts from many, still remains the poor relation of the curriculum. And citizenship, where will that go? It is too early to say, but unless it is assessed and leads to league table positions we suggest that it will simply become part of PSHE.

So yes, some of the issues surrounding the understanding of risk, its assessment and management, can and should be addressed in subject areas outside science. But the fact remains that, for the majority of our pupils, risk is likely to be one of a very small number of areas of science that will affect them when they leave school. However, as we have pointed out above, the current science curriculum has little space for such material. This leads to the argument that there is something fundamentally wrong with the National Curriculum for science: it is failing our children.

The continued criticism of science in the National Curriculum can be juxtaposed with the praise for innovative material post-16. AQA's AS *Science for public understanding* seems to have been well received by teachers and students. The syllabus contains a section devoted to 'Risk and risk assessment'. According to the syllabus, students should, *inter alia*: understand different ways of expressing the size of a risk, be aware of the range of factors that can influence people's willingness to accept specific risks, and be aware of the contribution of risk assessment to decisions about the management of risk (AQA, 1999). This knowledge and awareness is developed through a study of a health risk caused by a dietary or environmental factor (e.g. 'sick building' syndrome, smoking and lung cancer, aluminium and Alzheimer's). We would argue that serious consideration should be given to how similar knowledge and understanding could be made part of the curriculum throughout the 14–19 range.

Conclusion

In the world of the early 21st century the effects of the scientific enterprise on environmental and health issues are of rapidly increasing importance to the individual (see, for example, Beck, 1992). An understanding of how risk is dealt with is vital so we, the public, can make informed judgements that will have political effect. Without the necessary knowledge, society is left with the precautionary principle in its simplest form which, as we have argued, could lead to the severe curtailment of scientific research. Few science teachers want that, but to avoid that scenario there needs to be a radical reappraisal of the school science curriculum.

Acknowledgements

We would like to thank colleagues who attended our session on this topic at the Annual Meeting of the ASE at the University of Surrey in January 2001. The positive feedback and thoughtful questions that they generated encouraged us to write this article. The comments of three anonymous referees were also very helpful in finalising this version of the article.

References

AQA (1999) *Science for public understanding.* Advanced Subsidiary 5401 Specification. Manchester: Assessment and Qualifications Alliance (NEAB).
Beck, U. (1992) *Risk society. Towards a new modernity.* London: Sage.

Doll, R. and Peto, R. (1981) The causes of cancer: quantitative estimates of avoidable risks of cancer in the United States today. *Journal of the National Cancer Institute*, **66**(6), 1191–1308.

Environment Agency (1997) *Burning cattle cull waste in power stations—Environment Agency says risks are negligible.* Press Release 063/97, 25 June 1997. Bristol: Environment Agency.

ESRC Global Environmental Change Programme (2000) *Risky choices, soft disasters: environmental decision making under uncertainty.* Brighton: University of Sussex. Available from: www.gecko.ac.uk.

Fischhoff, B., Lichtenstein, S., Slovic, P., Derby, S. L. and Keeney, R. L. (1981) *Acceptable risk.* Cambridge: Cambridge University Press.

Gerrard, S. and Petts, J. (1998) Isolation or integration? The relationship between risk assessment and risk management. In *Risk assessment and risk management,* ed. Hester, R. E. and Harrison, R. M. pp. 1–19. Cambridge: Royal Society of Chemistry.

Harrison, R. M. (1998) Setting health-based air quality standards. In *Air pollution and health*, ed. Hester, R. E. and Harrison, R. M. pp. 57–73. Cambridge: Royal Society of Chemistry.

Hrudey, S. E. (1998) Quantitative cancer risk assessment—pitfalls and progress. In *Risk assessment and risk management,* ed. Hester, R. E. and Harrison, R. M. pp. 57–90. Cambridge: Royal Society of Chemistry.

Lash, S. and Wynne, B. (1992) Introduction. In Beck, U. *Risk society.* pp. 1–8. London: Sage.

Levinson, R. (2001) Should controversial issues in science be taught through the humanities? *School Science Review*, **82**(300), 97–102.

McQuaid, J. and le Guen, J-M. (1998) The use of risk assessment in government. In *Risk assessment and risk management,* ed. Hester, R. E. and Harrison, R. M. pp. 21–36. Cambridge: Royal Society of Chemistry.

Millar, R. and Osborne, J. F. ed. (1998) *Beyond 2000: Science education for the future.* London: King's College London.

National Cancer Institute of Canada (1997) *Canadian cancer statistics.* Toronto: Canadian Cancer Society.

Osborne, J. and Collins, S. (2000) *Pupils' and parents' views of the school science curriculum.* London: King's College London.

Slovic, P., Fischhoff, B. and Lichtenstein, S. (1980) Facts and fears: understanding perceived risk. In *Societal risk assessment: how safe is safe enough,* ed. Schwing, R. C. and Albers, W. A. New York: Plenum Press.

16 Approaching 'Soft Disasters' in the Classroom

Teaching about Controversial Issues in Science, Technology, Society, and Environment Education

Justin Dillon

Dillon, J. (2009). Approaching 'soft disasters' in the classroom: teaching about controversial issues in science, technology, society, and environment education. In, A. Jones, & M. de Vries (eds), *International Handbook of Research and Development in Technology Education*. Rotterdam: Sense, pp. 297–306.

Introduction

Today's citizens are confronted by a range of what are called 'soft disasters'—"environmental and political crises that emerge only slowly but at high cost to society, not least the erosion of public confidence and legitimacy" (ESRC Global Environmental Change Programme, 2000, p. 3). Such soft disasters include socio-scientific issues such as BSE, the GM food debate, HIV-AIDS and global climate change. To understand such issues, the public needs an education that is broad and deep enough to unravel the complex inter-relationships among science, technology, society and the environment (STSE). However, many of these topics are controversial and teachers face many challenges in attempting to address them in their classrooms.

Not surprisingly, the teaching of controversial STSE issues is itself somewhat controversial. In 2004, I gave a talk entitled 'Teaching Controversial Issues' as part of a session on 'Teaching Science for Responsibility' at the British Association Festival of Science in Exeter. My aim was to disseminate the findings of research I had been involved in with colleagues from two other universities (Oulton, Day, Dillon, & Grace, 2001). The press release issued by King's College to publicise the event contained a comment from me that summarised one of the key points emerging from our research:

> Teachers should be open with their own biases rather than pretend to be neutral and students should be asked to take any bias into account when making up their minds on a topic. Taking a neutral stance is not a good strategy for teaching children how society works.

The press release contained another statement attributed to me:

> The traditional approach to leading a discussion on a controversial science subject is for the teacher to take a neutral role. . . . We believe that this strategy is wrong and that it is unethical to pretend to pupils that teachers have no opinion.

The story was picked up by Mark Crow (2004) of the *Guardian,* one of the UK's more thoughtful newspapers. Crow opened his story by stating that "New research has lambasted science teaching in the UK as 'unethical' for its failure to acknowledge the importance of personal bias" (first paragraph). Crow had sought the views of a pressure group, Save British Science, whose spokesperson, Rosemary Davies, was quoted as saying: "Whilst it may be inappropriate for teachers to claim to be dispassionate about controversial science issues, students should still be trained to aspire towards objectivity". 'Aspire' maybe, but—as I will attempt to demonstrate later—students need to be aware that values play a key role in many of the decisions made by themselves, by politicians, by scientists, and by many others. Science and technology education that does not leave students with an awareness of the complexity of the STSE issues facing society is not fit for its purpose. Davies was also quoted in the article as saying:

> Just as subjects within science such as biology or physics are best taught by fully trained specialists, 'science for citizenship' type courses also need to be taught by those with adequate training and a thorough understanding of policy making and ethics. We can't invest in the skills of our pupils without investing in the people who are going to teach them.
>
> (Crow, 2004, final paragraph)

I would agree with the sentiment Davies expresses here. One of the findings of the research on which I was reporting is that science teachers often state that they are not trained to teach about socioscientific issues. Davies' quote masks a concern examined by Cross and Price (1996) about teachers who present science as unproblematic and characterised by content and certainty. As Kibble (1998) points out, there is a danger of an "over-simplistic presentation of moral dilemmas" (p. 54). Reporting on their work with pre-service and in-service teachers, Cross and Price found that science teachers were very loyal to their subject discipline, that is to say, a loyalty to science rather than science education. Partly as a result of this divided loyalty, Cross and Price reported that students lacked opportunities to discuss the nature of science or the construction of knowledge. A similar point has been made by Camino and Calcagno (1995):

> The most delicate matter still has to be faced, that is encouraging [science] teachers to change their ways of thinking. This implies, first, trying to

persuade teachers to abandon the safety of viewing science as an objective and neutral discipline in favour of the idea that knowledge is 'a-disciplinary', transitory and loaded with values.

(p. 72)

The British Association talk was also picked up by Anna Salleh (2004) of the Australian Broadcasting Corporation, whose online piece was entitled 'Teachers told to spice up lessons'. Salleh had contacted Deborah Crossing, Executive Director of the Australian Science Teachers Association, who said that "controversies such as immunisation, Australia's relationship to the Kyoto Protocol and genetic engineering were regularly discussed in Australian schools". She added that "teachers in Australia were probably not encouraged to express their own point of view when teaching controversies, and this might be because of concerns about litigation. . . . Instead," she said, "they were encouraged to facilitate debate". Crossing continued by providing what has become the orthodox view of teacher as neutral chair: "A good teacher will engage debate, provide the students with different points of scientific view, get them to question, get them to test and get them to come up with their own solution." She went on to make a point with which I wholeheartedly disagree: "It's kind of irrelevant whether they put their point of view". I would, however, agree with her subsequent statement that: "The important thing is that they are encouraging their students to come up with their point of view based on good scientific evidence and [showing them] how to assess whether it's good scientific evidence." Interestingly, Crossing seemed to contradict her support for teachers adopting a neutral stance when she commented that "students often did not question their teachers' views and this was counterproductive". Unless teachers give their views, how can students question them?

The Nature of Controversy and Controversial Issues

Controversy seems to be a common feature of many cultures. New 'soft disasters' appear daily somewhere in the world. Over time, controversies also disappear; what might be thought controversial one year might not be so the next. Issues are not intrinsically controversial—if it takes two to tango then it takes two or more for something to be controversial. Indeed, I am using the phrase 'controversial issues' in this chapter to mean that "significant numbers of people argue about them without reaching a conclusion" (Oulton, Dillon, & Grace, 2004, p. 411). Dearden (1981), writing more than 25 years ago, opined that:

A matter is controversial if contrary views can be held on it without those views being contrary to reason. This can be the case, for example, where insufficient evidence is held in order to decide the controversy, or, where the outcomes depend on future events that cannot be predicted with certainty, and where judgement about the issue depends on how to weigh or give value to the various information that is known about the issue.

(p. 38)

Using this conceptualisation, one can see how the increase in the use of mobile phones (see Chapter 25), which led to some people raising concerns about their safety, can be considered controversial because there was insufficient evidence about links between phone use and users' health. The point here is that people, when presented with the same evidence, whether it be scientific or otherwise, can come to radically different conclusions depending on their personal values (Lynch & McKenna, 1990). So Crossing's point, referred to above, that the "important thing is that [teachers] are encouraging their students to come up with their point of view based on good scientific evidence" seems oblivious to the impact of students' values on their interpretation of 'good scientific evidence'.

Writing a few years later than Dearden, Stradling (1985) defined controversial issues as "those issues on which our society is clearly divided and significant groups within society advocate conflicting explanations or solutions based on alternative values" (p. 9). Thus, nanotechnologies can be seen as offering great value or great peril depending on the values of people debating their use. Of course there are also degrees of being controversial, and what might be controversial in one situation is not controversial in another. For as long as I can remember, the use of nuclear power has been far more controversial in, say, the UK than it has been in France.

Teaching about Controversial Issues

> Education should not attempt to shelter our nation's children from even the harsher controversies of adult life, but should prepare them to deal with such controversies knowledgeably, sensibly, tolerantly and morally.
> (Qualifications and Curriculum Authority [QCA], 1998, p. 56)

Thus, the UK Qualifications and Curriculum Authority sees teachers as having a duty to prepare students to "deal with" controversial issues. Such an argument is supported by writers such as Dewhurst (1992), who argued that teachers have "to help their students to handle questions of value, to learn to make judgements which are truly their own" (p. 153); and Leib (1998), who argued that "it is [the educators'] duty not to shy away from tackling controversial topics" (p. 230). In order to carry out this "duty", Geddis (1991) asserts that teachers need to teach in such a way that students:

> . . . uncover how particular knowledge claims may serve the interests of different claimants. If they are to be able to take other points of view into account in developing their own positions on issues, they need

to attempt to unravel the interplay of interests that underlie these other points of view.

(p. 171)

The point is that controversial issues cannot be taught in a value-free way. Extending Geddis' points further, Oulton et al. (2004) argue that teachers need to teach about controversial issues in such a way that the following points are made:

- Groups within society hold differing views about them;
- groups base their views on either different sets of information or they interpret the same information in different ways;
- the interpretations may occur because of the different ways in which individuals or groups understand or 'see' the world (i.e. their worldview);
- differing worldviews can occur because the individuals adhere to different value systems;
- controversial issues cannot always be resolved by recourse to reason, logic or experiment; and
- controversial issues may be resolved as more information becomes available.

A point here is that students may well benefit from studying the nature of controversy, as well as developing better understanding of particular controversial issues. Science and technology teachers are, *potentially*, in a strong position to teach about the nature of controversy in their subject areas.

Research into pedagogical approaches to the teaching of controversial issues has developed relatively recently. Oulton et al. (2004, p. 415), summarising recent research, note that teachers often suggest that their pedagogy is underpinned by their stance on three issues:

- The focus should be on rationality, reasoning and sticking to the facts;
- a balanced view on the issue must be presented; and
- the teacher should remain neutral.

The neutral stance is a pedagogic strategy that has been advocated by, among others, Lawrence Stenhouse (1983). 'Procedural neutrality', as Stenhouse called it, is easier to advocate than to sustain in the classroom. Indeed, Stradling (1985) reports that teachers find procedural neutrality difficult to sustain because it threatens the rapport they have built up with the students and because the approach appears to cast doubt on their personal credibility. More realistically, Kelly (1986) suggests that teachers show 'committed impartiality'. With this approach, teachers attempt to provide different sides of an argument but do, at some stage, share their own views with the class.

In 1998, the UK Qualifications and Curriculum Authority concluded that teachers need to be flexible rather than show allegiance to one particular stance. Noting that it was not their "business or intention to try to tell teachers how to teach", they did, however, comment in a disturbingly vague manner that "a 'common sense' approach has much to recommend it" (QCA, 1998, p. 60). Three years later, the UK Department for Education and Employment (DfEE) published a non-statutory framework for Personal, Social and Health Education (PSHE) and Citizenship to accompany the recently implemented statutory National Curriculum for secondary (high) schools. Interestingly, the Citizenship documentation (DfEE, 1999) implied that certain values were to be promoted, for example:

> Education for sustainable development, through developing pupils' skills in, and commitment to, effective participation in the democratic and other decision making processes that affect the quality, structure and health of environments and society and exploring values that determine people's actions within society, the economy and the environment.
>
> (pp. 7–8)

So what might teachers do to promote such values should they so wish? Oulton et al. (2004) argue that there is a need for pedagogical approaches that:

- focus on the nature of controversy and controversial issues—that is, that people disagree and have different worldviews; values and limitations of science; political understanding; power; etc.;
- motivate pupils to recognize that a person's stance on an issue will be affected by their worldview;
- emphasize the importance of teachers and learners reflecting critically on their own stance and recognize the need to avoid the prejudice that comes from a lack of critical reflection;
- give pupils the skills and abilities to identify bias for themselves, encouraging them to take a critical stance towards claims of neutrality, a lack of bias and claims to offer a balanced view;
- promote open mindedness, a thirst for more information and more sources of information and a willingness to change one's view as appropriate, and avoid strategies that encourage pupils to actually make up their minds on an issue too hastily; and
- motivate teachers, as much as possible, to share their views with pupils and make explicit the way in which they arrive at their own stance on an issue. (p. 420)

Such a pedagogy, however, is rarely found in classrooms. Indeed, as suggested by Deborah Crossing, teachers might be reluctant to express their own point of view because of the risk of litigation (in Salleh, 2004, referred to above).

In recent years, curriculum changes in the UK have encouraged teachers to develop their students' awareness and understanding of controversial issues. For example, ever since the introduction of the National Curriculum in England, teachers have been encouraged by statute to devote some time to the nature of science. At its best, such an approach to science teaching can illuminate the values that underpin science and the use of science by society. In its latest incarnation, the science curriculum for 14–16-year-olds indicates that pupils should be taught about 'How Science Works'. Under the heading 'Applications and implications of science', teachers are instructed that pupils should be taught:

- about the use of contemporary scientific and technological developments and their benefits, drawbacks and risks
- to consider how and why decisions about science and technology are made, including those that raise ethical issues, and about the social, economic and environmental effects of such decisions
- how uncertainties in scientific knowledge and scientific ideas change over time and about the role of the scientific community in validating these changes. (QCA, 2007, p. 4)

A short series of 'Explanatory Notes' expands on what the learning outcomes should be:

> During key stage 4, pupils learn about the way science and scientists work within society. They consider the relationships between data, evidence, theories and explanations, and develop their practical, problem-solving and enquiry skills, working individually and in groups. They evaluate enquiry methods and conclusions both qualitatively and quantitatively, and communicate their ideas with clarity and precision.
>
> All pupils develop their ability to relate their understanding of science to their own and others' decisions about lifestyles, and to scientific and technological developments in society.
>
> (ibid.)

Few teachers have substantial experience of working as a scientist other than during their own education. As a result, they depend for their knowledge of how science works on a range of secondary sources, including television and textbooks. Some teachers arrange for scientists to come into the classroom, but they are in a minority. Perhaps teachers should spend some time shadowing scientists and technologists for a short time.

Recent Research in the Teaching of Controversial Issues

Although teaching about STSE issues has been advocated for many years (see, for example, Aikenhead, 2003; Solomon, 1993; van Weelie & Wals, 2002), it is only relatively recently that it has generated much research interest. For example, a growing body of research has examined how teachers deal with socioscientific issues. Albe (2008) carried out a study in a French 11th-grade science class (n=12) in the context of vocational secondary education. Albe's participants were 16–18-year-olds specialising in sciences and technologies for agronomy and the environment. The students were divided into two groups to debate whether or not mobile phones were dangerous to human health. Both groups contained equal numbers of boys and girls. The teaching intervention was based on a module designed by science education research-ers and teachers (Hind, Leach, Ryder, & Prideaux, 2001). The purpose of the lesson was to develop the students' understanding of how to assess the quality of scientific data.

> In the first part of the lesson, students examined the issues raised by the controversy, the technological and scientific concepts involved and were trained to evaluate the validity and reliability of research results. Students were then asked to review a number of pieces of research and to select those that support a particular point of view. To do so, students in group discussions studied seven research extracts on mobile phone effects. Each extract was an approximately 10-lines of text describing the research protocol and the results obtained by a team of scientists. They are related to the occurrence of illnesses in animals which had been subjected to experiments but also on epidemiological surveys and memory tests.
>
> (Albe, 2008, p. 69)

The teacher then set up a court case scenario. Students took on the roles of expert witnesses in a trial in which an employer was taken to court by an employee whose health was poor. The employee's argument was that it was excessive mobile phone use that had led to health problems. Students had to present coherent arguments to support the employer or the employee. Having presented their case, they then had to question the arguments put forward by the other group. The teacher's role was that of a magistrate presiding over the case. After the role playing was over, the students were given an opportunity to reflect on their opinions about the controversy. Albe used a "discourse analytic approach grounded in sociolinguistics and supported by an ethno-graphic perspective" (p. 86), reporting that:

> The activity of reviewing research results and elaborating arguments in a discussion group, with a view to participating in a role-play on the controversial issue of the danger of using cellular telephones, is very demanding in social and epistemological terms.
>
> (p. 85)

In addition:

> It has been observed that collaborative argumentation was developed within a group when a student acts as a helper collaborating with a

student assuming the leadership of group discussions. Students' requests addressed to others to explain their claims or to develop justifications create opportunities for collaborative argumentation. These results tend to indicate that group discussions on a socio-scientific issue should be organised to regulate students' interactions on the social plane. Teachers should attempt to arrange group discussions to avoid student conflicts.

(p. 86)

Following on from Grace (2005), who suggested ground rules for discussions of socioscientific arguments, Albe (2008) suggests that "guides could be provided to students within group discussions to explain their views and consider others. A discussion at the end of the activity may also be valuable to enable students to reflect on their views and appreciate the value of group discussion" (p. 86). Importantly, Albe argues that as well as looking at controversial issues themselves, students should also examine:

> . . . the way in which scientific knowledge is produced within a community and, in particular, the role of controversy in this process. For instance, the inclusion of considerations of the nature and limitations of science, the status, role and limits of evidence, the interests involved and the ways science communities operate to establish knowledge are very important when dealing with contemporary science controversies.
>
> (p. 86)

One wonders how many teachers are *au fait* with 'the ways science communities operate to establish knowledge'. (See also Chapter 25 for a discussion of the importance of the nature of science in relation to socioscientific issues.) Albe recognises that teachers need help, noting:

> Teachers can be uncomfortable with such issues that could question their own epistemology and teaching practices. A way of assisting teacher professional development can be the cooperative design of teaching sequences on socio-scientific issues. This can be an occasion to explore with teachers ideas about the nature of controversy and elaboration modes of scientific knowledge, issues about teaching approaches appropriate to socio-scientific issues and the question of teacher position in class.
>
> (p. 86)

Writing two years earlier, Levinson (2006) pointed out that "the teaching of controversial issues needs a stronger theoretical base" (p. 1201) and proposed a theoretical framework for the teaching of socio-scientific controversial issues:

> Drawing on a liberal democratic conception of possible sources of conflict, three strands are developed that provide a framework for teachers

when teaching socio-scientific issues: these are categories of reasonable disagreement, the communicative virtues, and modes of thought.

(ibid.)

Levinson also provides 'an epistemological model of controversy' that, he argues, is: "a framework for approaching the teaching of controversial issues in schools and a tool for analysing the ways in which elements of a controversial issue are deployed in the classroom" (p. 1217):

- Categories of reasonable disagreement.
- Communicative virtues when conducting a discussion with conflicting views
- Patience
- Tolerance
- Respect for differences
- Attentive and thoughtful listening
- Openness
- Honest self-expression
- Adherence to agreed procedures
- Freedom of expression
- Equality
- Modes of thought
- Narrative mode which involves the voices of the participants. For lay people addressing socioscientific issues the substantive science is often reconstructed, marginalized or inert.
- Logico-scientific mode, based on scientific evidence. (ibid.)

Drawing on various authors' work (for example, Rawls, 1983), Levinson provides a secure theoretical framework that complements and develops other work in the field of argumentation (e.g., Newton, Driver, & Osborne, 1999; Zeidler, Osborne, Erduran, Simon, & Monk, 2003; and Chapter 36 by Osborne).

With regard to research specifically focused on the teaching of controversial environmental issues, Cotton (2006) noted that "there has been little research into the strategies that teachers adopt" (p. 225). Adding that much existing research involved large-scale surveys of teachers' attitudes, students' attitudes, or classroom interventions aimed at attitudinal change, Cotton argues that they have provided little insight into teachers' beliefs and normal classroom behaviours:

The few recent studies which have investigated teachers' beliefs about teaching controversial issues in other curriculum areas have largely neglected to incorporate classroom observation, being reliant instead on post hoc reports of behaviour.

(p. 225)

Cotton's study "aimed to provide an alternative perspective by investigating the beliefs and practices of three experienced teachers in English secondary schools, drawing upon both interview data and transcripts of classroom interaction" (ibid.). Cotton focused on a geography course for students aged 16–19 (n=11–18) and studied three cases, each involving observations over a 5–6 week period at three different schools. The topics covered were: indigenous people's land rights in the rainforest, the role of NGOs in governing Antarctica, and reconciling the needs of conservation and tourism in National Parks. Cotton reported that all the teachers adapted their teaching strategies "to encourage student involvement in discussions and producing lively and interesting lessons" (p. 237). However, he went on to note:

> All of the teachers studied experienced great difficulty in implementing their beliefs about balance and neutrality, and the classroom data suggest that the influence of the teachers' own environmental attitudes was greater than they either intended or, in all probability, realized. Moreover, in line with the findings of Geddis (1991), there were occasions where the teachers' intent appeared to be persuasion rather than instruction. Their desire not to express their own views frequently led to the situation where these views were expressed indirectly in the form of questions, or by control of students' turns in discussion. Whilst these strategies enabled the teachers to avoid explicitly stating their views, such an indirect expression of attitudes may have been harder for the students to challenge than a direct argument presented by the teacher.
>
> (ibid.)

Cotton's findings seem to provide support for the more transparent approach advocated by Oulton et al. (2004) (see above).

In a similar vein, Tal and Kedmi (2006), reporting on the implementation of an innovative Science and Technology in Society (STiS) curriculum in Israel, described a study involving observations of six classes and interviews with four teachers and 20 grade 10 and 11 students. The unit taught was entitled 'Treasures in the Sea: Use and Abuse'. Noting that the "use of relevant and authentic topics that encourage and allow open discussions within and between communities reinforces the idea of citizens-science" (p. 640), they caution that "teachers cannot dramatically change their practice unless they go through a longer process of reviewing their own perceptions of science and teaching philosophy, and gradually practicing different teaching and assessment culture in class" (p. 641).

Concluding Comments

The long-haul journey to broaden the school curriculum in order to address STSE issues has been a bumpy ride. While many countries now promote the discussion of such issues, the evidence of any substantial impact on teachers' practices and students' understandings, attitudes, or behaviours remains weak. It is easy to advocate their inclusion in the curriculum; it is much more difficult to impact on the dominant pedagogies of science and technology education in an assessment climate that values rote learning of atomised concepts. This raises fundamental questions about how science teachers see science and technology and, thus, education in these areas. I worry that many science teachers see themselves as scientists first and teachers second when, in actuality, they have little experience of contemporary science in the making. Worse still, few science teachers appear to have had an adequate training in teaching socioscientific issues and so find it difficult to shift their pedagogical practices.

The focus of recent international comparisons, such as PISA, towards a focus on scientific literacy, might promote the cause of STSE education but it will take a major shift in teacher understanding and pedagogy before schools adequately prepare students for the complexity of life in the 21st Century.

Acknowledgements

I wish to thank two colleagues, Alex Manning and Melissa Glackin, for their insightful comments on an earlier draft of this chapter. I also acknowledge the contribution to my thinking about the teaching of controversial issues made by Marcus Grace (University of Southampton) and Chris Oulton (formerly of the University of Gloucester).

References

Aikenhead, G. (2003). STS education: A rose by any other name. In R. Cross (Ed.), *A vision for science education: Responding to the work of Peter Fensham* (pp. 59–75). London: RoutledgeFalmer.

Albe, V. (2008). When scientific knowledge, daily life experience, epistemological and social considerations intersect: Students' argumentation in group discussions on a socio-scientific issue. *Research in Science Education, 38*(1), 67–90.

Camino, E., & Calcagno, C. (1995). An interactive methodology for 'empowering' students to deal with controversial environmental problems. *Environmental Education Research, 1,* 59–74.

Cotton, D. R. E. (2006). Teaching controversial environmental issues: Neutrality and balance in the reality of the classroom. *Educational Research, 48*(2), 223–241.

Cross, R., & Price, R. (1996). Science teachers' social conscience and the role of teaching controversial issues in the teaching of science. *Journal of Research in Science Teaching, 33*(3), 319–333.

Crow, M. (2004). Call for a new approach to science teaching. *Guardian.* Retrieved from http://guardian.co.uk/education

Dearden, R. F. (1981). Controversial issues in the curriculum. *Journal of Curriculum Studies, 13,* 37–44.

Department for Education and Employment [DfEE]. (1999). *The national curriculum for England: Citizenship.* London: Department for Education and Employment and the Qualifications and Curriculum Authority.

Dewhurst, D. (1992). The teaching of controversial issues. *Journal of Philosophy of Education, 26*(2), 153–163.

ESRC Global Environmental Change Programme. (2000). *Risky choices, soft disasters: Environmental decision making under uncertainty.* Brighton, Brighton and Hove: University of Sussex. Retrieved from www.gecko.ac.uk

Geddis, A. (1991). Improving the quality of science classroom discourse on controversial issues. *Science Education, 75*(2), 169–183.

Grace, M. M. (2005, August/September). *Developing personal values and argumentation skills through decision-making discussions about biological conservation.* Paper presented at the European Science Education Research Association (ESERA) conference, Barcelona, Spain.

Hind, A., Leach, J., Ryder, J., & Prideaux, N. (2001). *Teaching about the nature of scientific knowledge and investigation on AS/A level science courses.* Leeds, UK: Centre for Studies in Science and Mathematics Education.

Kelly, T. (1986). Discussing controversial issues: Four perspectives on the teacher's role. *Theory and Research in Social Education, 14*(2), 113–138.

Kibble, D. (1998). Moral education dilemmas for the teacher. *Curriculum Journal, 9,* 51–61.

Leib, J. (1998). Teaching controversial topics: Iconography and the Confederate Battle Flag in the South. *Journal of Geography, 97,* 229–240.

Levinson, R. (2006). Towards a theoretical framework for teaching controversial socio-scientific Issues. *International Journal of Science Education, 28*(10), 1201–1224.

Lynch, D., & McKenna, M. (1990). Teaching controversial material: New issues for teachers. *Social Education, 54,* 317–319.

Newton, P., Driver, R., & Osborne, J. (1999). The place of argumentation in the pedagogy of school science. *International Journal of Science Education, 21*(5), 553–576.

Oulton, C. R., Day, V., Dillon, J., & Grace, M. (2001). *Unlocking controversial issues: A report to the Countryside Foundation for Education.* Worcester, UK: University College Worcester.

Oulton, C., Dillon, J., & Grace, M. (2004). Reconceptualizing the teaching of controversial issues. *International Journal of Science Education, 26*(4), 411–423.

Qualifications and Curriculum Authority [QCA]. (1998). Education for citizenship and the teaching of democracy in schools. London: Author.

QCA. (2007). Science. *Programme of study for key stage 4.* London: Author.

Rawls, J. (1993). *Political liberalism.* New York: Columbia University Press.

Salleh, A. (2004). Teachers told to spice up lessons. *ABC Science Online.* Retrieved from September 13. Retrieved from http://www.abc.net.au/science/news/

Solomon, J. (1993). *Teaching science, technology and society.* Buckingham, UK: Open University Press.

Stenhouse, L. (1983). *Authority, education and emancipation.* London: Heinemann.

Stradling, B. (1985). Controversial issues in the curriculum. *Bulletin of Environmental Education, 170,* 9–13.

Tal, T., & Kedmi, Y. (2006). Teaching socioscientific issues: Classroom culture and students' performances. *Cultural Studies of Science Education, 1*(4), 615–644.

Van Weelie, D., & Wals, A. (2002). Making biodiversity meaningful through environmental education. *International Journal of Science Education, 24,* 1143–1156.

Zeidler, D. L., Osborne, J., Erduran, S., Simon, S., & Monk, M. (2003). The role of argument during discourse about socioscientific issues. In D. L. Zeidler (Ed.), *The role of moral reasoning on socioscientific issues and discourse in science education* (pp. 97–116). Boston: Kluwer.

17 On Scientific Literacy and Curriculum Reform

Justin Dillon

Dillon, J. (2009). On scientific literacy and curriculum reform. *International Journal of Environmental and Science Education*, 4(3), 201–213.

Introduction

Since the first use of 'scientific literacy' in the late 1950s, science educators and policy makers have gradually reconceptualised the term to such an extent that one author remarked relatively recently that "scientific literacy is an ill-defined and diffuse concept" (Laugksch, 2000, p. 71). Despite this perceived imprecision, scientific literacy appears to underpin the curriculum standards of many countries and is at the heart of international comparisons of student attainment (and thus of education systems) including the Organisation for Economic Co-operation and Development's (OECD) Programme for International Student Assessment (PISA) study. But why is something so slippery—so hard to define—treated as if it is the Holy Grail of science education? The answer, it would appear, is that its slipperiness is the key to its longevity.

Whatever your opinion about the utility of the term 'scientific literacy', it would be difficult to argue with McEneaney's (2003) claim that it has a 'worldwide cachet'. In this paper I argue that the term is likely to be part of the discourse of science education for many years to come and that as science educators we need to identify how best to work with it. My analysis is set within attempts to improve science education within a number of European countries and towards the end of the paper, I reflect on how developments in Europe relate to those in other continents.

Underpinning the analysis presented here is a concern that science education frequently benefits the minority of students who go on to become scientists at the cost of those who do not (Osborne & Dillon, 2008). A related concern is that science education is frequently critiqued for its lack of imagination and relevance, and its focus on abstract concepts beyond the imagination and interest of most students. These concerns are not new nor are they easily addressed. Members of the 2007 Linné Scientific Literacy Symposium articulated the issues clearly in their "Statement of Concern":

> Science education, perhaps because of the sheer depth and volume of the knowledge base of modern science, has isolated that knowledge from its

historical origins and hence students are not made aware of the dynamic and evolving character of scientific knowledge, or of science's current frontiers. There is little flavour in school science of the importance that creativity, ingenuity, intuition or persistence have played in the scientific enterprise. Nor is there any real sense of any meaningful exploration of issues that relate ethical and personal accountability to modern scientific activity. Indeed, the existence of human enterprise that makes science possible is almost ignored in science education. Curricula and assessment need to support teachers' being able to share the excitement of the human dramas that lie behind the topics in school science with their students.

(Members of the Linné Scientific Literacy Symposium, 2007, p. 7)

However, their position does not represent those of all science educators. The recent introduction of a new science curriculum aimed at promoting scientific literacy drew criticism from "Leading educationalists [who] criticised [it] for being more 'fit for the pub' than the school room" (BBC News, 2006, 1). The longevity of the term scientific literacy relies on its ability to be seen as an umbrella for radically different philosophies of science education. However, the evidence suggests that when attempts are made to effect curriculum change to promote 'scientific literacy' the unreconciled philosophical clashes hinder progress.

What Do We Know about Scientific Literacy?

Writing at the turn of the century, Laugksch argues that "Scientific literacy has become an internationally well-recognized educational slogan, buzzword, catchphrase, and contemporary educational goal" (Laugksch, 2000, p. 71). Laugksch's paper is a thoughtful conceptual overview and in it he notes that the conceptualization of scientific literacy that he has described hides a range of meanings and interpretations, and he notes that a view has emerged "that scientific literacy is an ill-defined and diffuse concept (e.g., Champagne & Lovitts, 1989)" (Laugksch, 2000, p. 71).

"Ill-defined and diffuse" though 'scientific literacy' may be, the term has survived for half a century, and looks set to outlast 'science for all', and it seems unlikely that another phrase will knock it from its position at the heart of science curriculum policy making (even the rather clumsy "scientific literacy for all" seems to have come and gone (Lee, 1998)). As science educators, we will probably have to live with the phrase no matter how unsatisfactory we find it. It would be more accurate, perhaps, to refer to 'scientific literacies' or 'dimensions of scientific literacy', but more of that later.

Vision I and Vision II: Looking Inward and Outward

In his chapter on scientific literacy in the recent *Handbook of Research on Science Education,* Roberts (2007a) makes an elegant distinction between two ways of looking at the aims and purposes of science education. He identifies two 'visions' for generating conceptions of scientific literacy: Vision 1 and Vision II. Vision I, according to Roberts, "looks inward at science itself—its products such as laws and theories, and its processes such as hypothesizing and experimenting," whereas Vision II "looks outward at situations in which science has a role, such as decision-making about socioscientific issues" (Roberts, 2007b, p. 9).

These visions are underpinned by different philosophies and, at their most extreme, reflect competing interests that have and continue to influence the content of the science curriculum (see Hodson, 2008 for an in-depth study of these issues). At one extreme, there are those whose major preoccupation is the place of scientific content in the curriculum. At the other extreme are those who see science education as being primarily about criticizing the assumptions underpinning science as a cultural activity (see, e.g., Roth & Barton, 2004). These visions, which are philosophically irreconcilable, have established themselves under the increasingly broad umbrella of 'scientific literacy'.

Roth and Barton (2004) powerfully illustrate one dimension of the Vision II philosophy when they argue that:

> Conventional approaches to scientific literacy, knowing, and learning are based on an untenable, individualistic (neo-liberal) ideology that does not account for the fundamental relationships between individual and society, knowledge and power, or science, economics, and politics.
>
> (p. 3)

I am broadly in sympathy with that analysis but wherever you stand on the issues, I would argue that scientific literacy (or literacies) will be part of the discourse of science education for a long time. It suits all parties to use 'scientific literacy' as a weasel word.

The job of policy makers and curriculum planners is often to attempt to reconcile these conflicting visions. As a result, they can be put under substantial pressure to promote one vision over another (Blades, 1997; Fensham, 1998). A critical issue for contemporary curriculum makers, though, is can there be a balance between Vision I and Vision II, and do potential scientists need a different balance than everybody else? As Roberts puts it, "Everyone agrees that students can't become scientifically literate without knowing some science, and everyone agrees that the concept needs to include some other types of understanding *about* science [original emphasis]" (Roberts, 2007b, p. 11). It is this logic that allows the competing visions to co-exist under the same banner. In effect, 'scientific literacy' is what Stables (1996) referred to as a 'paradoxical' compound policy slogan. The conflicting visions of scientific literacy are sustained by a series of dubious rationales and questionable assumptions some of which are outlined in the following section.

Rationales for Promoting Scientific Literacy

Drivers for Change

There are several rationales for moving towards scientific literacy as a fundamental goal of science education, and they have all been used by politicians and policy makers in Europe and elsewhere at some point in recent years. To some extent, the drivers are inseparable from the rationales for science education itself. Laugksch (2000) groups the common arguments into two categories which he labels 'macro' and 'micro'. One macro-argument is that "national wealth depends on competing successfully in international markets" and that to compete, a nation must have a strong research and development base. Taking the assumption one step further, a steady stream of home-grown scientists is, therefore, essential to keep the R&D base strong. Such a rational was exemplified recently by Lord Grayson, the UK's new science minister who was quoted as saying that, "Science is fundamental to this country. As we go into this global downturn the importance of maintaining our investment in science has never been greater" (BBC News, December 5, 2008, ¶6).

The opening paragraph of the widely circulated and oft-quoted 'Rocard Report', *Science Now: A Renewed Pedagogy for the Future of Europe,* makes several macro-level assertions such as this one on the opening page:

> In recent years, many studies have highlighted an alarming decline in young people's interest for key science studies and mathematics. Despite the numerous projects and actions that are being implemented to reverse this trend, the signs of improvement are still modest. Unless more effective action is taken, Europe's longer term capacity to innovate, and the quality of its research will also decline.
>
> (High Level Group on Science Education, 2007, p. 2)

Another macro-level argument has probably been around even longer than the economic argument. For well over a century, scientists have argued that the public support of science is critical to the continuation of scientific research (Shamos, 1995; Waterman, 1960). Public attitudes towards science and technology have been the subject of repeated study. The 2006 PISA report of a survey of more than 400,000 15-year-old students from 57 countries notes that "92% [of respondents] said that advances in science and technology usually improved people's living conditions" (OECD, 2007, p. 6), and noting that:

> A strong acceptance by students that science is important for understanding nature and improving living conditions extends across all countries in the survey. However, this was mirrored to a much lesser extent in students' responses to the wider socio-economic benefits of science.

On average across OECD countries, 25% of students (and over 40% in Iceland and Denmark) did not agree with the statement "advances in science usually bring social benefits".

(OECD, 2007, p. 6)

Laugksch also identifies another macro-level argument for scientific literacy which is that the more the public understand how science works and what it can do for them, the more likely they are to support scientific and technological endeavour. Such an argument should cause scientists and their supporters in Iceland and Denmark, in particular, to advocate increasing scientific literacy as a matter of some urgency. However, such a view is based on an assumption that has never really been tested.

Another rationale identified by Laugksch at the macro-level is one that has undergone substantial discussion in the UK. The Public Understanding of Science movement, which was criticized as being based on a deficit model ("if only people knew more science, they'd be more positive about science") has been superseded by a Public Engagement with Science and Technology model that ostensibly encourages public discussion of issues such as xenotransplantation, GM foods and nanotechnology. The espoused argument for increased public engagement with scientists and policy makers is that increased transparency of decision making will lead to greater public confidence in the final decisions made on controversial issues (Royal Society, 1985). Again, this is an assumption that has yet to be rigorously tested.

The final argument listed by Laugksch relates to C. P. Snow's identification, in 1959, of a division between the 'two cultures' of science and literary intellectuals (Snow, 1962). Laugksch (2000) notes that the danger of science being seen as outside mainstream culture is that the public will fail to grasp what science is which might, in turn see them responding to science "with a mixture of adulation and fear" (p. 85). Increasing the public's scientific literacy might help to dispel the perceived 'cult' image of science and scientists.

While the macro-level arguments promote collective economic well-being, democracy and societal coherence, micro-level arguments focus on the benefits of scientific literacy to the individual. To some extent, the micro-level benefits are consequences of the macro-level benefits. The micro-level benefits might include increased economic prosperity and job opportunities, wiser health decisions, increased confidence in science and technology and reduced personal risk. An example of a micro-level rationale is provided by both Shortland (1988) and Snow (1962) who argue that an individual cannot be considered properly educated without a working knowledge of modern science which is one of the major successes of Western civilisation. Such an argument begs the question 'Who is to say what a 'proper' education looks like?'

The Rocard Report intertwines both macro- and micro-level benefits of scientific literacy. In a section headed 'Providing all citizens with both science literacy and positive attitude (*sic*) towards science' the report is unequivocal about the value of scientific literacy (which it never defines):

> There is *obviously* a need to prepare young people for a future that will require good scientific knowledge and an understanding of technology. Science literacy is important for understanding environmental, medical, economic and other issues that confront modern societies, which rely heavily on technological and scientific advances of increasing complexity.
> (High Level Group on Science Education, 2007,
> p. 6, emphasis added)

But the report argues that the 'key point' is:

> Equipping every citizen with the skills needed to live and work in the knowledge society by giving them the opportunity to develop critical thinking and scientific reasoning that will enable them to make well informed choices. Science education helps fighting misjudgements and reinforcing our common culture based on rational thinking.
> (High Level Group on Science Education, 2007, p. 6)

The idea that Europe has a common culture and that it is based on rational thinking might be seen as somewhat simplistic. The strength of the macro- and micro-level arguments for promoting scientific literacy may not be convincing and may rest on some false assumptions. What is important to realise, though, is that the arguments are as much philosophical as they are educational and that, as such, they can drive policy even if there is little evidence to support their veracity (whither 'rational thinking'?).

Some European Attempts to Make Science Education More Fit for Purpose

Turning now to some European attempts to address the problems of science education in terms of curriculum and pedagogy, I will examine some initiatives in individual nations and some cross-European projects. While the coverage of these initiatives is necessarily brief, I hope to convey a flavour of what has been tried and how it has, or has not, been received.

Britain: Twenty First Century Science

In Britain, the publishing of *Beyond 2000: Science Education for the Future* (Millar & Osborne, 1998) initiated a debate that led to a range of innovations aimed at making the science curriculum fit for a broader purpose than had been the case until then. Developments started in England and Wales with the piloting of an optional course, *Science for Public Understanding* aimed at 17–18 year olds. Subsequently, the University of York and the Nuffield Curriculum Centre developed a course for 14–16 year olds, *Twenty First Century Science*.

Twenty First Century Science consists of three components—a core curriculum that explores both the major explanatory themes of science and a set of 'ideas-about-science' that all students study. These components are then followed by an additional course of academic science which is for those who wish to pursue the study of science at a later stage. Alternatively, students with a more vocational inclination can take a course in Applied Science. The authors of *Twenty First Century Science* state that they would expect a scientifically literate person to be able to:

- appreciate and understand the impact of science and technology on everyday life;
- take informed personal decisions about things that involve science, such as health, diet, use of energy resources;
- read and understand the essential points of media reports about matters that involve science;
- reflect critically on the information included in, and (often more important) omitted from, such reports; and
- take part confidently in discussions with others about issues involving science. (*Twenty First Century Science*, 2008)

This conceptualisation of what it means to be scientifically literate reads almost as a meta-level set of aims. Underpinning these outcomes (what one might term separate 'literacies') are two different features: knowledge of how science works and knowledge of science explanations. In terms of the former, students need to "be able to reflect on scientific knowledge itself, including":

- the practices that have produced it;
- the kinds of reasoning that are used in developing a scientific argument; and
- the issues that arise when scientific knowledge is put to practical use.

In terms of knowledge of science explanations:

No one can be said to be 'scientifically literate' unless they understand some science. But what matters is to have a broad understanding of the main scientific explanations that give us a framework for making sense of the world around us.

(*Twenty First Century Science*, 2008)

The main scientific explanations in the scheme include: Chemical change; The interdependence of living things; The gene theory of inheritance; The theory of evolution by natural selection; Energy sources and use; Radiation; The Universe (*Twenty First Century Science*, 2008).

Approximately 10% of the project budget was set aside for an evaluation of the pilot. Three separate studies were commissioned into separate outcomes: knowledge and understanding; attitudes to science, and changes in classroom practice. One of the evaluation team, summarising the findings of the three studies, commented that:

> A clear majority of teachers indicated that they enjoyed teaching Core Science more than other courses. Given that nearly three-quarters also found it more demanding to teach, this clearly represents a major achievement.
>
> (Burden et al., 2007, p. 25)

The rather cautious summing up continued thus:

> Perhaps the single most positive feature in the findings of these studies is that the project has succeeded in persuading a majority of the pilot teachers that it is more professionally rewarding than standard Double Award science. It has thus, in my view, engaged to a measurable degree with the desire of science teachers to offer a curriculum, particularly at Key Stage 4, which is livelier and more engaging than what has traditionally been available. Moreover it cannot be assumed that the teachers involved are all volunteers, or sympathetic to the aims of the project. This achievement also needs to be set against the well- established view that producing educational change of any kind is difficult: the institutional, professional and political barriers are large. All this is a significant achievement.
>
> (Burden et al., 2007, p. 29)

The arrival of *Twenty First Century Science* was not greeted with universal acclaim as was indicated by the "being more 'fit for the pub' than the school room" quote mentioned above.

Netherlands—General Natural Sciences and the Junior College Utrecht

Innovative approaches have been tried in other parts of Europe. The Dutch minister of education set up an advisory committee in 1994 which proposed the introduction of an entirely new subject in the upper secondary curriculum of all students:

> The new subject, called *Algemene Natuurwetenschappen* ('General Natural Sciences'), or ANW, was to occupy well over 10% of the available time in Grade 10 [ages 16–17]. Its introduction, alongside the traditional science subjects that are optional in Grade 10, was part of a far-reaching innovation in upper secondary education.
>
> (De Vos & Reiding, 1999, p. 711)

Osborne and Dillon (2008, p. 21) report that "the course has been contentious and gone through some transformation" since its introduction. De Vos and Reiding (1999), the course evaluators, noted that it was "extremely difficult" for teachers "to escape from the shadows of the science teaching tradition" (p. 718). They reported that even after the introduction of the course, science teachers' pedagogy was still dominated by a focus on content rather than developing an understanding of science (see, also, Bartholomew, Osborne & Ratcliffe, 2004). As Osborne and Dillon (2008) point out that "the teaching of science is an established cultural practice passed on from one science teacher to another" (p. 22) and it is difficult to change that culture quickly or easily.

Another Dutch initiative which has been piloted recently was aimed at more able students who can become de-motivated at school. Junior College Utrecht is a specialized science-enriched secondary school (van der Valk & Eijkelhof, 2007):

> Entrance to this school is competitive and seen as high-status. Students are taught at an accelerated pace with students left to learn minor material independently. In addition, there is a greater research focus and a significantly enhanced curriculum in which university specialists teach specific modules. Students reported that they enjoyed the challenge, the enriched elements and working with their intellectual equals. Such a mechanism—essentially one of making the study of science a high status subject—is one means of attracting more able students.
>
> (Osborne & Dillon, 2008, p. 22)

Both these Dutch initiatives aim to address the issue of differentiating the curriculum offer so that students who want to continue into a science and/or technology career can be satisfied while, at the same time, the majority of students learn something potentially more relevant (though see Chapter 1 of Roth & Barton (2004) for a critical analysis of the relevance of contemporary science education).

Turkey: A New Science and Technology Curriculum

Turkey initiated a major primary school curriculum reform in 2003. Science was one of five subjects chosen for reform and a new curriculum for grades 1–8 has been implemented. Koc Isiksal and Bulut (2007), commenting on the rationale for the curriculum reform note that:

> One of the major motivations for this curriculum improvement is to reach ideal international standards of education implemented in Europe, North America and East Asia. For instance, the new curriculum aims at creating learning environments, where students can share their ideas and actively participate, relating various disciplines to each other, and using different teaching methods within the enriched environment.
>
> (p. 31)

Science is compulsory in Turkey from grade 4 (ages 9–10) through to grade 8 (ages 13–14). The seven learning areas in the new science and technology curriculum are: Physical Processes; Life and Living Beings; Matter and Change; The Earth and the Universe; Science Process Skills (SPS); Science-Technology-Society-Environment (STSE), and, Attitudes and Values (AV) (Taşar & Atasoy, 2006, p. 4).

What is particularly interesting about this curriculum, is the predicted outcomes of the Science-Technology-Society-Environment learning area. For example, at the end of Grades 6–8, a student should understand:

> That many sources of knowledge are utilized in developing technological products such as imagination, creative thinking, culture and traditions, mathematical knowledge, knowledge obtained through science about how nature functions, as well as the human capabilities of realizing and from whatever source bringing together the knowledge, facts, and materials that initially seem to be unrelated in order to make a technological product.
>
> (Taşar & Atasoy, 2006, p. 6)

Such an outcome would be unusual in a Western curriculum document. Indeed, at the beginning of this paper I noted the concern of members of the 2007 Linné Scientific Literacy Symposium: "There is little flavour in school science of the importance that creativity, ingenuity, intuition or persistence have played in the scientific enterprise" (Linder et al., 2007, p. 7). There is an irony here that 2009 is the 'European Year of Creativity and Innovation'.

Equally unusual to a Western European eye is one of the outcomes of the Attitudes and Values learning area ('Developing a life style') for the same three grades:

> (Development of a life style through the control of the value system over a long period of time)

- Continuously checks on her/himself and the environment
- Continues the habits for a healthy life
- Realizes that everything is for the service of love, peace, and happiness
- Self-disciplined (Self-controlled, prompt, self-evaluating, sincere, consistent)
- Takes safety measures for her/himself and the environment. (Taşar & Atasoy, 2006, p. 9)

Although Taşar and Atasoy do not specifically refer to scientific literacy in their description of the new science and technology curriculum, it would appear that the reforms, aimed at reaching the standards of more developed countries, could reasonably fall under the scientific literacy umbrella.

Cross European Projects: Pollen and Sinus

Osborne and Dillon (2008) identify another focus of development in Europe, that is, projects that have attempted to develop a more inquiry-based approach to the teaching of science:

> Notable amongst these are Pollen (www.pollen-europa.net) which is aimed at primary teachers in twelve European countries with an emphasis on teaching through inquiry; and Sinus and Sinus-Transfer which provide secondary school teachers in Germany with tools to change their pedagogical approach to science teaching in secondary school. The focus of these projects has been primarily on pedagogy and not on transforming the content itself. Such inquiry-based approaches are seen as providing children with: opportunities to use and develop a wider range of skills such as working in groups; more extended opportunities to explore their written and oral expression; and more open-ended, problem-solving experiences all in the belief that it will enhance student motivation and attainment. Some evidence does exist that these have been effective and it is these projects which are central to the recent report calling for a transformation in the pedagogy of science teaching in Europe.
>
> (p. 22)

The document that the authors refer to in their final sentence is the Rocard Report which, as we have seen, sets out an authoritative case for scientific literacy and radical change and yet promotes projects, originating in France and Germany that, in reality, offer little that teachers in the UK or the USA would regard as novel approaches to teaching science. Sometimes, though, it is necessary for every culture to reinvent the wheel.

Looking Beyond Europe

Although the focus of this brief overview has been Europe, there are some interesting parallels with what has been happening in North America, Australasia and Africa. For example, in Canada, which has seen many initiatives aimed at promoting Science-Technology-Society (STS) education for many years, a nationwide framework (CMEC, 1997) led to a series of province-based science curriculum revisions. The framework is based on a premise of science for all and scientific literacy is defined as "an evolving combination of the science-related attitudes, skills, and knowledge students need to develop inquiry, problem-solving, and decision-making abilities, to become lifelong learners, and to maintain a sense of wonder about the world around them" (Council of Ministers of Education, Canada [CMEC], 1997, p. 4). The Canadian approach differs from that taken in England and Wales and in the Netherlands in that it is an attempt to add a dimension to the curriculum rather than to create a special course. However, the approach used is to mandate that students do particular units which focus on problem- solving and on making decisions.

The approach taken in Australia (see Goodrum et al. 2000; Rennie et al. 2001) and the USA could be seen as more of an infusion in which frameworks and standards are used to impact across the curriculum. Roberts comments that frameworks such as *NSES* and the Australian example "typically reflect elements of both Vision I and Vision II, just because they are broad, idealized, multi-purpose, and intended to be enabling and facilitating" (2007a, p. 770).

In terms of issues arising from the Australian and the Canadian experiences, Roberts reviews what happens "whenever curriculum arrangements do not mandate Vision II outcomes" (2007a, p. 771). Fensham (1998) describes three instances where scientists have defeated proposals to implement courses based on Vision II approaches to scientific literacy. Blades (1997) identifies a similar situation in Canada. Roberts notes that these cases illustrate "the retreat from Vision II to Vision I" which, he continues, "occurred as a result of power politics within curriculum committees" (Roberts, 2007a, p. 771). De Vos and Reiding (1999) describe how a retreat from a Vision II approach can actually take place during the implementation of a new curriculum.

Moving beyond the more economically-developed countries, in South Africa, the Department of Education science curriculum policy actively promotes scientific literacy. Specifically, it does this by expecting:

- the development and use of science process skills in a variety of settings;
- the development and application of scientific knowledge and under- standing; and
- appreciation of the relationships and responsibilities between science, soci- ety and the environment. (Department of Education, Pretoria, 2002, p. 4)

The policy is unusual in that it aims to make the science curriculum "dis- tinctively South African" (Department of Education, Pretoria, 2002, p. 10). Roberts notes that the policy is worthy of "special interest because it includes attention to relationships between science, on one hand, and traditional prac- tices and technologies as these relate to traditional wisdom and knowledge systems, on the other" (2007a, p. 773). The policy explains the challenges that students face:

> One can assume that learners in the Natural Sciences Learning Area think in terms of more than one world-view. Several times a week they cross from the culture of home, over the border into the culture of science, and then back again. How does this fact influence their understanding of science and their progress in the Learning Area? Is it a hindrance to teaching or is it an opportunity for more meaningful learning and a curriculum which tries to understand both the culture of science and the cultures of home?
>
> (Department of Education, Pretoria, 2002, p. 12)

More details of the rationale for the curriculum and the impact of its implementation can be obtained from Ogunniyi (2007). The idea of indigenous science education has less resonance in Europe than it has in Canada, South Africa and Australasia. Perhaps as a consequence European science education policy makers and researchers have yet to critique the curriculum in terms of its cultural appropriateness to the same extent that has happened elsewhere.

The Dimensions of Scientific Literacy

McEneaney (2003) is correct to claim that 'scientific literacy' has a 'worldwide cachet'. It has become a rallying cry for a range of diverse interest groups all wanting to influence what science is taught in schools. The call can be heard in many parts of the world from Australasia to the Americas, and from Asia to Africa and Europe. However, as soon as words turn into action, the philosophical chasm that exists between Vision I and Vision II becomes a barrier to the major shift required in the culture of science education.

In Europe there are signs that, as we move towards the second decade of the 21st century, science educators and policy makers are rising to the challenge to provide a more appropriate science education for its citizens. However, when one reads the Rocard Report, with its clichéd rationales and overbearing assumptions, one is left feeling that we have a long way to go. We have yet to address, adequately, what is the purpose of science education? And we are nowhere near addressing the issues raised by Roth and Barton (2004) about the value of contemporary science education. So we plod on, looking over our shoulder at the monolith that is the content of science as students of all ages desert the sinking ship that is science education as we know it now.

Rather than wring our hands at the inadequacies of the term 'scientific literacy' we have to accept that it will have some considerable currency for years to come. We have to find ways to work with the term and then find ways to disrupt the hegemony that it holds over curriculum reform and assessment regimes. One way to do that might be to focus on the different dimensions of scientific literacy as identified by authors such as Shen who conceptualised three distinct but not mutually exclusive, categories of scientific literacy: *practical, civic,* and *cultural.* Practical scientific literacy is the "possession of the kind of scientific knowledge that can be used to help solve practical problems" (Shen, 1975, p. 46). Civic scientific literacy refers to the level of scientific knowledge and understanding necessary for informed public debate and sound policy-making. Cultural scientific literacy "is motivated by a desire to know something about science as a major human achievement" (Shen, 1975, p. 49). By breaking down scientific literacy into bite-sized chunks—scientific literacies, we can begin to see a way to organize the curriculum to meet the needs of different students throughout their time in and out of school. In so doing, we might begin to address the philosophical tensions between Vision I and Vision II conceptions of scientific literacy rather than pretend that they do not really matter.

References

Bartholomew, J., Osborne, J., & Ratcliffe, M. (2004). Teaching students 'Ideas-About-Science': Five dimensions of effective practice. *Science Education, 88,* 655–682.

BBC News (2006). *Q&A: Science GCSE controversy.* Retrieved 7 December 2008 from, http://news.bbc.co.uk/1/hi/education/6039950.stm

BBC News (2008). *Minister checks on science exams.* Retrieved 7 December 2008 from, http://news.bbc.co.uk/go/pr/fr/-/1/hi/education/7765769.stm

Blades, D. (1997). *Procedures of power and curriculum change: Foucault and the quest for possibilities in science education.* New York: Peter Lang.

Burden, J., Campbell, P., Hunt, A., Millar, R., Scott, P., Ametller, J., Hall, K., Leach, J., Lewis, J., Ryder, J., Bennett, J., Hogart, S., Ratcliffe, M., Hanley, P., Osborne, J., & Donnelly, J. 2007). *Twenty First Century Science Pilot Evaluation Report.* York/London: UYSEF/Nuffield Foundation. Retrieved 7 December 2008 from, http://www.21stcenturyscience.org/data/files/c21-evaln-rpt-feb07-10101.pdf.

Council of Ministers of Education, Canada (CMEC). (1997). *Common framework of science learning outcomes K to 12: Pan-Canadian protocol for collaboration on school curriculum for use by curriculum developers.* Toronto, ON, Canada: Author.

Champagne, A. B., & Lovitts, B. E. (1989). Scientific literacy: A concept in search of definition. In A. B. Champagne, B. E. Lovitts & B. J. Callinger (Eds.), *This year in school science: Scientific literacy* (pp. 1–14). Washington, DC: AAAS.

Department of Education, Pretoria. (2002). *Revised national curriculum statement for grades R-9 (schools): Natural sciences.* Pretoria; South Africa.

De Vos, W., & Reiding, J. (1999). Public understanding of science as a separate subject in secondary schools in the Netherlands. *International Journal of Science Education, 21,* 711–719.

Fensham, P. J. (1998). The politics of legitimating and marginalizing companion meanings: Three Australian case stories. In D. A. Roberts & L. Östman (Eds.), *Problems of meaning in science curriculum* (pp. 178–192). New York: Teachers College Press.

Goodrum, D., Hackling, M., & Rennie, L. (2000). *The status and quality of teaching and learning of science in Australian schools: A research report.* Canberra: Department of Education, Training and Youth Affairs.

High Level Group on Science Education. (2007). *Science Now: A renewed pedagogy for the future of Europe.* Brussels: European Union.

Hodson, D. (2008). *Towards scientific literacy. A teachers' guide to the history, philosophy and sociology of science.* Rotterdam: Sense.

Koc, Y., Isiksal, M., & Bulut, S. (2007). Elementary school curriculum reform in Turkey. *International Education Journal, 8*(1), 30–39.

Laugksch, R. C. (2000). Scientific literacy: A conceptual overview. *Science Education, 84*(1), 71–94.

Lee, O. (1998). Guest Editorial: Scientific literacy for all: What is it, and how can we achieve it? *Journal of Research in Science Teaching, 34*(3), 219–222.

McEneaney, E. H. (2003). The worldwide cachet of scientific literacy. *Comparative Education Review, 47(2),* 217–237.

Members of the Linné Scientific Literacy Symposium (2007). Statement of Concern. In C. Linder, L. Östman, & P.-O. Wickman (Eds.), *Promoting scientific literacy: Science education research in transaction* (pp. 7–8). Uppsala: Uppsala University.

Millar, R., & Osborne, J. (Eds.). (1998). *Beyond 2000: Science education for the future.* King's College, School of Education.

Ogunniyi, M. B. (2007). Teachers' stances and practical arguments regarding a science-indigenous knowledge curriculum: Part 2. *International Journal of Science Education, 29*(10), 1189–1207.

Organisation for Economic Cooperation and Development. (2007). *Executive Summary PISA 2006: Science competencies for tomorrow's world.* Paris: OECD.

Osborne, J., & Dillon, J. (2008). *Science education in Europe: Critical reflections.* London: The Nuffield Foundation.

Rennie, L., Goodrum, D., & Hackling, M. (2001). Science teaching and learning in Australian schools: Results of a national study. *Research in Science Education, 31,* 455–498.

Roberts, D. A. (2007a). Scientific literacy/Science literacy. In S. K. Abell & N. G. Lederman, (Eds.), *Handbook of research on science education* (pp. 729–780). Mahwah, NJ: Lawrence Erlbaum.

Roberts, D. A. (2007b). Opening remarks. In C. Linder, L. Östman, & P.-O. Wickman, (Eds.), Promoting scientific literacy: Science education research in transaction. *Proceedings of the Linnaeus Tercentenary Symposium* (pp. 9–17). Uppsala: Uppsala University.

Roth, W.-M., & Barton, A. C. (2004). *Rethinking scientific literacy.* New York, NY: RoutledgeFalmer.

Royal Society. (1985). *The public understanding of science.* London: Royal Society.

Shamos, M. H. (1995). *The myth of scientific literacy.* New Brunswick, NJ: Rutgers University Press.

Shen, B. S. P. (1975a). Scientific literacy and the public understanding of science. In S. B. Day (Eds.), *Communication of scientific information* (pp. 44–52). Basel: Karger.

Shortland, M. (1988). Advocating science: Literacy and public understanding. *Impact of Science on Society, 38*(4), 305–316.

Snow, C. P. (1962). *The two cultures and the scientific revolution.* Cambridge, UK: Cambridge University Press.

Stables, A. (1996). Paradox in compound educational policy slogans: Evaluating equal opportunities in subject choice. *British Journal of Educational Studies, 44*(2), 159–167.

Taşar, M. F., & Atasoy, B. (2006, November). *Turkish educational system and the recent reform efforts: The example of the new science and technology curriculum for grade 4–8.* Paper presented at the meeting of Asia Pacific Educational Research Association, Tai Po, New Territories, Hong Kong.

Twenty First Century Science (2008). Retrieved 7 December 2008 from, http://www.21stcenturyscience.org/

van der Valk, T., & Eijkelhof, H. (2007). Junior College Utrecht: Challenging motivated upper secondary science students. *School Science Review, 88,* 63–71.

Waterman, A. T. (1960). National Science Foundation: A ten-year résumé. *Science, 131*(3410), 1341–1354.

18 Science, Environment and Health Education

Towards a Reconceptualisation of Their Mutual Interdependences

Justin Dillon

Dillon, J. (2012). Science, environment and health education: towards a reconceptualisation of their mutual interdependences. In, A. Zeyer, & R. Kyburz-Graber (eds), *Science | Environment | Health—Towards a Renewed Pedagogy for Science Education*. Dordrecht: Springer, pp. 87–101.

Introduction

The search for causality, while not exclusive to science, is certainly one of its fundamental characteristics. In Victorian times, as new technologies allowed scientists to widen the boundaries of their knowledge and understanding, the environmental causes of human diseases became increasingly clear. In 1854, a major outbreak of cholera killed 616 people in the Soho area of London. Using biological and chemical testing, the physician John Snow identified the likely source of the disease as a public well on Broad Street. This early epidemiological study challenged the miasmatic theory that held that disease was carried in air polluted by particles from decomposing matter—a theory that had held sway since Roman times. Even well-educated people believed the miasmatic theory in Victorian times because it seemed to explain their everyday experiences. New techniques and theories allowed scientists to offer more compelling explanations.

More than 150 years later, the links between the environment and health have been well researched and our understanding has changed out of all recognition. That is not to say, however, that the public understanding of these links is particularly high. One reason for that poor state of affairs is that whereas science education is widely regarded as a core subject in the curriculum, health and environmental education are more likely to be seen as cross-cutting themes if they appear anywhere. Most current science curricula have relatively little health or environmental education in them and that is partly due to content over-load which has been a feature of science education for decades.

However, for a number of reasons, the situation is changing. One reason is that in recent times, significant efforts have been put into calculating the cost to society of a range of conditions from environmental pollution to alcohol abuse and obesity. For example:

> Scotland's obesity epidemic is costing the country around £450 million a year, according to a new Government study that predicts the bill could

soar to £3 billion by 2030 if there is no change in the nation's attitude to food and exercise.

(Gordon, 2010)

Scotland has a population of just over 5 million people which puts the scale of the problem into some perspective. The sheer scale of the cost of these health-related issues has led policy-makers to focus on strategies to change attitudes and behaviours. One of many policy responses was a pilot project, 'The Big Eat In', in Glasgow that involved pupils in 8 schools being encouraged to stay at school during lunchtime, eating healthily and taking part in activities (GCPH, 2010). The success of the year-long project led to more schools getting involved in the scheme.

Schools are one of the main vehicles through which public attitudes and behaviours can be influenced although some might argue that legislation is more effective. The counter-argument might be that without an educated public, it would be easier for opponents to regulation to thwart new laws. The recent UK Government Education White Paper notes that 'Good schools play a vital role as promoters of health and wellbeing in the local community' (DfE, 2010:28), adding:

> Children can benefit enormously from high-quality Personal Social Health and Economic (PSHE) education. Good PSHE supports individual young people to make safe and informed choices. It can help tackle public health issues such as substance misuse and support young people with the financial decisions they must make.
>
> (p. 46)

However, teacher training for PSHE, as it is labeled in the UK, is relatively brief and, consequently, not a good preparation for teaching about health issues (Walsh and Tilford, 1998). To have any lasting impact on young people, the links between science, health and the environment need to be reflected in the core of the curriculum. Such a repositioning would require an overhaul of science education as we know it, however there is a growing sense of frustration with the existing curriculum in many countries and resistance to change might be less of an issue than it has been in the past.

There is another reason why links between science, health and the environment are increasingly drivers for policy reform. The actual and potential impacts of climate change have caused a reconfiguration of policy agendas across the world. For many governments, the need for climate change mitigation and adaptation has already led to a range of new policies being implemented. Research into climate change education is increasing, and in 2009 NASA announced that it would be spending up to $8 million funding 'projects designed to educate students, teachers and lifelong learners about global climate change' (NASA, 2009). Climate change education will become a fixture in many education systems and, again, its most likely home is in the already crowded science curriculum.

So far, I have suggested that it is likely that links between science, health and the environment will increasingly be made in the school curriculum and that the most likely place for this to happen will be in school science education. In the next section some reasons why this change is particularly timely will be explored.

The Problems with School Science Education

In 2008, the Nuffield Foundation published a report entitled 'Science Education in Europe: Critical reflections' (Osborne and Dillon, 2008). The report emerged from two seminars, held in 2006, involving more than a dozen science educators and education researchers from a number of European institutions. The authors of the report identified why the seminars had been set up:

> Many countries are experiencing significant problems with engaging students with the advanced study of physical sciences. Where this is the case, it is a source of significant concern. However, this pattern is not universal across Europe and appears to be strongly correlated with the level of economic advancement in any given country.
>
> (Osborne and Dillon, 2008:13)

They noted, moreover, that 'one area [. . .] in which there is a common trend is in the decline of student attitudes to science' (p.11). This opinion is supported by data from the Relevance of Science Education (ROSE) survey which reported a '0.92 negative correlation between students' attitude towards school science and the UN index of Human Development' (p.11). This negative association between student attainment and student attitude towards science also emerges from the Third International Mathematics and Science Study (TIMSS) which carried out a major comparison of students' attainment and attitudes across the world in 1999 (Martin *et al.*, 2000).

The Nuffield report contained a series of specific criticisms of science education which were thought to be common to many European countries. These included: 'a lack of perceived relevance'; 'a failure to generate a sense of anticipation that accompanies an unfolding narrative'; 'a pedagogy that lacks variety'; 'a less engaging quality of teaching in comparison to other school subjects'; 'content which is too male-orientated'; and, 'an assessment system that encourages rote and performance learning rather than mastery learning for understanding' (Osborne & Dillon, 2008: adapted from p. 15).

In terms of the lack of relevance of science education, the authors noted that:

> School science is often presented as a set of stepping-stones across the scientific landscape and lacks sufficient exemplars that illustrate the

application of science to the contemporary world that surrounds the young person.

(p. 15)

What, then, might young people think are relevant topics? One answer to that question comes from an analysis of the English ROSE data carried out by Jenkins and Nelson (2005). Students in the ROSE survey were given a list of 108 science topics and asked to rate their level of interest on a scale of 1 ('not at all') to 4 ('very interested'). The results for the boys and girls were significantly different, as can be seen from Table 18.1 which lists the top five topics for each gender.

Whereas the boys indicate an interest in topics involving weapons and outer space, the girls' interests are predominantly focused on health topics. The implications of this divide is that if health and environmental topics are to have a bigger role in the science curriculum, then girls will be interested but ways will need to be found to engage boys.

Osborne and Dillon concluded their report by noting that secondary science education was not fit for purpose:

> The irony of the current situation is that somehow we have managed to transform a school subject which engages nearly all young people in primary schools, and which many would argue is the crowning intellectual achievement of European society, into one which the majority find alienating by the time they leave school. In such a context, to do nothing is not an option.

Europe is not alone in possessing an inadequate science education system if the response to the publication of the Programme for International Student

Table 18.1 The most highly-rated science topic chosen by English boys and girls in the Relevance of Science Education survey (Jenkins and Nelson, 2005)

Boys	Girls
Explosive chemicals	Why we dream when we are sleeping and what the dreams might mean
How it feels to be weightless in space	Cancer—what we know and how we can treat it
How the atom bomb functions	How to perform first aid and use basic medical equipment
Biological and chemical weapons and what they do to the human body	How to exercise the body to keep fit and strong
Black holes, supernovae and other spectacular objects in outer space	Sexually transmitted diseases and how to be protected against them

Assessment (PISA) 2009 study is anything to go by (see, for example, Tse, 2010). So, for many countries, doing nothing about the science curriculum is, as Osborne and Dillon argue, clearly not an option. The question, then, is—what could and should be done?

A New Mutualism?

The issue of the relationship between science and environmental education has been discussed by Annette Gough (2002) among others. She notes that in the early debates about the issue, some authors (for example, Fensham and May, 1979 argued for the two subjects to be brought closer together, while others have provided counter arguments for their separateness (for example, Lucas, 1980). The issue has continued to be debated and Gough refers to Webster's comment that:

> Science, like economics, has been reformed through the promotion of investigative science and the contextualisation of science. The contexts are often social, utilitarian concerns: health, science in everyday life, a nod to environment, and industry. Content still dominates, as does experimentation. As in economics, the hidden values and assumptions about the way the world works remain largely unexplored.
>
> (1996:82)

She concludes that it is timely to revisit the issue:

> If we are to achieve sustainable development then science education must have a role in encouraging ecological thinking (instead of being kept at a distance) and environmental education must move on from the insecure relationships that accompany the abstract arguments for it to adopt 'a holistic approach, rooted in a broad interdisciplinary base' (UNESCO 1978:24).
>
> (Gough, 2002:1203)

Sustainable development is a highly contested term but whether the desired goal is sustainability or survival, Gough's point is well made. Arguing for a new mutualism, and noting science education's insecurities, she argues that 'science education needs environmental education to reassert itself in the curriculum' noting that it could do this 'by making science seem appropriate to a wider range of students and making it more culturally and socially relevant' (p.1210). While, at the same time, she is of the opinion that 'environmental education needs science education to underpin the achievement of its objectives' and that it should 'provide it [environmental education] with a legitimate space in the curriculum to meet its goals because they are very unlikely to be achieved from the margins' (2002:1210–11). Not surprisingly, Gough suggests that this will not be an easy process.

So, extending Gough's argument to include the health dimension, how might science, environmental and health education come together in new ways that would add value to students' experiences in school rather than simply overwhelm them? A new curriculum would need to show that science is inherently political in terms of how it is funded and subject to commercial interest in terms of what research is favoured. Cancer, the second favourite topic of the girls in the ROSE survey mentioned above, provides an opportunity to look at how investments in prevention might be more effective than providing treatments. Cancer also provides an opportunity to examine issues of risk and probability, topics that do not often find themselves in traditional science curricula.

Recent events in Japan and their subsequent reporting in the media provide graphic examples of the need to educate people about topics such as the accuracy of predicting geological processes and the relative costs and dangers of different methods of power production. The engineering causes and the environmental, health and economic consequences of the Deepwater Horizon oil spill in 2010 provide other examples where a combination of knowledge and understanding is needed to make sense of major events in the world and of their implications for society.

One of the challenges that Gough identifies that any reformulated science education would need to take into account is 'critiques of traditional science education from cultural and constructivist perspectives' (p. 1211). In terms of identifying how science education might develop, it is instructive to consider what Aikenhead, writing from a cultural perspective, considers might constitute indicators of quality science teaching:

1   Acknowledgement of the degrees of cultural differences between students' cultural self-identities and the culture of their science classroom, and recognition that each student needs help when negotiating this cross-cultural classroom environment.
2   An enacted curriculum predominantly comprised of relevant science content outside the category of wish-they-knew science, but not ignoring that category.
3   An emphasis on the outcome: Teaching students *how to learn and use* science as the need arises in specific contexts.
4   Student assessment formulated in terms of monitoring students' learning how to learn and how to use science and technology as needed. (2011, p. 122)

As the main criterion for determining what might be taught in any future science curriculum, Aikenhead advocates 'educational soundness and relevancy' rather then political expediency (122–3). Such criteria would open the door for further inclusion of environmental and health education to be incorporated in science curricula.

The need for a less homogeneous version of school science curricula has also been identified by Jenkins (1999) who argued that 'curricula in different countries will show a greater degree of variety than is presently the case.' Jenkins uses the example of Bovine Spongiform Encephalopathy (BSE) in the United Kingdom to illustrate his point that:

> Not all science-related issues are global, and if the school science curriculum is to be sensitive to the interests of students, regional or other in-country variations will need to be accommodated.
>
> (p.708)

This is an interesting point and although much of the content of science—the laws and phenomena, for example—are clearly universal, their application is not. The question now is, what should be the purpose of this new curriculum? To answer that question, it might be useful to consider another of Gough's factors that any reconstructed science curriculum would need to attend to: calls for increasing the scientific literacy of the general public. Would this new vision of environmental | health | science education fit within a vision of scientific literacy?

Scientific Literacy

Although teachers might not use the term frequently, scientific literacy is relatively common in the lexicon of science education. Tracing its origins to the 1950s, McEneaney (2003) describes it as having achieved a 'worldwide cachet' and the concept underpins the Organisation for Economic Co-operation and Development's (OECD) Programme for International Student Assessment (PISA) study. However, the term is treated with scorn and distrust by some writers. Laugksch notes that 'scientific literacy is an ill-defined and diffuse concept' (2000:71). According to Dillon:

> The longevity of the term scientific literacy relies on its ability to be seen as an umbrella for radically different philosophies of science education. However, the evidence suggests that when attempts are made to effect curriculum change to promote 'scientific literacy' the unreconciled philosophical clashes hinder progress.
>
> (2009:202)

Some indication of the degree to which the philosophies clash can be gauged from this quote from Roth and Barton (2004):

> Conventional approaches to scientific literacy, knowing, and learning are based on an untenable, individualistic (neo-liberal) ideology that does not account for the fundamental relationships between individual and society, knowledge and power, or science, economics, and politics.
>
> (p.3)

In an attempt to clarify what is meant by scientific literacy, Roberts (2007) identifies two ways of conceptualizing science education's aims and purposes. He describes two 'visions' for generating conceptions of scientific literacy: Vision I and Vision II. Vision I 'looks inward at science itself—its products such as laws and theories, and its processes such as hypothesizing and experimenting' whereas Vision II 'looks outward at situations in which science has a role, such as decision-making about socioscientific issues' (Roberts, 2007:9).

Could the same set of visions illuminate what kind of literacies might be developed under the aegis of science, health and environmental education? Vision I would focus on a range of issues and topics such as climate change, environmental causes of cancers, and growth and reproduction. Vision II might focus on ethical issues concerned with stem cell research, how climate change scientists work and at the role of pharmaceutical industry in drug research.

Grace and Ratcliffe (2002) note that environmental issues affecting society tend to be underpinned by value judgements. Such approaches would require teachers to focus on teaching about the values underpinning science, health, the environment and society. Again, the challenge of such a pedagogical shift must not be underestimated. There are many science teachers who might find it challenging to teach such topics. Science teachers cannot shirk their responsibility to teach about the issues that fundamentally affect people's health and the environment—to do so would be intellectually bankrupt and morally indefensible.

Gayford (2002) makes the point that teaching about climate change in school science might be problematic because teachers' understanding of such complex issues might be inadequate. While this change would make the new curricula more interesting and relevant they might be difficult to teach because the border between 'scientific statements' and 'value statements' is often hard to see. Oulton *et al.* (2004) found a serious lack of preparation to teach about controversial issues among science teachers in England resulting in some reluctance to use them in the classroom. One of Lucas's earlier concerns about teaching environmental education through science education was whether science teachers' 'worldviews as empirical experimenters [would] seriously distort the nature of historical understanding and aesthetic judgement?' (1980:21). It is clear that any radical change to broaden the science curriculum would necessitate changes to initial and in-service teacher training as well as new resources for classroom and out-of-classroom use.

Health and Environmental Literacies

If Roberts's Vision I and Vision II for scientific literacy help us to reconcile some of the clashing philosophies that might impede curriculum change, do the notions of health and environmental literacy offer any opportunities to develop the mutual relationship between science, health and the environment? Nutbeam (2000) notes that health literacy is a relatively new concept and for a long time it has tended to refer to the ability of patients to read medical information including labels on medicine bottles. However, as Nutbeam points out, this is a very narrow conceptualization of literacy and it ignores the growth in the study of literacy and literacies. Broader interpretations of the term do exist, and the World Health Organisation, for example, notes that:

> Health literacy means more than being able to read pamphlets and successfully make appointments. By improving people's access to health information and their capacity to use it effectively, health literacy is critical to empowerment.
>
> (Nutbeam, 1998:264)

Nutbeam himself derives a model of three levels of health literacy: formal, interactive and critical. The highest level, critical health literacy:

> reflects the cognitive and skills development outcomes which are oriented towards supporting effective social and political action, as well as individual action.
>
> (2000:265)

This approach to health education, he argues, can focus more on 'achieving change in the social, economic and environmental determinants of health which may benefit the health of whole populations . . .' (p.265). Tones (2002:289) argues that adequate theoretical frameworks already exist and that expanding the meaning of 'health literacy' is redundant. An example of such a theoretical framework that includes social capital and action competence, can be found in Jensen *et al.* (2002). They advocate the development of:

> pupils' abilities to act at the personal and at the societal level [. . .] If pupils have to contribute to the solution of today's health problems, it follows [. . .] that they have to identify personal and structural causes behind the health problems and to develop their own possibilities to influence and change these conditions.
>
> (Jensen, 1995:6)

So, whether one takes the broader view of health literacy proposed by Nutbeam or sides with Tomes's view that critical approaches to health and environmental education already exist without recourse to the notion of literacy, then a new mutualism between science, health and environmental education should promote an educated citizenry able to critically examine issues of local importance and global significance in ways that they currently do not.

At this point I should make clear that I have been eliding between two ideas. The first is that the science curriculum (strictly speaking, the science curricula) should be reconstructed to include more health and environmental education (I am avoiding using acronyms such as SHE and HES here—the failure of STEM has taught me to be wary of them). I am not advocating that health and environmental education should be swallowed up by science education (hence the use of the term science | environment | health). There is a role for both beyond a reconstructed science education. What I am doing in this chapter is to focus on what a reconstructed science curriculum might look like.

One aspect of environmental and health issues that is poorly addressed in any part of the curriculum is risk. This is a fundamental problem because we are increasingly being confronted by a range of 'soft disasters'— 'environmental and political crises that emerge only slowly but at high cost to society, not least the erosion of public confidence and legitimacy' (ESRC Global Environmental Change Programme, 2000:3). Soft disasters include socio-scientific issues such as BSE, the GM food debate, HIV-AIDS and global climate change. These would seem to be just the sort of topics that might be studied in the science | environmental | health curriculum. To make sense of these complex issues, the public needs at least a basic understanding of risk assessment and management and some understanding of probability (Dillon & Gill, 2001; Jenkins, 2003).

What Might the Student Experience Look Like?

Some hypothetical topics for possible study in a science | environment | health curriculum were identified earlier. Examples of projects that might offer lessons about how to approach teaching the new inter-disciplinary curriculum can be found in countries with particularly democratic education systems such as Denmark. Jensen describes Danish educational activities that he identifies as being action oriented:

> Such activities may consist of physical, chemical and biological investigations of a polluted lake or they may embrace social science oriented activities such as interviews or document-analysis. Such activities are obviously valuable and productive to the extent that they facilitate motivation and the acquisition of knowledge. But in order to be characterised as actions, they must be targeted at effecting real change regarding the environmental problem that is being worked on.
>
> (Jensen, 1995:326)

Such projects might not seem particularly novel given the long history of water quality monitoring projects, however, the focus on developing students' action competence and on empowering them to take political and social actions rather than simply learning content might be seen as a more radical approach to education. However, given that we do not have examples of the science|environment|health curriculum in operation, we can imagine that if they were realised then there would be almost limitless opportunities for local and regional projects on a wide range of topics.

Another dimension to the curriculum would be a commitment to:

> focus on helping learners deal with the sheer complexity and splendor of the environment as well as looking to use the local environment as a vehicle for developing understanding of the more mundane aspects of the science curriculum.
>
> (Dillon and Scott 2002:1112)

In terms of outcomes, current science education tends to focus on a relatively small number, mainly related to knowledge and skills. However, the new curriculum might useful take a much broader look at what benefits might emerge from a range of pedagogical approaches. We already know, particularly from primary education, that environmental projects can have a wide range of benefits to students, to teachers, to schools and to the wider community. For example, Maller (2005) identified a number of aims for engaging children in hands-on contact with nature:

> to meet sustainability education, environmental education or science learning objectives. However, other reasons cited for the recent growth in these types of activities include beautification of school grounds, habitat restoration, and fostering qualities of stewardship and nurturing in children.
>
> (p. 16)

Maller's study showed that it is possible to identify science, environmental and health outcomes which were mutually reinforcing:

> The take-home message from this research is that hands-on contact with nature experienced via sustainability education is not only essential for protecting the environment, but it also appears to be a means of cultivating community and enhancing the mental health and wellbeing of children and adults alike.
>
> (pp. 21–22)

Other strategies to promote deeper understanding of science|environment| health links include public participation in scientific research (PPSR), sometimes known as Citizen Science. PPSR offers opportunities to develop a greater sense of how science works [Roberts' Vision II] in students and can encourage them and their schools to work collaboratively. In a review of PSSR projects, Bonney *et al.* noted that:

> Participants in many PPSR projects also gain knowledge of the process of science. Indeed, this is one area where PPSR projects have the potential to yield major impacts, particularly Collaborative and Co-created projects, which engage participants in project design and data interpretation to a significant degree.
>
> (Bonney *et al.*, 2009:12)

Such projects might involve the collection of environmental data, for example, of bird migration. Cowell and Watkins reported on a project that involved plants rather than animals. *Spring Bulbs for Schools,* a museum outreach programme, was set up in Wales in 2006. The project involved establishing 160 monitoring sites across the country. The project proved to be very successful as Cowell and Watkins report:

> Working with crocuses and daffodils made [participants] aware of the importance of bulbs in the life cycle of some plants. On a more general level, they become aware of the world around them and the idea that human activity can have noticeable effects, even on a local scale in the school garden.
>
> (p. 27)

Again, the scheme demonstrated a range of environmental and science outcomes. The authors noted that 'the project enabled them to undertake pattern-seeking and observational activities—aspects of scientific enquiry that are often underdeveloped throughout the science curriculum' (p. 28). What might be a next step would be to focus on the health dimension of growing flowers outdoors.

One could argue that the curriculum should focus on the types of experiences that students should have during their schooling: museum visits; long-term experiments; visits to the countryside in all the seasons; visits to a farm; an opportunity to care for animals and plants over an extended period; visits to a hospital; time to discuss with scientists about what they do, etc. Such a curriculum might provide opportunities for children to have individual responses and personal outcomes rather then be pushed into the homogeneity of contemporary education.

Values and Controversy

A new science | environment | health curriculum as described above, would necessarily involve teaching about values. Values are just one dimension of controversial issues such as growing GM crops or nuclear power. Traditionally, teachers have been recommended to adopt a neutral chair approach when teaching about controversial issue, however, Oulton *et al.* suggest that such an approach is unethical in that all pedagogic decisions would reflect the teachers' own position in some way and that it is better for them to be open about their position. Oulton *et al.* argue that teachers need to teach about controversial issues in such a way that the following points are made:

- groups within society hold differing views about them;
- groups base their views on either different sets of information or they interpret the same information in different ways;
- the interpretations may occur because of the different ways in which individuals or groups understand or 'see' the world (i.e. their worldview),
- differing worldviews can occur because the individuals adhere to different value systems;
- controversial issues cannot always be resolved by recourse to reason, logic or experiment; and
- controversial issues may be resolved as more information becomes available. (2004:420)

Theories of Learning

Underpinning the pedagogical approaches that would facilitate the new curriculum, there must be some theories of learning (Dillon, 2003). Vosniadou's (2001) review of research provides a good starting point for a discussion of what we know about learning and thus allows us to see how it might be used to inform a new pedagogy to support a new curriculum and new assessment. Table 18.1 provides an overview of the key points:

Learning requires the active, constructive involvement of the learner.

Learning is primarily a social activity and participation in the social life of the school is central for learning to occur.

People learn best when they participate in activities that are perceived to be useful in real life and are culturally relevant.

New knowledge is constructed on the basis of what is already understood and believed.

People learn by employing effective and flexible strategies that help them to understand, reason, memorize and solve problems.

Learners must know how to plan and monitor their learning, how to set their own learning goals and how to correct errors.

Sometimes prior knowledge can stand in the way of learning something new. Students must learn how to solve internal inconsistencies and restructure existing conceptions when necessary.

Learning is better when material is organized around general principles and explanations, rather than when it is based on the memorization of isolated facts and procedures.

Learning becomes more meaningful when the lessons are applied to real-life situations.

Learning is a complex cognitive activity that cannot be rushed. It requires considerable time and periods of practice to start building expertise in an area.

Children learn best when their individual differences are taken into consideration.

Learning is critically influenced by learner motivation. Teachers can help students become more motivated learners by their behaviour and the statements they make.

How children learn (adapted from Vosniadou, 2001)

Social constructivist theories of learning, based on the works of Piaget and Vygotsky in particular, would suggest that an effective pedagogy would involve the following characteristics:

1   Eliciting students' ideas about concepts and topics rather than assuming that they know nothing.
2   The provision of concrete experiences supported by appropriate vocabulary so that learners become familiar with the subject matter.
3   Choice of activities, so that they feel in control of aspects of their learning.
4   Cognitive challenge, so that learners are presented with something which is challenging without being overwhelming.
5   Plenty of time to discuss ideas with their peers and with adults.
6   Feedback on their performance so that they know how to improve their work.
7   Opportunities to practice operations so that they become confident in their skills.
8   Time to engage with activities, so that they have an opportunity to think about problems without feeling too pressured.

As before, there are some teachers who might feel that the list above describes their existing pedagogy. If so, fine, it would show that it can be done within the constraints of current curriculum and assessment regimes. Nevertheless, for all teachers to be able to use this approach would require support in the form of pre-service and in-service training.

More radical approaches to learning are outlined by Wals and Dillon (forthcoming). They note that we can learn from nature itself about learning process and about sustainability. Eco-systems, they argue, provide evidence of resilience and systems thinking allows us to examine how communities depend on each other to survive and to develop in the face of challenging circumstances.

They conclude that:

> Learning in the context of environment and sustainability then becomes a means for working towards a 'learning system' in which people learn *from* and *with* one another and collectively become more capable of withstanding setbacks and dealing with insecurity, complexity and risks.
> (Wals and Dillon, forthcoming)

Such a model of learning has substantial implications and would require a major shift in thinking about teaching students *how* to learn as individuals and groups rather than focusing on *what* they should learn.

Summary

The general sense of dissatisfaction with the existing science curriculum in many countries provides an opportunity to consider a radical reform based on Aikenhead's maxim that change must be based on 'educational soundness and relevancy' rather then political expediency (122–3). Gough's point that science and environmental education need each other and that there should be a new mutualism between the two disciples can be extended to include a third partner, health education.

The outcomes of the new curriculum should be diverse and more personalized and local than is currently the case. Students should be empowered rather than drilled to absorb information for the purpose of testing. In particular, students should develop an understanding of risk and probability and learn to appreciate the values implicit in a range of scientific, environmental and health issues.

Teachers need a pedagogy based on sound theories of learning and need to find out what students know, design activities to challenge students, provide opportunities for discussion, and provide formative feedback. They will need to develop their skills and knowledge and they will need to be able to teach about values and about controversial issues openly.

There have been many calls for radical change to the way that the curriculum is organised. Now, however, the health and environmental challenges to society are of such a magnitude that we must rise to them otherwise we will condemned to repeat the failures of the past.

References

Aikenhead, G. (2011, forthcoming). Towards a cultural view of quality science teaching. In D. Corrigan, J. Dillon, & R. Gunstone (Eds), *The professional knowledge base of science teaching* (107–127). Dordrecht: Springer.

Cowell, D., & Watkins, R. (2007). Get out of the classroom to study climate change—the "Spring Bulbs for Schools" project. *Primary Science Review, 97,* 25–28.

Department for Education (DfE) (2010). *The importance of teaching: Schools White Paper.* London: DfE. Available at: http://publications.education.gov.uk/eOrderingDownload/ CM-7980.pdf.

Dillon, J. (2003). On learners and learning in environmental education: Missing theories, Ignored communities. *Environmental Education Research, 9*(2), 215–226.

Dillon, J. (2009). On scientific literacy and curriculum reform. *International Journal of Environmental and Science Education, 4*(3), 201–213.

Dillon, J., & Gill, P. (2001). Risk, environment and health: aspects of policy and practice. *School Science Review, 83*(303), 65–73.

Dillon, J., & Scott, W. (2002). Perspectives on environmental education-related research in science education. *International Journal of Science Education, 24,* 1111–1117.

ESRC Global Environmental Change Programme (2000). *Risky choices, soft disasters: Environmental decision making under uncertainty.* Brighton: University of Sussex. Retrieved from www.gecko.ac.uk.

Fensham P. J., & May, J. B. (1979). Servant not master—a new role for science in a core of environmental education. *Australian Science Teachers' Journal, 25,* 15–24.

Gayford, C. (2002). Controversial environmental issues: a case study for the professional development of science teachers. *International Journal of Science Education, 24,* 1191–1200.

Glasgow Centre for Population Health (GCPH). *Developing capacity for effective action to tackle health inequalities.* Available at: http://www.gcph.co.uk/work_programmes/ local_authority_role/healthy_school_food_policy

Gordon, T. (2010). Cost of obesity could reach £3bn a year and hurt economic growth, *heraldscotland* (February 21, 2010). Available at: http://www.heraldscotland.com/news/health/ cost-of-obesity-could-reach-3bn-a-year-and-hurt-economic-growth-1.1008165.

Gough, A. (2002). Mutualism: A different agenda for environmental and science education. *International Journal of Science Education, 24*(11), 1201–1215.

Grace, M. M., & Ratcliffe, M. (2002). The science and values that young people draw upon to make decisions about biological conservation issues. *International Journal of Science Education, 24,* 1157–1169.

Jenkins, E. W. (1999). School science, citizenship and the public understanding of science. *International Journal of Science Education, 21*(7), 703–710.

Jenkins, E. W. (2003). Environmental education and the public understanding of science. *Frontiers in Ecology and the Environment, 1,* 437–443.

Jenkins, E., & Nelson, N. W. (2005). Important but not for me: students' attitudes toward secondary school science in England. *Research in Science & Technological Education, 23,* 41–57.

Jensen, B. B. (1995). Teaching for and with democracy, In D. Colquhoun, K. Goltz, M. Sheehan, & B. Marshall (Eds) *The Proceedings of the Inaugural National Health Promoting Schools Conference.* Geelong: Deakin University.

Jensen B. B. (2002). Knowledge, action and pro-environmental behaviour. *Environmental Education Research, 8*(3), 325–334.

Jensen, B. B., Schnack, K., & Simovska, V. (2002). *Critical environmental and health education research issues and challenges.* Copenhagen: Research Centre for Environmental and Health Education, University of Education.

Laugksch, R. C. (2000). Scientific literacy: A conceptual overview. *Science Education, 84*(1), 71–94.

Lucas, A. M. (1980). Science and environmental education: Pious hopes, self praise and disciplinary chauvinism. *Studies in Science Education, 7*, 1–26.

Maller, C. (2005). Hands–on contact with nature in primary schools as a catalyst for developing a sense of community and cultivating mental health and wellbeing. *Eingana, 28*, 16–21.

Martin, M. O., Mullis, I. V. S., Gonzales, E. J., Gregory, K. D., Smith, T. A., Chrostowski, S. J., Garden, R. A., & O'Connor, K. M. (2000). *TIMSS 1999 International Science Report: Findings from IEA's Repeat of the Third International Mathematics and Science Study at the eighth grade*. Chestnut Hill, MA: Boston College. National Institute for Educational Research: Tokyo.

McEneaney, E. H. (2003). The worldwide cachet of scientific literacy. *Comparative Education Review, 47*(2), 217–237.

NASA (2009). NASA Announces Climate Change Education Funding Opportunity. Press release 09–131. Available at: http://www.nasa.gov/home/hqnews/2009/jun/HQ_09-131_Edu_Climate_Opp.html.

Nutbeam, D. (1998). Health promotion glossary. *Health Promotion International, 13*, 349–364.

Nutbeam, D. (2000). Health literacy as a public health goal: a challenge for contemporary health education and communication strategies in the 21st century. *Health Promotion International, 15*(3), 259–267.

Osborne, J., & Dillon, J. (2008). *Science education in Europe: Critical reflections*. London: Nuffield Foundation.

Oulton, C., Day, V., Dillon, J., & Grace, M. (2004). Controversial issues—teachers' attitudes and practices in the context of citizenship education. *Oxford Review of Education, 30*, 489–507.

Roberts, D. A. (2007). Opening remarks. In C. Linder, L. Östman, & P.-O. Wickman, (Eds), Promoting scientific literacy: Science education research in transaction. *Proceedings of the Linnaeus Tercentenary Symposium* (9–17). Uppsala: Uppsala University.

Roth, W.-M., & Barton, A. C. (2004). *Rethinking scientific literacy*. New York, NY: RoutledgeFalmer.

Sjøberg, S., & Schreiner, C. (2005). How do learners in different cultures relate to science and technology? Results and perspectives from the project ROSE. *Asia Pacific Forum on Science Learning and Teaching, 6*, 1–16.

Tones, K. (2002). Health literacy: new wine in old bottles? *Health Education Research, 17*(3): 287–290.

Tse, V. (2010). Swedish pupils slide in new global ranking. *The Local* (December 7, 2010). Available at: http://www.thelocal.se/30668/20101207/

UNESCO (1978) Intergovernmental Conference on Environmental Education: Tbilisi (USSR), 14–26 October 1977. Final Report. Paris: UNESCO.

Vosniadou, S. (2001). *How children learn*. Brussels: International Academy of Education.

Walsh, S., & Tilford, S. (1998). Health education in initial teacher training at secondary phase in England and Wales: current provision and the impact of the 1992 government reforms. *Health Education Journal, 57*(4), 360–373.

Webster, K. (1996) The secondary years. In J. Huckle, & S. Sterling (Eds), *Education for Sustainability* (72–85). London: Earthscan.

Section 6

Science Engagement and Communication

This section illustrates how I have moved into the realm of science engagement and communication. 'Science communication—a UK perspective' is a contribution to the first edition of Part B of the *International Journal of Science Education* which I co-edit. '"If the public knew better, they would act better": the pervasive power of the myth of the ignorant public' is a critique of traditional linear theories of knowledge, attitude and behaviour. My co-author, Elin Kelsey, was one of my doctoral students and was a constant source of energy and insight. The book in which the chapter was published emerged from a symposium that Bob Stevenson, then at the University of Buffalo and I took part in at an American Educational Research Association conference.

'Communicating global climate change: issues and dilemmas' is an empirical study of what effective science engagement might look like. The chapter was written with Marie Hobson who was then working at the Science Museum in London. Co-authoring with museum practitioners is something that my research group at King's tried to do but never succeeded in systematically. This chapter was published in a book edited by John Gilbert, Bruce Lewenstein and Sue Stocklmayer. John has been a constant source of support throughout the last third of my career. When it became clear that he was about to retire from Reading University, he offered himself to the research group at King's and I secured him a position as a Visiting Professor. In that unpaid role John was exemplary—happy to lighten the load of colleagues and to offer constructive and positive support to doctoral students. He is one of the giants of science education (he is another contributor to this series) and continues to edit the *International Journal of Science Education*.

19 Science Communication— A UK Perspective

Justin Dillon

Dillon, J. (2011). Science communication—a UK perspective. *International Journal of Science Education, Part B: Communication and Public Engagement*, 1(1), 5–8.

If the UK is not the home of science communication, it is a place where science communication, surely, feels at home. That is not to say, however, that science itself is unconditionally appreciated or that scientists are universally trusted. One rationale for the existence of a thriving science communication and engagement sector is that the public trust in science has been shaken by a series of what might be termed 'public relations disasters', including controversy over the MMR vaccine, the debate about GM foods, the BSE outbreak, and more recently, the leak of emails from British climate change scientists. Rightly or wrongly, scientists and science communicators perceive a significant anti-science bias in some quarters of the media and in sectors of British society such as the civil service. It is now just over 50 years ago since C. P. Snow gave a Rede Lecture entitled *The Two Cultures*, in which he articulated the emergence of a division between scientists and literary intellectuals. One wonders just how far have we come since then?

Despite the science crises, something seems to be happening to the public perception of science and scientists. In April 2010, *The Guardian* newspaper ran a feature entitled 'How science became cool' explaining that:

> The incredible ambition of the Large Hadron Collider has fired our imagination; physicists have become cult TV stars; dramatic new pictures from space grace a million computer screensavers. Is this a golden age of science?
>
> (*The Guardian*, 2010)

Even in an age when the plurality of TV channels and internet websites seems endless, blockbuster series featuring science, particularly nature or space, can generate huge audiences. The BBC, in particular, has introduced generations of viewers to habitats and species that few will ever encounter otherwise, and it was no surprise when the Natural History Museum announced that it was naming its premier venue in which, every day, members of the public can engage with working scientists from the Museum and elsewhere, the Attenborough Studio.

The UK has a relatively large number of science and discovery centres. The Association for Science and Discovery Centres (ASDC) represents over 100 science centres and discovery centres in museums, botanic gardens, aquariums, and zoos (ASDC, 2010). The total number of annual visitors is around 19 million (Frontier Economics, 2009). While smaller centres might attract 2,000 visitors per year, the Natural History Museum attracted more than 4 million visitors in 2009, putting it fourth in popularity behind the British Museum (5.6 million visitors), the National Gallery, and Tate Modern. Not everyone is happy, though, *The Guardian's* regular 'Bad Science' columnist, Ben Goldacre, is of the opinion that:

> The indulgent and well-financed 'public engagement with science' community has been worse than useless, because it too is obsessed with taking the message to everyone, rarely offering stimulating content to the people who are already interested.
>
> (Goldacre, 2008, p. 321)

The 2010 Royal Institution Christmas lectures, another permanent fixture in the UK's cultural calendar, were given by a colleague, Dr Mark Miodownik, a material scientist and engineer. Miodownik was a member of the Science for All Expert Group, an independent committee tasked by the government to write an action plan that would, *inter alia*, deliver a shift in cultural awareness, recognition, and support for science. The Science for All Expert Group's report focuses on public engagement and the term science communication barely appears—there are three references to 'science communication organisations' and a couple to the annual Science Communication conference organised by the British Science Association, originally known as the British Association for the Advancement of Science. Public engagement is another term whose meaning has shifted significantly in quite a short space of time. Originally the term referred to scientists and policy makers engaging with the public about the direction and focus of science research. Now, however, the term seems to be applied to any contact between scientists and the public including outreach and entertainment.

The report commissioned a mapping exercise which used a framework organised around four modes of public engagement: 'telling', 'sharing', 'involving', and 'consulting' (Science for All Expert Group, 2010, p. 30). What the report failed to do was to engage with the critical question of what counts as effective science engagement? Indeed, one might argue that the sector has suffered from a lack of clarity about what the purpose of science communication is and what constitutes good quality science communication. Some indication of the challenges facing the sector can be seen in the following extract from one of the seven reports commissioned by the Science for All Expert Group:

> *Public engagement remains counter-cultural* to the ethos of most public and educational institutions, the civil service and scientific research. Over the

last ten years, public engagement has been encouraged; yet the ethos of expert leadership and one-way communication still predominates.

(2010, p. 39, emphasis in original)

So, then, what role might Part B of *IJSE* play in this complex and shifting scene? There is much that we do not know about science communication and public engagement. The Science for All Expert Group noted that 'we have only partial knowledge of why the public engages, how engagement activities can be most effectively developed and delivered, and what the impact of these events actually is' (2010, p. 10). I would hope that Part B would report on studies that addressed these issues critically and informed by a wide range of theoretical frameworks.

One particular area that needs academic study is the impact of science communication and public engagement. The Millennium Commission funded 18 new science centres at the end of the last century. However, several of the centres have experienced severe financial pressures and two have since closed. Calls for the sector to receive government funding resulted in the House of Commons Science and Technology Committee setting up an inquiry. A micro-economics company, Frontier Economics, was commissioned to evaluate the impact of science centres in England on the Government's Science and Society agenda and to assess whether science centres represented 'good value for money in comparison with other STEM-related organisations' (Frontier Economics, 2009, p. 1). Despite undertaking a survey of 39 science centres and five case studies, the authors of the subsequent report noted that:

We have not been able to assess whether science centres are good value for money relative to other comparator programmes. This is because there is insufficient evidence on the long term outcomes of science centres or comparator programmes.

(Frontier Economics, 2009, p. 2)

The report recommended that the sector needed to systematically collect a greater range of data, including 'participants' satisfaction with the programme' and 'measures of the programmes' effectiveness, i.e. whether the objectives of these programmes are achieved' (Frontier Economics, 2009, p. 4). In a similar vein, the Science for All Expert Group recommended that the sector required:

A set of indicators which would act as a basket of measures of the health of the relationship between society and the scientific anµd policy communities, building on public attitudes surveys, the attitudes of scientists and policy makers, and including measures of diversity.

(2010, p. 14)

It is at this conjunction of science, the public, and policy that *IJSE* (Part B) can make a contribution. The science communication and public engagement sector in the UK is at a crossroads—it is fragmented, lacking in credible evidence of its success (in the eyes of its critics) and yet intuitively aware that it can make a major contribution to a vision of the world that has at its heart a knowledge that science is seen as society's greatest cultural asset of the twenty-first century.

References

ASDC (Association of Science and Discovery Centres). (2010). About the association of science and discovery centres. Retrieved from http://www.sciencecentres.org.uk/about/

Frontier Economics. (2009). *Assessing the impact of science centres in England: A report prepared for BIS.* London: Author. Retrieved from http://www.sciencecentres.org.uk/govreport/docs/impact_of_science_centres.pdf

Goldacre, B. (2008). *Bad science.* London: Fourth Estate.

Science for All Expert Group. (2010). *Science for all: Report and action plan from the Science for All Expert Group.* London: BIS. Retrieved from http://interactive.bis.gov.uk/scienceand society/site/science-for-all/

The Guardian. (2010, April 13). How science became cool. Retrieved from http://www.guardian.co.uk/science/2010/apr/13/science-cool

20 'If the Public Knew Better, They Would Act Better'

The Pervasive Power of the Myth of the Ignorant Public

Elin Kelsey and Justin Dillon

Kelsey, E., & Dillon, J. (2010). 'If the public knew better, they would act better': the pervasive power of the myth of the ignorant public. In, R. Stevenson, & J. Dillon (eds), *Engaging Environmental Education: Learning, Culture and Agency*. Rotterdam: Sense, pp. 99–110.

Introduction

Museums, aquariums, science centres, zoos and other informal science institutions (ISIs) are increasingly committed to engaging the public in issues connected to environmental conservation and sustainability. Although ISIs around the world may hold different views about what information should be shared with the public, they appear to share the belief that 'if the public knew better, they would act better'. They operate within a common authoritative discourse about the power of education to transmit information from those who are knowledgeable to those who are not (Kelsey, 2001).

In this chapter, we explore the implications of this particular discourse on environmental learning, participation and agency within informal science institutions. More specifically, we examine a case study of conversational learning between guests (visitors) and volunteer guides in the galleries of a major U.S. aquarium. This is a particularly timely topic, as the interaction between ISIs and their publics has undergone significant change in recent years. Throughout the 1980s and 1990s, environmental public participation programs operated in a type of 'decide-announce-defend' mode based on a 'one way' transfer of information from experts to the public (Davies *et al.*, 2009). Such programs echoed a deficit model of Public Understanding of Science (PUS) rhetoric, with its tacit assumption of public ignorance (Lehr *et al.*, 2007).

Today, a new emphasis on 'co-determined' decisions and 'two-way' exchanges between experts and the public of both information and values has emerged. As Lehr *et al.* (2007) noted, 'the deficit model has—in theory, at least—been firmly rejected in response to a series of crises in the public trust of science and the government in the 1990s (for example, the BSE and genetically modified foods controversies), and a 'new mood for dialogue' between scientists, policy-makers, and various publics has emerged as its replacement' (House of Lords Select Committee on Science and Technology, 2000, p. 44).

A major response from ISIs to this shift toward more authentic public participation has been the creation of 'dialogue events' (Lehr *et al.*, 2007). Lehr *et al.* define these as face-to-face, adult-focused forums that bring scientific and technical experts, social scientists, and policy-makers into discussion with members of the public about contemporary scientific and socio-scientific issues. A number of ISIs now host *Café Scientifiques* where members of the public are invited to informal gatherings to discuss current issues of science, environment and/or technology (McCallie *et al.*, 2007). The Dana Centre, which opened in 2003 at the London Science Museum, for example, is a purpose-built venue which describes itself as 'a place for adults to take part in exciting, informative and innovative debates about contemporary science, technology and culture' (Dana Centre, 2008).

Rather than focusing on special 'dialogue events', which occur in specialized areas and/or at scheduled times, this paper deals with another highly complementary locus for environmental learning, participation and agency in ISIs, that is, the interactions between volunteer guides and visitors in the public galleries of ISIs, and the potential they hold for learning through conversations.

The Value of Learning through Conversations

A growing body of research in out-of-school contexts recognizes that people learn in museums through conversations (Leinhardt, Crowley, & Knutson 2002). Indeed, much attention has been paid to research on conversations between visitors at recent annual conferences of the Visitor Studies Association, the Association of Science-Technology Centers, the National Association for Research in Science Teaching (NARST) and the American Educational Research Association (AERA) (see for example the AERA 2002 symposium entitled *Learning conversations for all: Explanation, reflective reasoning, thematic content and significant events*). There is also a well-articulated awareness within the research literature of the importance of conversation to enhancing and changing knowledge, attitudes and values (Baker, Jensen, & Kolb, 2002; Jickling, 2004; Laurillard, 1993; Lave & Wenger, 1991).

In terms of environmental learning, Rennie (2003) finds that conversation promotes engagement in environmental awareness or action projects. A number of authors highlight the role of conversation in increasing public participation in politics and in real-world issues (Bobbio, 1987; Bohman & Rehg, 1997; Chambers, 1996; Cohen, 1989; Elster, 1998; Fishkin & Luskin, 2005; Gutmann & Habermas, 1996; Keane, 1991; Public Conversations Project, 2008; Zeldin, 1998).

Much of the literature on the role of language in learning fits within a Vygotskian/sociocultural paradigm (as explicated by Wertsch, 1991) in which it is the social plane that is so critical to development, through interaction, primarily through talking. Indeed, a growing body of literature emerging from research in schools points to the key role of 'exploratory talk' (Barnes, 1976), discussion, dialogic teaching, dialogic inquiry, collaborative reasoning (Chinn and Anderson, 1998) and argumentation, as playing critical roles in developing conceptual understanding and changes in mood and emotion (Rojas-Drummond & Mercer, 2003; von Aufschnaiter *et al.*, 2008). At the heart of the debate is language which, as Halliday (1993) points out, 'has the power to shape our consciousness; and it does so for each human child, by providing the theory that he or she uses to interpret and manipulate their environment' (p. 107).

Mercer and Littleton (2007) characterise exploratory talk (Barnes, 1976) as being:

> dialogue which involves partners in a purposeful, critical and construc-
> tive engagement with each other's ideas. Statements and suggestions are
> offered for joint consideration. These may be challenged and counter-
> challenged, but challenges are justified and alternative hypotheses are
> offered. Partners all actively participate, and opinions are sought and
> considered before decisions are jointly made. 'Exploratory talk' has some
> similarities with the notions of 'accountable talk' (Resnick, 1999) and
> 'collaborative reasoning' (Chinn and Anderson, 1998).
>
> (Mercer, 2008, p. 357)

Now, many of these ideas may not be supported by rigorous empirical research but they act as powerful lenses through which to see how discussion and dialogue impact on learning in its broadest sense. In terms of dialogue, Mercer (2008) notes that:

> It is our natural habit to express our ideas in dialogue, to test our views
> against those of others, and to attempt to persuade other people to share
> the conceptual understandings that we believe are the best. It is of course
> also normal that we resist changing our minds, if the views we hold are
> bound up with aspects of our social identities. But, nevertheless, most
> of us proceed as if we believe that one of the most important ways of
> changing someone's mind is to talk with them.
>
> (p. 355)

Many environmental issues, as they impact on the interface between science and society, are controversial. To help make sense of these controversies, we would argue that the public would benefit from a deeper understanding of the ways in which scientific understanding develops. One of the ways in which young people may come to understand the nature and development of science is by engaging in the processes of argumentation, that is, building knowledge through purposeful weighing of evidence and analysis of warrants for 'truth'. We know something about how argumentation, a foundation of the ways in which science works can be taught to young people. Much of the work of Osborne and colleagues points to the fact that teachers can be taught to develop higher order argumentation given enough time (see, for example, Simon, Erduran, & Osborne, 2006).

Rather than teaching the 'neutral' skills of argumentation or dialogic talk, ISI programs, such as the Monterey Bay Aquarium's Seafood Watch initiative, endeavor to use these formats to engage visitors and persuade them to change their actions. But human decision-making is hard to affect, as Eiser and van der Pligt (1988) argue:

> [evidence suggests] that the conscious thought preceding a decision may be of a relatively simple nature, given the difficulty of processing complex information. People seem to rely on simple heuristics for making probability judgements and hardly seem to think about more complex combinations of probabilities and values or utilities involved in a decision [. . .] In other words, people's decision processes seem relatively inarticulated and are hardly compatible with the sort of rigorous, systematic thinking required by normative decision models.
>
> (p. 181)

So, given these limitations, how can ISIs play an active role in promoting conservation using conversations? The potential to do so is very high: every year, for example, more than 143 million people visit zoos and aquariums (Falk *et al.*, 2007). Furthermore, there is evidence that visitors seek opportunities to converse about issues of societal importance during their visits. Cameron (2003) found that 95% of people surveyed in an Australian sample wanted museums to provide *more* opportunities for visitors to have their say about topics; to converse; to exercise their democratic right to be heard in a publicly funded institution. Fortunately, volunteer educators (guides or docents) already exist as a well-established part of the operation of ISIs in many countries and rather than static exhibits, these volunteers represent a tremendous opportunity to engage the public in personally relevant conversations about current environmental issues.

The Prevalence of Mini-Scripts

Despite the potential value of conversational learning, dialogue and exploratory talk in theory, a multi-year study of a major USA aquarium reveals that such conversations are rarely observed in practice. Instead, guides typically default to a one-way transfer of information to visitors in the form of 'mini-scripts' (Kelsey, 2004). Kelsey defines mini-scripts as predetermined statements which guides tend to pair with specific animals or props. Although not officially scripted, these statements are repeated so frequently that they take on the appearance of standard scripts, and are sometimes shared across shifts and individuals. At the touch pools, for example, at least one tenth of the aquarium's 500 volunteer guides say 'sea cucumbers and chitons are the vacuum cleaners of the sea' and 'sea urchins feel like a hairbrush' even though no formal script actually exists. Rather than engage in conversations, guides tend to create longer engagement sequences by moving from one prop to another, stringing together mini-scripts for each specimen or piece of apparatus.

Though no other study of mini-scripts has yet been conducted, their presence appears to be well-recognized by professional educators working in ISIs. Each time they are mentioned at presentations at AERA, NARST and the North American Association for Environmental Education (NAAEE), professional colleagues have been quick to acknowledge their existence at their own host institutions. Sanders, for example, describes the tendency for Explainers at the Natural History Museum in London to favor particular entry points, or "opening gambits" in their interactions with students (personal communication).

In 2006, an opportunity to explore the tenacity of mini-scripts presented itself at the same aquarium where they were first identified. The aquarium had just established a new 'Take Action' temporary exhibit on marine protected areas (MPAs) to coincide with a major initiative to grant further conservation protection to MPAs along the California coast. The exhibit provided information on the issues and names and addresses of elected officials. It invited guests to become more engaged and explore issues in more detail by inviting them to sign up for an email listserve operated by the aquarium.

The exhibit was located in close proximity (approximately 5 metres) from a guide station called the 'Ocean Advocacy Station'. The station takes the form of a large cart equipped with props such as cans of seafood, a computer screen and 'Seafood Watch' cards. Guides use the station as a base from which to interact with visitors or "Guests" as they are referred to at the aquarium. Each guide shift was given specific training and enrichment sessions about the issue of MPAs and asked to engage guests in conversations about the issue. It is important to note that these training sessions deliberately used a conversation-based instruction style that invited guides to share their own thoughts about and experiences with MPAs in a facilitated group format. The MPA campaign was timely, local, clearly endorsed by the aquarium and supported by the new exhibit. It was hypothesized that guides would readily engage guests in conversations about MPAs as a result of:

- overt institutional support;
- specific guide training; and,
- specific gallery location (guide cart in close proximity to 'Take Action' exhibit)

However, during 15 half-hour observation sessions by one of the authors (EK), guides did *not* engage guests in conversations about MPAs even when presented with the opportunity. Instead, they stuck to their mini-scripts associated with the Ocean Advocacy Station.

Further analysis of the guides' actions revealed the following findings:

1) *Guides adhered to mini-scripts that were prop-driven.* Rather than discuss MPAs, the guides talked about Seafood Watch using the cards, video clips and cans of seafood on their cart.

2) *Guides were 'glued' to their cart, even in circumstances where there were no guests at the cart and there were guests at the Exhibit.* Only one guide in the 15 observation sessions left the guide cart to engage guests at the 'Take Action' exhibit.

3) *The 'Mini-scripts' used at the carts lacked context and created confusion.* Mini-scripts at the guide cart existed as a series of simplified sentences that were strung together and repeated frequently. The problem here is that many important concepts that served to make the complexity of the ideas understandable were lost in the repetition. For example, some guides were so eager to advocate the consumption of wild caught salmon that when a guest picked up a can of tuna and asked which kind of tuna is dolphin safe, the guide responded with a mini-script about the benefits of wild caught salmon.

4) *The 'Mini-scripts' used were not personalized.* In responding to visitors, the guides tended to draw on a number of favourite phrases and linked them together in response to guest questions or comments, creating the impression that a conversation was happening. However, the sentences themselves remained the same no matter who the guide was speaking with or what the guest asked/answered. For example, one guide asked every guest who stopped at the station if they had seen the movie *A Perfect Storm*. None of the guests answered in the affirmative, yet each time, the guide proceeded to explain how well the movie depicted a certain kind of fishing practice.

5) *The 'Mini-scripts' used were not age appropriate.* In the observation sessions, guides were very friendly to children but did not change their 'mini-scripts' when children were present. Thus, in a number of encounters preschool and early elementary school aged children were asked if they liked to eat tuna and then told about the dangers of high mercury levels or the entrapment of dolphins and their babies during tuna purse seine fishing. The lack of age sensitivity on the part of the guides is worrying, not least because there is mounting evidence from researchers such as Sobel (1996, 2008) of the dangers of presenting children with examples of environmental problems before they are emotionally and developmentally (around age eight years old) equipped to deal with them.

Subsequent discussions with a focus group of eight guides revealed that they had a different perception of their interactions with visitors in that when asked specifically how they transferred the experiences of their MPA training to their conversations with guests, most guides answered decisively that this readily occurred. As one guide expressed: 'Many, probably most (guides), are highly educated here so they are able to integrate that information into the other stuff that they know and to present it at different locations.' Yet, from observations, the transfer did not happen. Guides did *not* share information they learned at the MPA enrichments with guests at the guide cart during any of the observation sessions. Instead, they adhered to the familiar mini-scripts associated with the Seafood Watch program.

So Why are Mini-Scripts So Pervasive?

The evidence above demonstrates the pervasiveness of mini-scripts in these volunteers' practice. It appears that the guides' own experiences of schooling and, perhaps, the traditional lecture approach become so ingrained that they are hard to shake despite a training format that modelled conversational learning and an explicit request to engage the guests in conversation. It suggests that this transmissionist model on learning, teaching, and communicating is the default, despite interest in the field (both researcher and practitioners) and the aquarium leadership to act and believe otherwise.

We see an answer, partly, in the work of Mortimer and Scott who have researched the difficulties faced by schoolteachers trying to move from what they call authoritative teaching to more dialogic teaching. Mortimer and Scott have identified 'Four Classes of the Communicative Approach' (Scott, Mortimer, & Aguiar, 2006):

a   *Interactive/dialogic:* Teacher and students consider a range of ideas. If the level of interanimation is high, they pose genuine questions as they explore and work on different points of view. If the level of interanimation is low, the different ideas are simply made available.

b   *Noninteractive/dialogic:* Teacher revisits and summarizes different points of view, either simply listing them (low interanimation) or exploring similarities and differences (high interanimation).

c   *Interactive/authoritative:* Teacher focuses on one specific point of view and leads students through a question and answer routine with the aim of establishing and consolidating that point of view.

d   *Noninteractive/authoritative:* Teacher presents a specific point of view.

Table 20.1 indicates the four classes of communicative approach model schematically.

What the model points to is a tension between the talk associated with authoritative science knowledge and the kind of knowledge built by students engaged in dialogic activity. While Mortimer and Scott suggest a need to have a balance of approaches, they argue that interactive dialogic is preferable for exploring ideas and facilitating their engagement. Wells (1997)

Table 20.1 Four classes of communicative approach

	INTERACTIVE	NON-INTERACTIVE
AUTHORITATIVE	*interactive/authoritative* (e.g., teacher-led discussion)	*non-interactive/authoritative* (e.g., teacher lecture)
DIALOGIC	*interactive/dialogic* (e.g., teacher/student collaboration)	*non-interactive/dialogic* (teacher summarises students' views)

Source: (Adapted from Mortimer & Scott, 2003, p.35)

argues that dialogic discourse does not necessarily mean an equal discourse. From this perspective the guides would be seen to have a responsibility to shape the exchange. Yet the prevalence of mini-scripts suggests that the transition from authoritative communication to dialogic communication fails to occur.

Furthermore, McCallie (2008, personal communication) makes a distinction between teaching in contexts in which there is a preordained set of information to be learned—'learning for mastery'—as opposed to a more open learning agenda in which what is to be learned is not yet known or codified. In other words, what is to be learned is yet to be figured out. The question of whether or not to establish MPAs, for instance, is a socio-scientific issue that fits in the latter category. Guides could engage in discussions about what could be done, for example, 'The Aquarium thinks this, what do you think?'

Yet we believe that these teaching and learning considerations are only part of the answer. The idea of a common discourse that prevents ISIs from realizing their stated aim with respect to public engagement is supported by the notion of 'structure' as described by Sewell (1992) and Giddens (1991). According to Sewell (1992, p. 3) structure is an elusive and difficult to describe a notion that reflects 'something very important about social relations: the tendency of patterns of relations to be reproduced, even when actors engaging in the relations are unaware of the patterns or do not desire their reproduction.'

The degree of guide *agency* meanwhile, that is the capacity of individuals to act independently of the *structures* imposed by social systems, remains a question of debate. In his theory of structuration, for example, Giddens (1991) argues that it is a mistake to pose social systems and individual agency as separate from one another because neither exists except in relation to the other. In this sense, there is what Giddens calls a *duality of structure*, which is to say the structure of a system provides individual actors with what they need in order to produce that very structure as a result. Structures, says Giddens, are both the medium and the outcome of the practices that constitute social systems. Structures shape people's practices, just as people's practices constitute and reproduce structures.

One of the hallmarks of structures, according to Sewell (1992), is that it is often difficult for one engaged in a pattern to be aware of it. Thus, it is possible that ISIs operating within a common structure—in this case, a common discourse about scientific knowledge, the public and education—will be unaware of it even while their actions serve to sustain and reproduce it.

The power of discourses to shape public life, whether or not individuals engaged within these discourses are aware of them, forms the basis of Foucault's (1988) work on the connections between language, knowledge, power and social control. Foucault argues that language and knowledge form a basis for power in their role in the social construction of reality. The modern mode of domination, he claims, is based on a combination of scientific disciplines and professional and administrative practices which penetrate each and every socialised subject of society. How we talk and think about the world shapes how we behave and the kind of world we help to create. Discourses are powerful because they both define and limit the ways in which we conceptualise reality (Gee, 1999).

Blades (1997) provides an example of Foucault's theory in action in a formal education setting through his case study of curriculum change in secondary school science. According to Blades (1997, pp. 2–3):

> attempts to change secondary school science education curricula are defined and thus limited by the positivistic, technical-rational assumptions of the discourse of modernity. So en-framed, curriculum change seems destined to technicality, to a view of change as a problem to be solved once all the factors are elucidated; a search for the correct method and generalisable technique.

So, Where Next?

The prevalence of mini-scripts indicates a worrisome disconnect between the stated intention of ISIs to serve as sites for public engagement and the realities of the interactions between guides and guests. The use of mini-scripts has the unintentional effect of treating the visiting public as if they are ignorant and/ or as if they don't know what questions to ask or what information they need. This disconnect is further mirrored in volunteer guide training programs and enrichments which are typically structured as information dissemination sessions where guides are *told* by an expert staff member (often in a friendly though authoritative, non-interactive manner) how they should interact and engage in dialogue with visitors. Adding a few conversational learning sessions is not sufficient to help guides learn how to transition between authoritative information giver and dialogic discourse promoter.

A major issue with dialogue is that it is not 'secure' or 'consistent.' It is far more challenging than following mini-scripts. In order to engage in argument (or dialogue in general) one must have a much stronger command of the information in order to think and be flexible with it. It is the difference between 'knowing something' 'and being 'literate' in the sense of being able to apply information and skills in a variety of contexts.

Failure to create and model a conversational learning environment for guides serves to reinforce the status quo and to undermine the guides' participation and agency in engaging guests in conversational learning in the public galleries. Yet changing the structure of guide training sessions is not an easy task. Training sessions are a mainstay of most volunteer guide programs and both staff and volunteers have strong, well-established expectations about how training should be conducted. Volunteers at the aquarium described in this paper, for example, speak in proud terms about having survived 'training boot camp' wherein they mastered scientific names of marine invertebrates and challenging concepts of ocean geomorphology.

For many ISIs, including the one mentioned in this paper, the corps of volunteers is even more stable than its staff. These long-term, experienced guides serve an important 'gate-keeping' role in inspiring and maintaining a high level of professionalism. Though keenly committed to remaining 'cutting edge', a number of these individuals are of the opinion that the existing system of guide training and guest/guide encounters is working well and needs little change. The fact that experienced guides mentor new guides at the stations further perpetuates the traditional discourse of guide as information giver rather than dialogist.

Furthermore, many volunteers are seniors who attended school at a time when the teacher was the unquestioned source of expertise and authority. Beginning attempts to create training sessions that challenge this norm by facilitating learning through conversations have been met with enthusiasm by some volunteers but not-unexpectedly, with confusion and scepticism from others.

Nevertheless, this aquarium and a growing number of ISIs (see for example Osborne and Rodari's work regarding training programs for museum educators across Europe) are interested in improving and developing their training programs, especially with regards to learning literature.

Perhaps the greater barrier to progress is the institutional identity of ISIs. ISIs have a distinguished history as a 'trusted source of information' with respect to science (Astor-Jack et al., 2006). This identity as a purveyor of science authority and expertise further reinforces a transmission-based learning culture in both guide training programs, and the ways in which guides interact with the public in exhibit galleries. Yet as the past decades of conservation initiatives attest, the issues are rarely straightforward. Nor are they exclusively confined to problems answered by science. As Johnson et al. (2001) note, conservation issues are defined as much by socio-cultural values and political and economic factors as by the biophysical dimension. Indeed, the complexity of conservation issues is evidenced by the multiple roles that ISIs are increasingly adopting (information source, habitat protector, political advocate, role model, etc.) with respect to conservation action.

The ideas put forward in this paper recognize the importance of learning models that openly encourage and value multiple ideas and perspectives (Layton *et al.* 1993; Larochelle, Bednarz, & Garrison, 1998). Such models challenge the belief that facts speak for themselves and, instead, emphasize the active role of the learner and the contextual nature of learning. Clarifying messages and transmitting ISIs positions on key conservation issues is one important institutional role. Yet, the goal of engaging the public in conservation demands that ISIs continue to expand their identity as a scientific authority to more fully embrace their identities as a forum and facilitator of a conversational learning culture.

References

Astor-Jack, T., Balcerzak, P., & McCallie, E. (2006). Professional development and the historical tradition of informal science institutions: Views of four providers. *Canadian Journal of Science, Mathematics, & Technology Education, 6*(1), 67–81.

Baker, A., Jensen, P., & Kolb, D. (2002). *Conversational learning: An approach to knowledge creation.* Westport, CT: Quorum.

Barnes, D. *(1976). From communication to curriculum.* Harmondsworth: Penguin Books.

Blades, D. W. (1997). *Procedures of power and curriculum change: Foucault and the quest for possibilities in science education.* New York: Peter Lang.

Bobbio, N. (1987). *The future of democracy: A defence of the rules of the game.* Cambridge: Polity.

Bohman, J., & Rehg, W. (Eds.). (1997). *Deliberative democracy: Essays on reason and politics.* Boston: MIT Press.

Cameron, F. (2003). Transcending fear—engaging emotions and opinion—a case for museums in the 21st century. *Open Museum Journal, i,* 1–46. Retrieved December 24, 2008, from http://archive.amol.org.au/omj/volume6/cameron.pdf

Chambers, S. (1996). *Reasonable democracy: Jurgen Habermas and the politics of discourse.* Ithaca, NY: Cornell University Press.

Chinn, C. A., & Anderson, R. C. (1998). The structure of discussions that promote reasoning. *Teachers College Record, 100,* 315–368.

Cohen, J. (1989). Deliberation and democratic legitimacy. In A. Hamlin & P. Pettit (Eds.), *The good polity.* New York: Blackwell.

Dana Centre/Science Museum. (2008). *Dana centre: About us.* Retrieved December 24, 2008, from http://www.danacentre.org.uk/aboutus

Davies, S., McCallie, E., Simonsson, E., Lehr, J., & Duensing, S. (2009). Discussing dialogue: Perspectives on the value of science dialogue events that do not inform policy. *Public Understanding of Science, 18*(3), 338–353.

Eiser, J. R., & van der Pligt, J. (1988). *Attitudes and decisions.* Routledge: London.

Elster, J. (Ed.). (1998). *Deliberative democracy.* Cambridge: Cambridge University Press.

Falk, J. H., Reinhard, E. M., Vernon, C. L., Bronnenkant, K., Deans, N. L., & Heimlich, J. E. (2007). *Why zoos and aquariums matter: Assessing the impact of a visit.* Silver Spring, MD: Association of Zoos and Aquariums.

Fishkin, J. S., & Luskin, R. C. (2005). Experimenting with a democratic ideal: Deliberative polling and public opinion. *Acta Politica, 40*(3), 284–298.

Foucault, M. (1988). *Politics, philosophy, and culture: Interviews and other writings, 1977–1984.* In M. Morris & P. Patton (Eds.). New York: Routledge.

Gee, J. (1999). *An introduction to discourse analysis: Theory and method.* New York: Routledge.

Giddens, A. (1991). *Modernity and self-identity. Self and society in the late modern age.* Cambridge: Polity Press.

Gutmann, A., & Thompson, D. (1996). *Democracy and disagreement.* Cambridge, MA: Belknap Press of Harvard University.

Habermas, J. (1996). *Between facts and norms.* Boston: MIT Press.

Halliday, M. A. K. (1988). On the language of physical science. In M. Ghadessy (Ed.), *Registers of written English: Situational factors and linguistic features.* London: Frances Pinter.

House of Lords Select Committee on Science and Technology. (2000). *Third report: Science and society.* Retrieved December 24, 2008, from http://www.parliament.the-stationery-office.co.uk/pa/ld199900/ldselect/ldsctech/38/3801.htm

Jickling, B. (2004). Making ethics an everyday activity: How can we reduce the barriers? *Canadian Journal of Environmental Education, 9,* 11–30.

Keane, J. (1991). *The media and democracy.* Cambridge: Blackwell.

Kelsey, E. (2001). *Reconfiguring public involvement: Conceptions of 'education' and 'the public' in international environmental agreements.* Unpublished PhD thesis, King's College London, UK.

Kelsey, E. (2004, April 1–3). *From science learning to conversations about conservation: A study of guide training at the Monterey Bay Aquarium.* Paper presented at the annual meeting of the National Association for Research in Science Teaching, Vancouver, British Columbia, 2004.

Larochelle, M., Bednarz, N., & Garrison, J. (Eds.). (1998). *Constructivism and education.* Cambridge: Cambridge University Press.

Laurillard, D. (1993). *Rethinking university teaching: A framework for the effective use of educational technology.* London: Routledge.

Lave, J., & Wenger, E. (1991). *Situated learning. Legitimate peripheral participation.* Cambridge: Cambridge University Press.

Layton, D., Jenkins, E., Macgill, S., & Davey, A. (1993). *Inarticulate science? Perspectives on the public understanding of science and some implications for science education.* Nafferton: Studies in Education Ltd.

Lehr, J. L., McCallie, E., Davies, S., Caron, B. R., Gammon, B., & Duensing, S. (2007). The value of "dialogue events" as sites of learning: An exploration of research and evaluation frameworks. *International Journal of Science Education, 29*(12), 1467–1487.

Leinhardt, G., Crowley, K., & Knutson, K. (2002). *Learning conversations in museums.* Mahwah, NJ: Lawrence Erlbaum Associates.

McCallie, E., Kollmann, E. K., Simonsson, E., Chin, E., & Dillon, J. (2007). *Visitors and engagement: Findings from research and evaluation studies of discussion forums on controversial issues.* Paper presented at the 20th Annual Visitor Studies Association conference. Columbus, OH: Visitor Studies Association. Retrieved from http://www.informalscience.org/research/show/3574

Mercer, N. (2008). Changing our minds: A commentary on 'Conceptual change: A discussion of theoretical, methodological and practical challenges for science education' by D.F. Treagust and R. Duit. *Cultural Studies in Science Education, 3*(2), 351–362.

Mercer, N., & Littleton, K. (2007). *Dialogue and the development of children's thinking: A sociocultural approach.* London: Routledge.

Mortimer, E. F., & Scott, P. H. (2003). *Meaning making in secondary science classrooms.* Maidenhead, UK: Open University Press.

Public Conversations Project. (2008). *Public conversations project.* Retrieved December 24, 2008, from http://www.publicconversations.org/pcp/pcp.html

Rennie, L. (2003). *The Australian Science Teachers Association science awareness raising model: An evaluation report.* Canberra: Department of Education, Science and Training. Australian Government.

Resnick, L. B. (1999). Making America smarter. *Education Week, 18*(40), 38–40.

Rojas-Drummond, S., & Mercer, N. (2003). Scaffolding the development of effective collaboration and learning. *International Journal of Educational Research, 39,* 99–111.

Scott, P. H., Mortimer, E. F., & Aguiar, O. G. (2006). The tension between authoritative and dialogic discourse: A fundamental characteristic of meaning making interactions in high school science lessons. *Science Education, 90*(4), 605–631.

Sewell, W. F. (1992). A theory of structure: Duality, agency, and transformation. *American Journal of Sociology, 98*(1), 1–29.

Simon, S., Erduran, S., & Osborne, J. (2006). Learning to teach argumentation: Research and development in the science classroom. *International Journal of Science Education, 28*(2&3), 235–260.

Sobel, D. (1996). *Beyond ecophobia: Reclaiming the heart in nature education.* Great Barrington, MA: Orion Society.

Sobel, D. (2008). *Childhood and nature: Design principles for educators.* Portland, ME: Stenhouse Publishers.

von Aufschnaiter, C., Erduran, S., Osborne, J., & Simon, S. (2008). Arguing to learn and learning to argue: Case studies of how students' argumentation relates to their scientific knowledge. *Journal of Research in Science Teaching, 45*(1), 101–131.

Wells, G. (2008). Learning to use scientific concepts. *Cultural Studies in Science Education, 3*(2), 329–350.

Wertsch, J. V. (1991). A sociocultural approach to socially shared cognition. In L. B. Resnick, J. M. Levine, & S. D. Teasley (Eds.), *Perspectives on socially shared cognition.* Washington, DC: American Psychological Association.

Zeldin, T. (1998). *Conversation: How talk can change your life.* London: Harvill Press.

21 Communicating Global Climate Change

Issues and Dilemmas

Justin Dillon and Marie Hobson

Dillon, J., & Hobson, M. (2013). Communicating global climate change: issues and dilemmas. In, J. Gilbert, B. Lewenstein, & S. Stocklmayer (eds), *Communication for Engagement in Science and Technology*. New York: Routledge, pp. 215–228.

Introduction

> A campaign is being launched across Australia to promote respect for science. The campaign comes after it was announced last week that climate scientists in the country had received death threats, with the Australian National University increasing security around nine climate scientists and administrative staff. With national debate over climate change becoming increasingly heated owing to the government's carbon tax, scientists are having to battle against what Anna-Maria Arabia, chief executive of the Federation of Australian Science and Technological Societies, called "a noisy misinformation campaign by climate denialists".
>
> *(Times Higher Education, June 30, 2011, p. 16)*

This chapter critically examines the relationship between science, scientists, and science communication through the lens of the major environmental issue facing society, climate change. Many, if not all, of the issues are relevant to science communicators wherever they work. The long-term nature of the challenges thrown up by climate change means that developments in the relationship can be identified and, as the subject will not 'go away' quickly, it allows opportunities for speculation about future trends and possibilities for science communication. Climate change is an issue that can provoke strong responses among experts as well as the lay public. Increasingly, as the story from the *Times Higher Education* above suggests, it is already an issue of life and death.

A range of topics needs to be considered when discussing the communication of climate change. These include: the role of the media; public trust in scientists; public understanding of the science; and the nature of climate science, specifically, what counts as evidence.

So, this chapter will address the various conceptualizations of climate change/global warming and look at how the relationship between science, scientists, and the public has changed over recent years. We will also examine whether science communicators should promote the scientific consensus or encourage debate about the ways that science is carried out and at how scientists operate. Throughout the chapter, we will draw on the experience of the Science Museum, London and the planning that went into the design of a new gallery on climate change *(atmosphere . . . exploring climate science)* that was opened in 2010.

The first thing to note is that, as with many issues, climate change has taken on a political dimension. Writing in *The Guardian* several years ago, the columnist Polly Toynbee described the tendency of the right-wing to line up against scientific evidence in issues concerned with public health and the environment:

> Posing as hard-headed realists, those on the right are more prone to pit their ideology against the weight of science. Seat belts? Motorbike helmets? Chlorofluorocarbons and the ozone layer? Smoking bans? Advertising junk food to children? The science-based realos tend to be on the left, conviction fundis on the right.
>
> *(Toynbee, 2006)*

Five years on, Toynbee (2011), commenting on the influence of the US extreme right-wing on political debate in the UK, noted that 'a taste of the Tea Party arrives on these shores in the peculiar paranoia of the climate-change deniers'. Toynbee's position, which is mirrored in the public policies of the major UK political parties is that 'On matters of fact, those of us who are not scientists can only listen to what scientists say and trust such an overwhelming global consensus'. For many people, access to 'what scientists say' is moderated by the media and by science communicators on the Internet and in museums and science centres.

The question is, however, what does this moderation entail? Does it, for example, involve presenting the scientific consensus and looking at the range of predicted impacts of climate change? Or does it involve presenting climate change science as hotly contested? These, and other questions, are ones which faced the Science Museum when they decided to create a new gallery: *atmosphere . . . exploring climate science.*

The gallery was designed to achieve the following goals:

- To deliver an immersive, enjoyable and memorable (life-enhancing) experience that increases interest, deepens understanding and is robust against deeply held convictions;
- To be recognised and admired as *the* UK destination for clear, accurate, up-to-date information on climate science for the non-specialist.

The exhibition was targeted at independent (non-specialist) adult visitors; families with children aged 8+; and high school science and geography teachers and their students (aged 11–16).

The development of the gallery was heavily influenced by research carried out by the second author and her colleagues in the Audience Research and Advocacy Department. This research involved consultation, desk-reviews and prototype testing. Initial consultations with the target audiences involved focus groups and in-depth interviews to find out their prior knowledge, opinions, and 'misconceptions' about climate change and climate science. The consultation also identified potential visitors' expectations of what the gallery would look like and tried to identify any barriers to attendance or engagement. The research team carried out desk-reviews of academic papers, public opinion polls, and other climate change exhibitions (both at the Science Museum and elsewhere). The final stage of the research involved prototype testing of the interactive exhibits to reduce and remove barriers to usability, comprehension, and motivation.

A key part of this work was a survey of 30 adult visitors' familiarity and understanding of terms and concepts, such as 'greenhouse gases', 'remediation', and 'carbon footprint' to develop a mental model of what visitors considered to be the causes of and possible responses to climate change. The findings were presented to the exhibition team as a two-page diagram, with areas of misunderstanding clearly highlighted, so it could be easily referred to when discussing content, to ensure that the team defined terms where necessary and avoided reinforcing misconceptions. This work showed that visitors have pockets of knowledge about climate change which they struggle to link together accurately and it demonstrated the need for the audience to have the science behind climate change clearly explained. The findings and implications of the research is described more fully below.

The Science and Terminology of Climate Change

Our understanding of climate change as a scientific phenomenon has developed rapidly since the 1980s. Climate change refers to long-term changes in weather patterns over a region or across the planet. Global warming refers to the process by which the average temperature of the Earth's near-surface air and oceans has increased relatively recently and continues to increase. The greenhouse effect refers to the process in which thermal radiation from the Earth is absorbed by gases in the atmosphere and then re-radiated. This re-radiation increases the temperature of the Earth.

The evolution in our understanding of climate change, and the issues which are currently open to debate, are outlined by Steve Jones, Emeritus Professor of Genetics at University College London, in his independent report to the BBC Trust, which was carried out in 2010:

> In its early days, two decades ago, there was a genuine scientific debate about the reality of climate change (although that attracted rather little attention). Now, there is general agreement that warming is a fact even if there remain uncertainties about how fast, and how much, the temperature might rise. At present, the pessimists are in the ascendant and today's increase in floods and snow (as predicted for a warmer atmosphere which can take up more water) is on their side. A debate remains, and

it deserves to be reported with as much objectivity as would any other unresolved issue.

(BBC Trust, 2011, p. 68)

The scientific consensus on the key issues is outlined in the Third Assessment Report of the Intergovernmental Panel on Climate Change (IPCC 2001). The report's main conclusions are that the global average surface temperature has risen by 0.6 ± 0.2 °C since the late 19th century, and by 0.17 °C per decade since the 1970s, and that there is new and stronger evidence that most of the warming observed over the last 50 years is attributable to human activities. The Panel also concluded that if emissions of greenhouse gas continue, then the warming will also continue. Temperatures were projected to increase by 1.4 °C to 5.8 °C between 1990 and 2100. Accompanying this temperature increase will be increases in some types of extreme weather and a projected sea level rise of 9 cm to 88 cm.

The growing consensus about climate change science coalesces around findings such as these. Scientists, by their very nature, should be open to contrary evidence and to new ideas which might emerge with future research. Science has a habit of making new knowledge by falsifying old ideas. There is a debate about the degree to which climate change will impact on the planet—science can predict phenomena in the real world, but not always with 100 per cent accuracy—but the debate about what were once contested ideas has withered on the vine. The question is, though, to what extent have scientists and science communicators been successful in presenting these ideas and educating the public about some of the most important issues facing civilisation?

What Do the Public Know and Believe?

According to a large-scale survey carried out early in this century, most of the UK's population had heard of the terms 'climate change', 'global warming' or the 'greenhouse effect' (DEFRA, 2002). At first, the public seemed more familiar with 'global warming' than 'climate change' (DEFRA, 2002; Whitmarsh, 2009). However, during the late 2000s, it is thought that people have become equally aware of both terms (Upham, et al., 2009).

An increased familiarity with both 'climate change' and 'global warming' suggests that the public regard the terms as synonymous. Only four per cent of Whitmarsh's respondents explicitly differentiated between 'global warming' and 'climate change' without being prompted to do so (Whitmarsh, 2009, p. 410). This situation may have arisen because the terms are used interchangeably by journalists and scientists. Generally, however, the media prefers to use 'global warming', while scientists and policy writers prefer 'climate change' (Whitmarsh, 2009).

The Center for Research on Environmental Decisions (CRED) in the US (CRED, 2009) suggests that 'climate change' is a better choice than the term 'global warming' because 'climate change' better conveys broader changes in the earth's ecosystems and varying global temperatures from one year to the next (CRED, 2009, p. 2). Unlike 'global warming', the term 'climate change' avoids 'the misleading implications that every region of the world is warming uniformly' and 'the idea that the only concern with increased greenhouse gases is higher temperatures' (ibid).

Surveys of the public suggest that CRED is right to make this recommendation. It does seem that the public associate 'climate change' with a broader range of impacts on the climate and weather (for example, hot summers, wetter winters, rainfall, and drought) and a broader range of causes (that is, both natural and human) than 'global warming' (Whitmarsh, 2009, p. 410). Associated with the term 'global warming' are the notions of heat being 'trapped', increasing temperatures, human causes, and misconceptions surrounding the depletion of the ozone layer.

Atmospheric Science: The Controversy and Sources of Confusion

Atmospheric science entered the public consciousness some years before climate change grabbed the headlines. The discovery of a 'hole' in the ozone layer in 1985 attracted substantial media attention. Ozone depletion, a phenomenon that had been detected in the 1970s, refers to both the relatively steady decline in the total volume of ozone in the stratosphere, and to the seasonal decrease in the concentration of ozone above the polar regions. Ozone depletion has since been linked to the level of human use of substances, such as some refrigerants. The phenomenon persists and it has been associated with climate change although the nature of the link is unclear.

In the early 1990s, Boyes and Stanisstreet noted that global warming might present some significant challenges to educators. Their survey of undergraduate students' perceptions identified a range of incorrect ideas including a confusion between global warming and the depletion of the ozone layer (Boyes & Stanisstreet, 1992). Like many scientific phenomena, the key concepts are abstract and somewhat distant for students. It is difficult to carry out experiments or hands-on inquiries, so a lot of climate change education depends on data interpretation (with or without a computer), discussions, or watching videos.

Ironically, while the media coverage of the ozone hole was successful in raising public awareness of that issue, and prompting action to virtually wipe out the use of chlorofluorocarbons (CFCs), it is now presenting a barrier to the successful communication of climate change more broadly. In 2010, as part of the planning and preparation for the Science Museum's *atmosphere* gallery, the Audience Research and Advocacy team conducted a survey of visitors' knowledge of the science behind climate change (Science Museum, 2010a). One third of the participants thought that greenhouse gases caused the hole in the ozone layer. As one participant put it: 'They [greenhouse gases] are depleting our ozone layer—there are more rays coming through and we are heating up'.

The Science Museum adult visitor survey found that the analogy of a greenhouse fosters misconceptions, particularly relating to the ozone layer. Visitors in the survey thought that the greenhouse gases formed a physical layer in the atmosphere, like a ceiling, and that the ozone hole was in this layer. Consequently, in designing the gallery, Science Museum staff took great pains to avoid any graphical depictions of greenhouse gases forming 'ceilings', instead trying to depict varying concentrations of gases.

What and Who Do the Public Believe?

A person's level of understanding or mental model of climate change both affects, and is affected by, their level of belief. Altering either represents a huge challenge for climate change communicators as individuals are prone to 'confirmation bias', that is, a tendency to interpret information in such a way that it affirms their prior knowledge (CRED 2009). In the case of the ozone layer, sceptics may interpret the hole as letting hot air out to justify their belief that climate change is *not* happening, while a believer may interpret it as letting more hot air in to justify their belief that it *is* happening.

Public opinion about climate change varies significantly from country to country. It also varies within countries, according to various demographics. In the UK, believers in anthropogenic climate change tend to be children or adults aged between 18–40 and female, while sceptics/deniers are more commonly adults aged 40+ (particularly aged 60+) and male (Science Museum 2010b). Overall it seems most people believe climate change is happening, but not necessarily that it has anthropogenic causes. According to a survey conducted by the BBC in 2010, while 75 per cent believed climate change was occurring, only a third of them also thought it was caused by humans (BBC 2010). This finding resonates with Cardiff University research which found that 78 per cent of people surveyed believed in climate change, but only 31 per cent thought it was primarily the result of human activity (Spence, Venables, Pidgeon, Poortinga & Demski, 2010). In addition, the levels of belief in climate change appear to be decreasing over time. The Cardiff research revealed a swing of 13 per cent (down from 91 per cent to 78 per cent) from believers to disbelievers between their 2005 and 2010 surveys (Spence et al., 2010). The BBC identified a swing of eight per cent (down from 83 per cent to 75 per cent) between November 2009 and February 2010 (BBC 2010)—a period of a few months.

In 2010, George Mason University's Centre for Climate Change Communication (C4) published a report that compared US public opinion in November 2008 and June 2010 (Leiserowitz, Maibach, Roser-Renouf & Smith, 2010). The percentage of respondents classified as 'alarmed' or 'concerned' about climate change dropped from 51 per cent to 41 per cent during the period. The number of respondents classified as 'doubtful' or 'dismissive' rose from 18 per cent to 24 per cent. These data indicate the challenge facing climate change communicators, as well as the malleability of public opinion. The segmentation of the audience used in the analysis of the C4 report points to a need for different messages for different sectors of the public (or publics).

Issues of Trust

Earlier, we pointed to Polly Toynbee's statement that 'On matters of fact, those of us who are not scientists can only listen to what scientists say and trust such an overwhelming global consensus' (Toynbee, 2006). Trust in science and scientists underpins much of what passes for science education. At school, much of what is taught about science is the received wisdom of teachers and textbooks. What evidence do you have, for example, that the Earth spins on its axis? Trust, though, can be undermined, as recent events have all too clearly shown.

The 'Climategate' affair, in 2009, involving the release of 160 mb of emails and data from the University of East Anglia's Climate Research Unit, has changed the ways in which climate scientists operate and communicate with each other and with the public. The reporting of the affair may well be the major reason for the swing of eight per cent in the proportion of the population who believed that climate change was real, that the BBC identified as happening between November 2009 and February 2010.

Such was the potential damage to public trust in government climate change policy, that the UK political response to 'Climategate' involved then Prime Minister Gordon Brown being forced to comment that 'With only days to go before Copenhagen [Summit] we mustn't be distracted by the behind-the-times, anti-science, flat-earth climate skeptics'. Such language may not always help to build trust in the government's position as Leo Hickman (2011) noted in a *Guardian* blog. Hickman was concerned that a new term 'climate crank' had been added to the already long list of terms varying from relatively neutral to derogatory: 'sceptic, denier, contrarian, realist, dissenter, flat-earther, misinformer, and confusionist'. Characterising sections of society in such negative terms may backfire on those who use such strong terms.

The Public's Levels of Trust

In developing its exhibition, the Science Museum commissioned focus groups in April 2008, more than a year before Climategate, from TWResearch. The study revealed three factors which affect people's level of trust in sources of information about climate change: *hypocrisy* in that the public do not want to be told what to do without seeing any evidence of others taking action and 'practising what they preach'; *profit* in that there are two confusing dichotomies here: firstly, between the altruistic act of combating climate change being championed by businesses, such as energy companies encouraging home owners to switch to renewable energy sources, etc; secondly between 'independent' scientists being funded by the Government or businesses; and, *inconsistency* in that sources are expected to have a consistent stance, rather than, as the public perceive it, changing their views to suit the latest trend. As a result, the government, businesses, and the media were least trusted by the focus group participants, while scientists, charities, and non-profit public organisations (such as the Science Museum) were most trusted. Science communicators working in the media, then, would seem to be in a curious position—trusted because of their science credentials, but mistrusted because of their media employment.

The public's association of 'science' with 'truth' and 'facts' results in scientists being viewed primarily as independent truth-seekers (TWResearch, 2008). Confusion therefore arises over scientists' lack of agreement around climate change issues and this perceived 'inconsistency' is often cited as a reason for lack of belief in climate change. This observation suggests that, in order to communicate climate change effectively, more work needs to be done in communicating the nature of science to the public.

The Science Museum, as an educational, science-based institution, is considered a place to find reliable information based on evidence, rather than opinions. However, to overcome some of the potential barriers to trust in their exhibition, the Museum: employed a Sustainability Consultant; set up a Carbon Reduction Working Group, and succeeded in reducing its carbon footprint by 17 per cent between 2009 and 2010; retained editorial control from all sponsors; and focused on presenting the science behind climate change.

Teachers' Views of Climate Change

Teachers' understandings of global warming are critical in that many museum visits take place in the form of school visits. It is during the preparation and follow-up of visits that the messages of exhibitions can be enhanced and moderated. Science Museum focus groups with secondary school science and geography teachers revealed, however, that some of the science teachers were not convinced that climate change was caused by humans (TWResearch, 2010). Dove's (1996) survey of student teachers' understanding of the greenhouse effect, ozone layer depletion, and acid rain found similar confusions to those identified earlier by Boyes and Stanisstreet in university students. Dove, though, was puzzled as to why the prospective teachers understood the science behind the ozone layer but did not understand the greenhouse effect.

Following her survey of pre-service teachers, Dove hypothesized that while the link between CFCs and the depletion of the ozone layer was well-established, global warming was somewhat contentious. Another possible challenge might be that the science behind the greenhouse effect is more difficult than that behind ozone depletion. Dove noted that the difference in understandings of the different phenomena raised the question 'as to whether understanding would be improved by simply presenting the concepts involved, or if alternative teaching methods are needed to make the message clear' (p. 99). Although Dove's research was carried out in the mid-1990s, its findings are likely to be relevant today.

Mason and Santi (1998) advocate that teachers should use constructivist approaches including discussions about different interpretations of evidence for global warming. The Science Museum focus groups revealed that teachers found dealing with conflicting and ever-changing evidence difficult as this sense of uncertainty conflicts with their perception of their role as teaching the 'truth' (TWResearch 2010). This is quite a paradox for teachers as they also need to teach about the nature of science, such as the tentative nature of some scientific knowledge and the value of disagreement over explanations of phenomena in the natural world.

Changing Attitudes/Behaviours

As the scientific consensus about the human causes of climate change has strengthened, and the potential consequences of global warming have become more immediate, some educators have advocated a shift in the type of education that is offered to students. Uzzell (1999) criticized much environmental education as being top-down and from the centre to the periphery, and argued that it did not have a good track record of changing the attitudes and values of children to the environment. More recently, educators have advocated new models and strategies for climate change education such as harnessing the power of community action (Moser & Dilling, 2004).

Cordero, Todd and Abellera (2008) reported that to be effective, climate change education should emphasize the personal connection between the student, energy, and climate change using methods such as environmental footprint calculations. Such strategies, they argue, can improve students' understanding of the links between personal energy use and global warming, a point echoed by Devine-Wright, Devine-Wright and Fleming (2004).

Barriers to Communicating Climate Change

Climate change communication is a challenging activity. In addition to the lack of understanding and lack of trust discussed above, the Science Museum focus groups (TWResearch 2008 and 2009) revealed a range of barriers for people engaging in climate change, regardless of their opinion as to whether it is happening, human caused or threatening:

- *Boredom:* climate change is constantly in the media and, for children, it is a topic they encounter in multiple subjects throughout their school career;
- *Irritation:* the public do not want to be told what to do and how to live their lives, particularly when it involves foregoing activities they enjoy, such as travelling abroad;
- *Powerlessness:* the public feel that individual actions are futile and have no sense of collective impact; they feel there has been little change and

have a low awareness of international efforts, for example, the 2009 Copenhagen conference;

- *Fear:* the public do not know how bad the impacts will be, the effect it will have on themselves, or if it is even too late to act.

According to Roser-Renouf and Maibach (2010) action has been hampered by political partisanship and industry disinformation campaigns; principles of fairness in news coverage have given a far greater voice to the handful of skeptics than is merited by either their numbers or their evidence; and publication of their views has fostered a widespread perception in the public of scientific controversy, where none actually exists. Therefore, the issue remains a low policy priority for most people, and it is likely to remain so until the perception of controversy is overcome and people clearly understand both the dangers we face, and the actions we must take to avert these dangers.

The issue of impartiality is one that continues to vex science communicators. In his review of the BBCs science coverage, Steve Jones identified the issue as one that goes beyond climate change:

> A belief in alternative medicine or in astrology and a fear of vaccines or of GM food are symptoms of a deep mistrust in conventional wisdom. Such scepticism should be part of every scientist's, every journalist's or every politician's, armoury. However, mistrust can harden into denial. That faces the media with a problem for, in their desire to give an objective account of what appears to be an emerging controversy, they face the danger of being trapped into false balance; into giving equal coverage to the views of a determined but deluded minority and to those of a united but less insistent majority. Nowhere is the struggle to find the correct position better seen than in the issue of global warming.
>
> *(BBC Trust, 2011, p. 66)*

Towards Overcoming Barriers

To communicate climate change effectively, communicators need to continue to engage and educate the public with the evidence that it is happening and that it is caused primarily by humans. Through audience consultation, the Science Museum (TWResearch 2008) identified the following strategies to *engage* the public with the issue of climate change:

- Focusing on humans: The public seem to be interested in the human stories, particularly those relating to:
  - the UK: these are emotive and can make the issue personally relevant;
  - the class war: the sense of injustice is motivating;

- countries already experiencing the effects: that the effects are happening to people now helps make the issue seem less remote and more immediate;
- Personal relevance: Many visitors fail to relate to the global issue. They want to know how it will impact on them and in what time-frame or they can dismiss the issue as not relevant to them;
- Providing examples of possible adaptation and innovative solutions: Examples of action that could or have been taken can provide a message of hope in an otherwise gloomy picture. Visitors have low awareness of these broader solutions beyond re-using carrier bags and replacing incandescent light bulbs;
- Providing examples of solutions from other countries: Visitors are intrigued by what other countries have done and how it provides hope.

In order to educate, communicators need to present what the public considers to be evidence that anthropogenic climate change is happening. A crucial point to be borne in mind is that the public do not consider the effects of climate change (such as sea level rise), or statements about the consensus of scientific opinion, as evidence of climate change (even though they are frequently reported in the media)—they could be attributed to many different problems, not just climate change. Instead the public want to see why those impacts, responses, and opinions are related specifically to climate change. The Science Museum did this in the *atmosphere* exhibition by:

- Presenting graphs which demonstrate rising CO_2 or temperature levels, such as the Keeling Curve;
- Displaying objects which scientists use to work out how the climate has changed (for example, an ice core) or might change (for example, a weather balloon);
- Developing interactive exhibits which explain the carbon cycle and the greenhouse effect;
- Designing an exhibit which encourages visitors to compare natural and human causes of climate change to deduce which is the more likely cause of the current period of warming;
- Creating a whole zone of the exhibition on how scientists predict the future through climate modelling.

The C4 report noted that 'regardless of their beliefs about global warming, large numbers of Americans said they engage in energy conservation actions at home—turning off lights and electronics, reducing their use of heating and air conditioning, conserving water and replacing incandescent bulbs with compact fluorescents' (Leiserowitz et al., 2010, p. 6). This finding suggests that a range of factors are influencing public patterns of consumption, including economics and perceptions of social responsibility. Education is just one of the influences on public behaviour. Changes in consumer behaviour are themselves likely to influence members of the public so that new patterns of behaviour emerge. One strategy for climate change communicators is to spread awareness of changes in public behaviour such as recycling or installing energy-saving lighting (actually, energy-saving is a misnomer, a more accurate term would be 'fuel saving').

The Science Museum survey and subsequent focus groups established that visitors participated in such 'energy-saving' activities although they did not know what else they could do beyond that. They wanted to know what else they could do, but they did not want to be dictated to. Neither did they want to make major lifestyle changes, such as giving up flying, as that would have too negative an impact on their lives. They had limited knowledge of mitigation or remediation, and they could not see how actions such as writing to an MP, or events such as the Copenhagen conference, would help.

Conclusions

Climate change presents significant opportunities and challenges to science communicators. The scientific consensus grows year by year, and there seems to be no justification nowadays for science communicators presenting the arguments of sceptics and deniers, except as an example of how hard it is to convince some sections of society.

We know that public understanding of the issues is patchy and fragmented. The public, generally, trust scientists and their explanations. We suspect that part of the challenge facing science communicators is that the public do not fully appreciate the scale of the scientific consensus. Sometimes the uncertainty that scientists display, in terms of predicting what might happen in future decades, is seen as evidence that they disagree about whether or not climate change is caused by humans. Science communicators can help by showing people that uncertainty is part of how science works.

The London Science Museum's approach to researching public understanding and opinions about climate change provides a model for other institutions. The findings of the research provided a firm base for the exhibition developers. The realisation that visitors' often displayed serious misunderstandings, shifted the exhibition development team's focus towards explaining the science behind climate change, rather than, as was originally planned, on the broader impacts and issues raised by climate change.

What does the future hold? The sheer economic consequences of climate change will ensure that governments and industry takes action even if public opinion lags behind. Even in countries such as the US where there seems to be a perverse delight in ignoring scientific evidence, public opinion will eventually swing to accept that the human influence on the climate has been catastrophic. The challenge for science communicators is to maintain the public's interest in the topic while simultaneously showing that each individual can, and should, do their best to reduce their fuel consumption.

References

BBC (2010). *BBC Climate Change Poll.* Available at: http://news.bbc.co.uk/nol/shared/bsp/hi/pdfs/05_02_10climatechange.pdf. (Accessed on August 11, 2011).

BBC Trust (2011). *BBC Trust review of impartiality and accuracy of the BBC's coverage of science.* London: BBC Trust.

Boyes, E., & Stanisstreet, M. (1992). Students' perceptions of global warming. *International Journal of Environmental Studies*, 42(4), 287–300.

Center for Research on Environmental Decisions (CRED). (2009). *The Psychology of Climate Change Communication: A Guide for Scientists, Journalists, Educators, Political Aides, and the Interested Public.* CRED: New York.

Cordero, E. C., Todd, A. M. & Abellera, D. (2008). Climate change education and the ecological footprint, *Bulletin of the American Meteorological Society*, 89, 865–72.

Department for Environment, Food and Rural Affairs (DEFRA) (2002). *Survey of public attitudes to quality of life and to the environment—2001.* London: DEFRA.

Devine-Wright, P., Devine-Wright, H. and Fleming, P. (2004). Situational influences upon children's beliefs about global warming and energy. *Environmental Education Research*, 10(4), 493–506.

Dove, J. (1996). Student teacher understanding of the greenhouse effect, ozone layer depletion, and acid rain. *Environmental Education Research*, 2(1), 89–100.

Hickman, L. (2011). The need for caution when 'calling out the climate cranks'. *The Guardian*. Available at: http://www.guardian.co.uk/environment/blog/2011/feb/14/climate-cranks-caution-sceptics-protest [accessed on August 8, 2011].

Intergovernmental Panel on Climate Change (IPCC 2001). *Climate Change 2001. IPCC Third Assessment Report.* Cambridge: Cambridge University Press.

Leiserowitz, A., Maibach, E., Roser-Renouf, C. and Smith, N. (2010). *Global Warming's Six Americas, June 2010.* Yale University and George Mason University. New Haven, CT: Yale Project on Climate Change.

Mason, L., & Santi, M. (1998). Discussing the greenhouse effect: Children's collaborative discourse reasoning and conceptual change. *Environmental Education Research*, 4(1), 67–85.

Moser, S., & Dilling, L. (2004). Making climate hot: Communicating the urgency and challenge of global climate change. *Environment*, 46, 32–46.

Roser-Renouf, C., & Maibach, E. (2010). Communicating climate change. In S. Priest (Ed.), *The Encyclopedia of Science and Technology Communication*, Sage Publications.

Science Museum (2010a). *Visitors' Mental Model of Climate Change* (unpublished).

——(2010b). *Audience's Attitudes Table.* (unpublished).

Spence, A., Venables, D., Pidgeon, N., Poortinga, W., & Demski, C. (2010) *Public Perceptions of Climate Change and Energy Futures in Britain: Summary Findings of a Survey Conducted in January-March 2010.* Cardiff: Cardiff University.

Times Higher Education (2011). Science is fair dinkum. *Times Higher Education*, June 30, 2011, p. 16. Available at: http://www.timeshighereducation.co.uk/story.asp?storyCode=416621&sectioncode=26 [accessed on August 10, 2011].

Toynbee, P. (2006). The climate-change deniers have now gone nuclear. *The Guardian*. Available at: http://www.guardian.co.uk/commentisfree/2006/jul/18/comment.politics3 [accessed on August 8, 2011].

——(2011). Britain must resist Tea Party thinking. *The Guardian*. Available at: http://www.guardian.co.uk/commentisfree/2011/aug/01/britain-resist-tea-party-thinking [accessed on August 8, 2011].

TWResearch (2008). *A Climate Change Gallery at the Science Museum* (unpublished).

——(2009). *Developing the Climate Change Exhibition* (unpublished).

——(2010). *A Climate Change Toolkit for Teachers* (unpublished).

Upham, P., Whitmarsh, L., Poortinga, W., Purdam, K., Darton, A., McLachlan, C., & Devine-Wright, P. (2009). *Public attitudes to environmental change: a selective review of theory and practice.* Swindon: Economic and Social Research Council. Available at: http://www. esrc.ac.uk/_images/LWEC-research-synthesis-full-report_tcm8-6384.pdf. Accessed on August 10, 2001.

Uzzell, D. L. (1999). Education for environmental action in the community: New roles and relationships. *Cambridge Journal of Education,* 29(3), 397–413.

Whitmarsh, L. (2009). What's in a name? Commonalities and differences in public understanding of "climate change" and "global warming". *Public Understanding of Society,* 18(4), 401–20.

Section 7

Science, Environment and Sustainability

The final section sets out my position on the relationship between science, the environment and sustainability. '*Silent Spring*: science, the environment and society' is a critique of simplistic thinking about the relationship between science and environmental education. 'Education for sustainable development: opportunity or threat?' is a critique of education for sustainable development. 'Convergence between science and environmental education' lays out my current thinking on the need for a convergence of science and environmental education. This last paper was one of the most difficult to get published that I can remember. *Science* has exceptionally high standards and my co-authors and I lost count of the number of iterations that we went through to turn our initial ideas into the final paper. The experience, though, was very useful and we went on to edit a special section on Citizen Science for *Conservation Biology*, another highly regarded science journal.

22 *Silent Spring*

Science, the Environment and Society

Justin Dillon

Dillon, J. (2005). *Silent Spring*: science, the environment and society. *School Science Review*, 86(316), 113–118.

Recent events in south-east Asia have shown all too clearly that the environment plays a major role in our lives and the lives of the billions of people around the world who we will never meet, some of whom make our clothes, grow our food and drink and benefit from our overseas aid. Influenced by the 2002 Johannesburg Summit and the Earth Summit in Rio de Janeiro ten years earlier, the media frequently draw our attention to the links between the environment, poverty, development, politics and people. As science educators we take it for granted that our curricula enable (or mandate) us to teach these issues to today's young citizens in the hope that they can make a better job of sorting out the planet than we have done so far. In England, key stage 3 (11–14 year-old) pupils should be taught '*about ways in which living things and the environment can be protected, and the importance of sustainable development*'. An appreciation of ecology and an awareness of '*how the impact of humans on the environment depends on social and economic factors, including population size, industrial processes and levels of consumption and waste*' is a desired outcome of the curriculum at key stage 4 (14–16 year-olds). In Scotland, students visiting the heritage centre, the Robert Burns Experience, are encouraged to '*research and debate opinions on the use of fertilisers, pesticides and crop enhancers used in modern farming*' (see Websites) as a follow up to their visit. But it was not ever thus.

The rise in concern for the environment can be traced back to the 1950s and 1960s when many members of the public began to be aware that there was sometimes an uncomfortable price to pay for scientific and technological advance. Public concern for the environment has remained high throughout the last 30 years or so, although it is not often backed up with an adequate knowledge base. Only 3 in 10 people in a recent telephone survey in the US recognised 'biodiversity' and could describe what it meant (Elder, 2002, unpublished conference paper). Concern is sometimes triggered more by feelings and events than by inculcation with specific pieces of knowledge.

Rachel Carson's *Silent Spring,* which exposes the catastrophic effects of pesticide spraying in the United States in the 1950s, is probably the most influential book in the history of the environmental movement. Carson trained as a scientist in the days when few women did and was an award-winning writer before the publication of *Silent Spring*. The book celebrated its 40th birthday in 2002. It was first published on 27 September 1962 following a successful serialisation in the *New Yorker* earlier in the year. Since publication, the book has hardly been out of print. Indeed, Linda Lear, Carson's biographer, puts the book in the same category—'books that changed the world'—as *Das Capital, The wealth of nations, The origin of the species* and *Uncle Tom's cabin*. At a more individual level, many environmentalists pay homage to its influence on their lives, as exemplified by these two comments (solicited anonymously) from two generations of environmental educators in Australia:

> *I read* Silent Spring *for the first time as a 14-year-old teenager. At the time I was horrified, but vividly inspired by this text. It provided a doorway to the environmental movement and inherently inspired me to enter the debate. As a young teenager at the time, I immersed myself in literature such as Suzuki, Ehrlich, Weston, Evernden . . . Ten years have passed, and I am. still intrigued by* Silent Spring, *such that I now endeavour to lead a career in the environmental movement and live my life accordingly.* (PhD student)
>
> *My personal relationship with Rachel Carson is through a Peanuts (Lucy, of course!) cartoon that [my husband] gave me for my . . . office door. In it she says, referring to Rachel Carson, that* 'we girls need our heroes'. *I'm not sure which Peanuts book it came from, but I do like to think of Carson as one of my heroes. Of course I have only ever dipped into the book, but its significance goes without saying . . . and she is a hero because she was willing to stand up against the male science establishment and speak out. We need more courageous people like that!* (academic)

Rachel Carson: Writer, Scientist, and Ecologist

Born, Springdale, Pennsylvania, USA, 27 May 1907
Died, Silver Spring, Maryland, USA, 14 April 1964
1929 Graduated from Pennsylvania College for Women
1932 MA in zoology from Johns Hopkins University
1936 Began a 15-year career as a scientist and editor and rose to
        become Editor-in-Chief for the US Fish and Wildlife Service
1937 Publication of an article, 'Undersea', in *Atlantic Monthly*
1941 Publication of *Under the sea-wind*
1952 Publication of *The sea around us*
1955 Publication of *The edge of the sea*
1962 Publication of *Silent Spring*

If she were alive today, Rachel Carson would have celebrated her 97th birthday on 27 May 2004. However, she died in April 1964, less than two years after the publication of *Silent Spring*, after a catalogue of illnesses including a misdiagnosed breast cancer. Her interests in writing and in nature seem to have been permanent features of her life. She attributed both interests to her mother. At university she changed her major from literature to zoology and in 1936, with an MA in Zoology from Johns Hopkins University, she began a 15-year career as a scientist and editor for the US Fish and Wildlife Service. She rose to become Editor-in-Chief but still managed to publish articles and books independently (Lear, 1998)—see panel.

Silent Spring begins with an apocalyptic vision 'A fable for tomorrow' in which a town '*in the heart of America . . . lay in the midst of a checkerboard of prosperous farms, with fields of grain and hillsides of orchards where, in spring, white clouds of bloom drifted above the green fields*' (Carson, 1999: 21). But then '*a strange blight crept over the area and everything began to change. Some evil spell had settled on the community: mysterious maladies swept the flocks of chickens; the cattle and sheep sickened and died. Everywhere was a shadow of death*' (1999: 21). Carson goes on to catalogue the growth in the use of pesticides and the sometimes catastrophic impacts on flora and fauna other than the intended targets. The book is a populist account of the science of ecology: '*The chemical weedkillers are a bright new toy. They work in a spectacular way; they give a giddy sense of power over nature to those who wield them*' (1999: 73). As a finale, Carson points to a range of alternatives including the use of biological approaches to pests. '*Some*', she writes, '*are already in use and have achieved brilliant success. Still others are little more than ideas in the minds of imaginative scientists*' (1999: 240).

The true impact of *Silent Spring* is hard to measure but there is no doubt that it affected many people's lives permanently. Influenced by the book, President John F. Kennedy ordered the Presidential Science Advisory Committee to consider the issue of pesticides the year after the book was published. Rachel Carson gave evidence at its deliberations and she endured a period of intense vilification and relentless attacks from many in the pesticide industry and the media (Lear, 1998).

At the time the book was written, US public interest in the environment was high. Indeed, Carson's earlier work, *The sea around us,* spent 86 weeks on the *New York Times* bestseller list and made her enough money to be able to concentrate on writing full-time (Logan, 1992). With her interest primarily in marine matters, Carson was not immediately attracted to writing a book on pesticides, although she had mooted the idea of writing an article for a magazine as early as 1945. However, as she found out more about the size and scale of the problem, she devoted more time to researching the topic and produced the book that we know as *Silent Spring.*

In the 1950s and 1960s, interest in science and the environment was remarkably high. The post-war years were, politically, times when progress was expected to be driven by careful government making decisions with the best interests of people at heart. Progress was assumed to mean bigger, better, faster, as was evident in the almost obsessive quest for air, land and sea speed records and, the epitome of such competition, the space race. But technological progress was seen as requiring secrecy—from the prying eyes of the United States' enemies and competitors. Carson's book, which exposed a range of agencies and industries to the public gaze, while not actually an act of espionage in itself, was portrayed as, in some ways, 'un-American'.

Partly, I suspect, as a reaction against the attacks on Carson and on environmentalists in general, criticism of *Silent Spring* is relatively rare in the environmental movement. Linda Lear, writing for a US website designed as a resource for teachers (Lear, 2002), noted that:

> *Some historians argue that Carson took the middle ground in Silent Spring and did not go far enough in challenging the power of big business and the scientific establishment. Yaakov Garb presents the best argument that Carson's moderation had costs which underlie the pesticide triumph of today. He is joined by critics who point to her defense of the outmoded concept of a balance of nature as contributing to the political softness of her attack. These critics chide Carson for not naming names of chemical companies who were irresponsible polluters and therefore failing to give a full critique of the process by which science invents and then assesses its own results . . . Gregg Easterbrook is among those who find Carson an early scaremonger who did more harm than good.*

Science in *Silent Spring*

In his book *Playing safe: science and the environment,* Jonathan Porritt, the noted environmental advocate, invoked Carson's memory in lamenting the fact that '*a huge percentage of scientists are now paid by private or public sector employers who have little interest in open scientific debate*' (Porritt, 2000: 19–20). He continues: '*One can only surmise that Rachel Carson would be distraught at what is happening today, when not just cowardice but "science for sale" has become a familiar phenomenon*' (2000: 20).

Carson is highly critical of some aspects of science and of some scientists. For example, *Silent Spring's* final paragraph contains a strong condemnation of applied entomology:

> *The 'control of nature' is a phrase conceived in arrogance, born of the Neanderthal age of biology and philosophy, when it was supposed that nature exists for the convenience of man. The concepts and practices of applied entomology for the most part date from that stone age of science. It is our alarming misfortune that so primitive a science has armed itself with the most modern and terrible*

weapons, and that in turning them against the insects it has also turned them against the earth.

(1999: 257)

Carson quotes the biologist Carl Swanson, when she writes:

Any science may be likened to a river. It has its obscure and unpretentious beginning; its quiet stretches as well as its rapids; its periods of drought as well as of fullness. It gathers momentum with the work of many investigations and as it is fed by other streams of thought; it is deepened and broadened by the concepts and generalizations that are gradually evolved.

(1999: 241)

Silent Spring's final chapter, 'The other road', argues strongly that the alternative to chemical control of pests must be biological control. Carson writes of '*forging weapons from the insect's own life processes*' (1999: 247). She quotes a range of scientists in support of her case against chemical control. Towards the end of the final chapter, after describing a range of lethal strategies from X-raying male insects in order to sterilise them to using ultrasonic sound waves to kill mosquito larvae, she writes:

Through all these new, imaginative, and creative approaches to the problem of sharing our earth with other creatures there runs a constant theme. The awareness that we are dealing with life—with living populations and all their pressures and counter-pressures, their surges and recessions. Only by taking account of such life forces and by cautiously seeking to guide them into channels favourable to ourselves can we hope to achieve a reasonable accommodation between the insect hordes and ourselves.

(1999: 256)

Carson was a scientist who was highly critical of chemical methods used to control insect pests and of the influence of industry on scientific research and policy. However, unlike many other readings of the book, it seems to me that one interpretation of Carson's point of view is that she believes that it is an inalienable right of 'man' to wipe out any species that might be considered a pest. She would be quite happy to see science used to provide biological solutions that would see billions of insects slaughtered for the commercial and health benefits of humans. The use of the phrase, '*the problem of sharing our earth with other creatures*', suggests to me that Carson might not want to share '*our*' earth with quite as many species as many radical ecologists might tolerate. Science, and only science, can provide the '*new, imaginative, and creative approaches*' that allow us to produce mass annihilation techniques that are '*favourable to ourselves*' and cope with the '*insect hordes*'. It seems to me that Carson is saying that ecology, which purports to bring a more synthetic approach to scientific knowledge, could provide a more 'natural' approach to the problems than chemistry alone can offer.

Although Carson rails against the use of the phrase *'control of nature'* she seems to be willing to use nature to control itself on our behalf. The most we can hope for, she seems to be arguing, is to coexist with the *'hordes'*. In coming to terms with her perception of the problems and the possible solutions, science plays a dual role for Carson—as both friend and enemy—and it is to that duality that I turn next.

Science, the Environment and Society

Some commentators on the place of science in our cultures have recognised the dual-edged nature of the role of science. In *Understanding the present: science and the soul of modern man*, Brian Appleyard (1992) wrote that:

> *Concern for the environment is our age's mechanism for resolving the contradictions inherent in the two opposing aspects of science. Environmentalism is based on scientific insight, and yet it is violently opposed to the effects of most of the more obvious and spectacular achievements of science and technology. It is a way of turning science against itself, of rejecting the progressive ideals of economic growth by using scientific means to expose them as potentially suicidal. It is the single most successful popular solution to the terrible contrast between penicillin and atom bombs, air conditioning and concentration camps.*

Jonathan Porritt comments that he has *'met literally hundreds of scientists who have become adept at suppressing their own values and passions out of an implicit or explicit fear of their work being corrupted by such suspect tendencies'* (Porritt, 2000: 127). His solution is to develop a *'deeper understanding of the utter impossibility of either physical or psychic disconnection from the rest of life on earth . . . a licence to engage as citizen scientists, as alert to the social and ethical importance of the work they're doing as they are to its intellectual and technical significance'* (2000: 127).

Porritt admits that *Silent Spring* had a *'shattering effect'* (2000: 54) on him at an early age. One wonders what Porritt would have made of the following quotes from a speech made by a scientist in 1952:

> *The materials of science are the materials of life itself. Science is part of the reality of living; it is the what, the how, and the why of everything in our experience. It is impossible to understand man without understanding his environment and the forces that have molded him physically and mentally . . . The aim of science is to discover and illuminate truth.*
>
> (quoted in Lear 1998: 219)

He might have criticised the positivist view explicit in the use of terms such as *'discover'* and *'truth'*. He might even rail at the certainty of the scientist's views of the power of science to explain *'the what, the how, and the why of everything in our experience*'. The words are those of Rachel Carson as she accepted the National Book Award in New York.

Science, the Media and Public Trust

John Burnside, writing in the *Guardian* in May 2002, noted that the chemical industry learned much from the furore surrounding *Silent Spring*. 'Corporations' he writes, '*have become highly skilled in managing public opinion*' (2002: 2). Burnside charts the change in strategy that has taken place in the last 40 years:

> *In 1962, the field where battles were fought, in public at least, was scientific debate; the trick then was to have control over the nature, terms and extent of the debate. An unexpected bonus, in recent years, has come from public awareness of that control; now when the scientific organisation speaks, the voice we hear is too often that of the sustaining industry as the MMR [measles, mumps and rubella vaccine] scandal so clearly demonstrates. We do not know who to trust, and in such cases, we tend to hope that our leaders and elected representatives are still as well meaning as they seemed when we elected them.*

Indeed, *Silent Spring* is more about trust than about the environment or about scientific knowledge. Carson does not deny or devalue her scientific training. Indeed, without it, she would have been unable to write the book. Our trust in her comes, partly, from our knowledge that she is '*writer, scientist, and ecologist*' (Lear, 2004). The 42 pages of principal sources in *Silent Spring* are dominated by reports from scientific journals and correspondence with scientists. Without the evidence that science provides, her case would not be remotely credible.

Final Thoughts

Silent Spring, writes Burnside, is '*a call to a new way of thinking, a challenge to us all, to create, and live by, a radical philosophy of life*' (2002). However, Burnside notes that Jonathan Bate has pointed out that '*the two other radical movements that emerged in the 1960s, feminism and anti-racism, have been tolerated: gender and post-colonial studies are offered in most universities, for example*' whereas '*Radical ecology; a philosophy that challenges all the accepted social and economic models, lags far behind*' (2002: 1). Burnside cautions us when he writes '*Yet mystical and sentimental is exactly what ecology is not: these honours belong to the old religions of market values and objectivity*' (2002: 2). Carson's important message about the misuse of science, sustained by individual greed and market forces, has got lost.

Although the book was published more than 40 years ago, it still has much to offer students and teachers. The message now is as relevant as it was back in 1962. The language and scientific content probably put it outside the reach of most key stage 4 students but anyone studying A-level biology or chemistry or the AS-level science for public understanding would appreciate the relevance of the content and the accessible style of writing. Beyond that, Rachel Carson offers a role model of a woman scientist, ridiculed by a predominantly male scientific establishment and pesticide industry, who used her knowledge of science to make the public stop and think about what was being done in the name of progress.

Acknowledgements

This is an abridged version of an article published in the 2002 edition of the *Australian Journal of Environmental Education*. I am happy to acknowledge the helpful comments and encouragement that I received from Annette Gough and Noel Gough, two of the editorial collective of *AJEE*.

References

Appleyard, B. (1992) *Understanding the present: science and the soul of modern man.* New York: Doubleday.

Burnside, J. (2002) Reluctant crusader. *Guardian Saturday* Review, 18 May, pp 1–2.

Carson, R. (1999) *Silent Spring.* Originally published 1962. London: Penguin.

Lear, L. (1998) *Rachel Carson: witness for nature.* London: Allen Lane.

Lear, L. (2002) Rachel Carson and the awakening of environmental consciousness. Scholars Debate. http://www.nhc.rtp.nc.us:8080/tserve/nattrans/ntwilderness/essays/carsonf.htm Accessed 3 June 2002.

Lear, L. (2004) Biography. http://www.rachelcarson.org/ Accessed 12 September 2004.

Logan, J. (1992) *An introduction to Rachel Carson and her legacy: a resource for Maine educators to commemorate the 30th anniversary of the publication of Silent Spring.* Boothbay: Boothbay Region Land Trust.

Porritt, J. (2000) *Playing safe: science and the environment.* London: Thames and Hudson.

Further Reading

Freeman, M. ed. (1995) *Always, Rachel: the letters of Rachel Carson and Dorothy Freeman, 1952–1964—the story of a remarkable friendship.* Boston: Beacon Press.

Lear, L. ed. (1999) *Lost woods: the discovered writing of Rachel Carson.* Boston: Beacon Press.

Waddell, C. ed. (2000) *And no birds sing: rhetorical analyses of Silent Spring.* Carbondale, IL: Southern Illinois University Press.

Websites

Biography, bibliography, etc.: www.rachelcarson.org/

WWF toxic chemicals site: http://www.worldwildlife.org/toxics/

Cornell University's guide to biological control in North America: http://www.nysaes.cornell.edu/ent/biocontrol/

Association for the Study of Literature and the Environment: http://www.asle.umn.edu/

Robert Burns Experience: http://www.burnsheritagepark.com/farming.htm

23 Education for Sustainable Development

Opportunity or Threat?

Justin Dillon and Jing Huang

Dillon, J., & Huang, J. (2010). Education for sustainable development: opportunity or threat? *School Science Review*, 92(338), 39–44.

During a period of real pressure on public finances, government faces difficult choices in how to spend its limited resources in meeting its social goals. Whilst global issues such as climate change or international poverty remain a high priority, it is increasingly recognised that government cannot succeed in creating social change on its own. These issues require UK people to be engaged and take action ourselves, whether this be to buy fairtrade goods, donate to overseas development charities, or shift our behaviour towards sustainability.

(DEA, 2010: 3)

It is becoming increasingly clear that it is unrealistic to expect policy makers and politicians to solve the problems that face society in today's troubled times. To some extent, then, 'we' are all in this together. One of the implications of this state of affairs is that, as the DEA report quoted above goes on to suggest, *'Moving towards greater environmental sustainability requires people to rethink a variety of behaviours including around transport, energy use, purchasing and waste'* (p. 4). This shift in behaviour applies equally well at school as it does at home. Given that scientific knowledge underpins understanding of issues such as transport, energy use, waste and climate change, now might be a good time to take a close look at what science departments might consider as their role in encouraging behaviour change directly or indirectly. This contribution to the special issue focuses specifically on education for sustainable development (ESD), a term that may well become more familiar to science departments in the next few years. We consider whether it is ESD or environmental education that school science departments may feel most comfortable focusing their efforts on promoting.

It would be difficult to argue that ESD has a high profile in school science departments in the UK or elsewhere at the moment. Few teachers seem to know that we are in a 'Decade of ESD' and fewer still would be able to tell you who decided what it should be called (it was the United Nations) and when the decade began (2005). The UN chose UNESCO as the lead agency for the Decade, which might explain why it has slipped past almost unnoticed. UNESCO has a relatively low profile in the UK and was viewed with some suspicion by the previous Conservative government, which, following the USA's lead, withdrew from the organisation in 1985. Britain rejoined in 1997 after the change of government and the USA followed suit in 2003.

The Decade is related in policy terms to a series of initiatives that appeared at around the turn of the century and that shared a sadly unrealistic sense of optimism. For example, progress towards the Millennium Development Goals, established in 2000, has been, to say the least, uneven. The chances of achieving 'Goal 2'—universal primary education for all by 2015—in many of the signatory countries are very low. The use of goals to drive development policy has, as a result, been called into question by, among others, the Institute of Development Studies (2009).

However, the UK Government responded to the Decade for ESD by bringing together a panel of advisers—the Sustainable Development Education Panel (SDEP)—with a brief to ensure that *'pupils are fully-equipped to be active citizens for the new millennium"* (Ofsted, 2003). Speaking at the launch of the SDEP report in April 1999, the then deputy prime minister, John Prescott, said:

> *We have placed sustainable development at the heart of the Government's agenda. We all need to learn how we can be full citizens in the new millennium—how we can live our lives more sustainably at work, at home, and in our leisure time.*

> (DfEE, 1999)

Charles Clark, then parliamentary undersecretary for school standards and the Department for Education and Employment's green minister, added:

> *The report is a significant contribution to the debate on education and training about the environment and sustainable development . . . I intend to take a personal interest in taking this forward . . .*

> (DfEE, 1999)

Three months later, Mr Clark, who was widely regarded as one of the more popular education secretaries, was moved to the Home Office but, nevertheless, the message going out to lobbyists, educators and the wider world was that the Government took the issues seriously. However, the political support was and is for ideas conceived and moulded by another layer of policy makers. Civil servants and representatives from a range of non-governmental organisations had a hand in constructing the SDEP report, which, as a result, was not as radical as some environmental educators had hoped might be the case.

Origin of the Term 'Education for Sustainable Development'

As Ofsted pointed out in their 2003 report, *'ESD is not a new concept'*. By way of explanation, they added that:

> *It has evolved from a mixture of environmental as well as development education ideas and links to a number of related areas concerned with personal, social, economic and citizenship issues.*

> (Ofsted, 2003)

We will return to this highly contested term later but, when the English and Welsh National Curriculum was revised in 2000, the Government took the opportunity to focus schools' attention on issues of sustainability and the environment:

> *Pupils should . . . develop the knowledge, values and skills to participate in decisions about the way that we do things individually and collectively, both locally and globally, that will improve the quality of life now without damaging the planet for the future.*
>
> (Ofsted, 2003)

Guidance to schools on how to approach this challenging task followed in 2002 when the then Qualifications and Curriculum Authority (QCA) produced a report based on the seven key concepts identified by the SDEP that pupils should address (see SDEP, 1998):

- citizenship and stewardship;
- sustainable change;
- needs and rights of future generations;
- interdependence;
- diversity;
- uncertainty and precaution;
- quality of life, and equity and justice.

Further guidance, including case studies, was produced in 2009 by the QCA, although the term 'education for sustainable development' is used rather sparingly in the document (QCA, 2009).

As we enter the second half of the Decade, perhaps it is a good time to look back at what has happened and forward to what might be on the horizon. In terms of what is happening in schools, a number of reasonably independent sources of evidence are available to us, including a recent report from Ofsted (2009).

The View from Ofsted

For people with backgrounds in science and research, reading Ofsted reports can be somewhat frustrating. The reports are written for various audiences—in some cases parents and teachers, in other cases policy makers and the wider public—so it might be rather unfair to judge them by standards other than Ofsted's own. The report *Education for sustainable development: improving schools—improving lives,* published in 2009, illustrates the challenges Ofsted faces in terms of collecting data, making valid judgements and providing recommendations to schools.

Between September 2005 and December 2008, one of Her Majesty's Inspectors, representing Ofsted, visited eight primary schools, five secondary schools and one special school, three times, to establish to what extent each school was *'developing pupils 'experience and understanding of sustainable development'*. In addition, Ofsted looked at *'whether an increased commitment to sustainability had wider benefits'* (Ofsted, 2009: 4).

The schools had all volunteered to take part in the study and the Ofsted report notes that *'their focus on sustainability was sharper than might normally be found'* Nevertheless, the report's author argues that the schools' *'experiences should be a valuable resource for other schools wishing to improve provision in this area'* (p. 7). The approach to collecting data for the study is briefly described in the report:

> *The initial visit was used to establish a baseline against which to map progress over the following three years. During the visits, the inspector observed lessons across the range of the National Curriculum, scrutinised documentation, and attended assemblies and school council meetings. He held discussions with pupils, parents, governors and members of the community and met representatives of organisations supporting the schools in their work on sustainability.*

(p. 24)

Changes in Attitudes

It is not possible to assess how reliable and valid the Ofsted findings are. Claims such as *'In the first year of the survey, girls tended to show more interest than the boys in the topics being explored* are unsubstantiated and no data are provided to support the claim, nor the claim that *'By the third year, however, boys and girls displayed equal levels of interest'* (Ofsted, 2009: 14). It is not clear from the report how the inspector measured interest. What level of interest did boys and girls show? Had the girls' level of interest dropped to that of the boys? Or had the level of interest of both boys and girls risen by different amounts? Without that information, the significance of the claims for changed levels of interest cannot be deduced with any degree of certainty.

Changes in Behaviour

If, as was suggested earlier, the objectives of ESD are long-term changes in behaviour, then the following claim is critical:

> *Over the course of the survey, there was increasing evidence that pupils who were committed to sustainability at school were leading more sustainable lives at home and influencing those around them.*

(Ofsted, 2009: 23)

The evidence for this statement is given in the penultimate paragraph of the report:

> *Responses to a questionnaire produced by one school showed that, influenced by their children, many families were now recycling, composting and installing low*

energy light bulbs and were supporting the 'cycle to school' initiative. Parents were also responding to their children's requests for healthier snacks and the amount of packaging used for school lunches had been cut by 60%.

(p. 24)

More evidence to support these claims for the impact of ESD can be found in the following statement:

The parents and governors were very positive about the way in which the school was promoting sustainability and acknowledged the influence that it was having on their own as well as their children's lives.

(p. 24)

Finally, Ofsted noted that the parents and governors commented on:

how the children now looked healthier and how the incidence of obesity in the school had fallen.

(p. 24)

Some more robust evidence would make the statements more reliable.

Changes in Knowledge and Understanding

Some of the limitations of the Ofsted (2009) study are identified by its author. In terms of the impact of ESD on learning, the report states quite categorically that '*It would be impossible, without more detailed research, to make a strong link between education for sustainable development and improved attainment*' but goes on to state that '*nine of the headteachers cited examples of work on sustainability leading to higher levels of commitment and engagement and to improved performance on the part of their pupils*' (p. 20). The report describes how a year 9 class study on water supply in geography led, eventually, to the local council changing its policy on recycling plastic bottles.

Despite the author's caveats about the need for more detailed research, the report comes to some firm conclusions, one of which is that schools '*should develop a whole-school approach to education for sustainability in the curriculum to enable it to become firmly embedded in teaching and learning*' (p. 6). It is difficult to see how this conclusion is supported by the evidence in the report. The 14 schools are atypical and the evidence base is relatively weak. To conclude that schools '*should*' do anything as a result might be somewhat premature.

Problematising Education for Sustainable Development

We noted above that, in an earlier report, Ofsted (2003) found that ESD had '*evolved from a mixture of environmental as well as development education ideas*' What they failed to note was that the term was, and is, highly contested.

The underlying philosophy of the UK policy brand of ESD is clear from the SDEP key principles:

- *Sustainable development is the responsibility of everyone;*
- *Education for sustainable development needs to pervade every aspect of life;*
- *UK prosperity in the long term depends on our capacity to learn about sustainable development.* (SDEP, 1998)

The Panel does not see ESD as conceptually problematic:

> *Education for sustainable development is not new ... In the past decade these approaches have increasingly found commonality under the label of 'education for sustainable development' and there is a strengthening consensus about the meaning and implications of this approach for education as a whole.*

(SDEP, 1998)

Indeed, it takes for granted that the purpose of ESD is uncontroversial:

> *The Panel has deliberately chosen not to repeat in any detail the arguments for education for sustainable development. The case for an education which enables young people to participate in efforts to achieve a more sustainable future is largely understood and endorsed by policy makers and teachers, business and the community.*

(SDEP, 1998)

Evidence for their position is not immediately apparent. Cross's survey (1998), for example, found that, while teachers took the language of sustainable development at *'face value'*, they were *'inhibited by a lack of knowledge of the complexities of the issues and how their teaching might contribute'* (p. 50). The situation might have changed since then but our feeling is that ESD is still poorly understood, partly because it is based on the idea of sustainable development, which is itself highly contested.

Sustainable development is, as Pawley (2000) defines it, a *'political fudge'* combining opposing positions by proposing a third. *'Sustainable development calls for the conservation of development, not for the conservation of nature'* (Sachs, 1990: 34). Arguing along similar lines, Jickling (2000) relates the contradictions inherent in sustainable development to Orwell's *'double think'*, in that ordinary citizens increasingly accept contradictory meanings for the same term and accept both. Stables (1996) refers to terms such as sustainable development as *'paradoxical compound policy slogans'*, although, as Scott and Gough (2003) argue, having distinctive perspectives can bring distinct learning opportunities provided that educators and others do not foreclose the issue by focusing on their preferred view to the exclusion of others.

Arthur Lucas (1991) noted a tendency that began in the 1970s to label educational interventions as being 'for' the environment. This philosophical stance has framed much policy formation and curriculum planning since then. During the 1990s, the appropriateness of the 'for' position was challenged from a number of perspectives, for example by Jickling and Spork (1998). When sustainable development becomes the 'something' that education should be 'for', as a number of environmental educators have argued (see Tilbury, 1995), more debate is stimulated. Foster, for example, argues that:

The relation between education and sustainability cannot be an external, still less an instrumental one . . . Learning to understand the natural world and the human place in it can only be an active process through which our sense of what counts as going with the grain of nature is continually constituted and recreated. This process cannot have its agenda set to subserve sustainability criteria which it actually makes meaningful.

(Foster, 2001: 153)

That is to say, how can education be 'for' a particular view of economics and development? Jickling (2000) is highly critical of Hopkins' 'determinist' position outlined in the proceedings of a major conference on ESD held in Thessalomki in 1997 and hosted by UNESCO and the Government of Greece. Hopkins asserted that:

education should be able to cope with determining and implanting these broad guiding principles [of sustainability] at the heart of ESD.

(Hopkins, 1998: 172)

Jickling retorted that:

When highlighted in this way, most educators find such statements a staggering misrepresentation of their task. Teachers understand that sustainable development, and even sustainability, are normative concepts representing the views of only segments of our society. And, teachers know that their job is primarily to teach students how to think, not what to think.

(Jickling, 2000: 469)

What, then, is the role of science education in such a charged context? In the final section of this article, we consider what science departments might wish to consider further in terms of education, sustainability and the environment.

Where Next?

From a bio-centric position, as opposed to an anthropo-centric one, we could argue that the best way to ensure the continuity of life on earth would be to allow the human species to expand so much that it forced itself into extinction by destroying the features of the earth that supported it: then another evolutionary explosion such as the one that followed the extinction of the dinosaurs would occur, with new life forms evolving that exploited the new niches created by the circumstances that drove Homo sapiens and many other species to extinction.

(Lucas, 1995: 2)

Lucas's observation might be awkwardly prescient, particularly as this is the International Year of Biodiversity. What the role of science education would be for such a scenario to be enacted is impossible to say. A more plausible option has been put forward by another Australian academic, Annette Gough, who notes that:

Science education needs environmental education to reassert itself in the curriculum by making science seem appropriate to a wider range of students and making it more culturally and socially relevant. Environmental education needs science education to underpin the achievement of its objectives and to provide it with a legitimate space in the curriculum to meet its goals because they are very unlikely to be achieved from the margins.

(Gough, 2002: 1210)

While ESD might be a bridge too far for science departments to consider—perhaps because it is a contested term or a political fudge, or because research evidence for its effectiveness is, as yet, minimal—environmental education might provide a more productive next step. Indeed, many schools actively promote environmental education, both in and out of school. Despite also being a contested term, environmental education is a less slippery beast than ESD!

To embrace environmental education more strongly would not, however, be at the expense of education *about* sustainability. In effect, most of the schools cited in the Ofsted and QCA/QCDA reports are doing education *about* sustainability rather than education *for* sustainability. We already know that science departments can do much to teach students about what sustainability involves and why it should be at the heart of government and always at the back of the minds of individual members of the public but, as Jickling (2000: 470) points out, *'if there is to be a future for sustainability within education, we must begin to recognise the educational limitations of the deterministic manifestations of the sustainability agenda'*. Our purpose in contributing to this special issue is to reinforce Jickling's point that:

Unfortunately, the mantra of sustainability has conditioned many to believe that this term carries unconditional or positive values. Yet critical thought depends on transient elements in ordinary language, the words and ideas that reveal assumptions and worldviews, and the tools to mediate differences between contesting value systems.

(p. 472)

So what might science departments do if they are as sceptical about ESD as we are? It would be no bad thing if students left school knowing how to grow some of their own food, understanding how to calculate their carbon footprint and being able to identify how their individual consumer behaviours impacted on the lives of other people and other species. Few subjects in the curriculum have greater potential to contribute to environmental education than science. To ignore that potential is no longer an option.

References

Cross, R. T. (1998) Teachers' views about what to do about sustainable development. *Environmental Education Research,* **4**(1), 41–52.

DEA (2010) *The Impact of global learning on public attitudes and behaviours towards international development and sustainability.* London: DEA.

Department for Education and Employment (DfEE) (1999) *Green learning for the new century—government receives think tank blueprint.* Available at: www.dcsf.gov.uk/pns/Display PN.cgi?pn_id=1999_0176.

Foster, J. (2001) Education *as* sustainability. *Environmental Education Research,* **7**(2), 153–165.

Gough, A. (2002) Mutualism: a different agenda for environmental and science education. *International Journal of Science Education,* **24**(11), 1201–1215.

Hopkins, C. (1998) The content of education for sustainable development. In *Environment and society: education and public awareness for sustainability. Proceedings of the Thessaloniki International Conference,* ed. Scoullos, M. J. pp. 169–172. Paris: UNESCO.

Institute of Development Studies (2009) *After 2015: rethinking pro-poor policy.* Focus Policy Brief 9.1. Brighton: Institute of Development Studies.

Jickling, B. (2000) A future for sustainability? *Water, Air, and Soil Pollution,* **123**, 467–476.

Jickling, B. and Spork, H. (1998) Education for the environment: a critique. *Environmental Education Research,* **4**(3), 309–327.

Lucas, A. M. (1991) Environmental education: what is it, for whom, for what purpose and how. In *Conceptual issues in environmental education,* ed. Keiny, S. and Zoller, U. pp. 25–45. New York: Peter Lang Publishing.

Lucas, A. M. (1995) *Beware of slogans.* Paper presented at the Opening Session of the British Council Seminar 'Environmental Education: From Policy to Practice', held at King's College London, March 1995.

Ofsted (2003) *Taking the first step forward . . . towards an education for sustainable development. Good practice in primary and secondary schools.* London: Ofsted.

Ofsted (2009) *Education for sustainable development: improving schools—improving lives.* London: Ofsted.

Pawley, M. (2000) Sustainability: a big word with little meaning. *The Independent,* 11 July 2000. Available at: www.independent.co.uk/opinion/commentators/sustainability-a-big-word-with-little-meaning-709649.html.

Qualifications and Curriculum Authority (QCA) (2009) *Sustainable development in action. A curriculum planning guide for schools.* London: QCA.

Sachs, W. (1997) No sustainability without development. *The Aisling Magazine,* issue 21, Lúghnasa 1997. Available at: www.aislingmagazine.com/aislingmagazine/articles/TAM21/Sustainability.html.

Scott, W. and Gough, S. (2003) *Sustainable development and learning: framing the issues.* London: RoutledgeFalmer.

Stables, A. (1996) Paradox in compound educational policy slogans: evaluating equal opportunities in subject choice. *British Journal of Educational Studies,* **44**(2), 159–167.

Sustainable Development Education Panel (SDEP) (1998) *First annual report 1998: Annex 4.* Available at: webarchive.nationalarchives.gov.uk/20080305115859/http:/www.defra. gov.uk/environment/sustainable/educpanel/1998ar/ann4.htm.

Tilbury, D. (1995) Environmental education for sustainability: defining the new focus of environmental education in the 1990s. *Environmental Education Research,* **1**(2), 195–212.

24 Convergence Between Science and Environmental Education

Arjen E. J. Wals, Michael Brody, Justin Dillon, and Robert B. Stevenson

Wals, A. E. J, Brody, M., Dillon, J, & Stevenson, R. B. (2014). Convergence between science and environmental education. *Science*, 344, 583–4.

Urgent issues such as climate change, food scarcity, malnutrition, and loss of biodiversity are highly complex and contested in both science and society (*1*). To address them, environmental educators and science educators seek to engage people in what are commonly referred to as sustainability challenges. Regrettably, science education (SE), which focuses primarily on teaching knowledge and skills, and environmental education (EE), which also stresses the incorporation of values and changing behaviors, have become increasingly distant. The relationship between SE and EE has been characterized as "distant, competitive, predatorprey and host-parasite" (*2*). We examine the potential for a convergence of EE and SE that might engage people in addressing fundamental socioecological challenges.

Since the end of World War II, SE has been driven primarily by a need to develop a sufficient pool of science and engineering talent to accelerate innovation and to remain competitive. EE emerged in the early 1960s out of a need to respond to emergent environmental crises. It tried to do so by developing the ecological and environmental literacy required to understand the sociopolitical, value-laden, place-based, and emotional contexts in which environmental issues arise and need to be resolved. An example of the difference between early SE and EE is that, while the former might teach students how to monitor water quality, identify pollutants, and understand technologies that can reduce pollution, EE would involve an analysis of circumstances and behaviors that caused the pollution, as well as identifying ways to clean up a river involving the local community, policy-makers, and industry.

The complex nature of current sustainability challenges, and the need for competent citizens who can adequately respond to them, is such that EE and SE need to develop a mature symbiotic relationship. The recent *International Handbook of Research on Environmental Education* (*3*) describes a trend in favor of such convergence, which, in combination with increased interest in citizen science supported by information and communications technology (ICT), may make education more responsive to current global challenges.

Figure 24.1 **Science education and environmental education do not have to clash.** The Harlem Success Garden in New York City is a living classroom where students grow their own produce.

Research in Environmental Education

Initially, much research in EE (especially in the United States) focused on the effectiveness of EE activities in changing individual environmental behaviors. This approach contributed to the persistent but ill-founded assumption that there is a simple linear relationship between knowledge, awareness, attitude, and environmental behavior. Research, most notably from social psychology, has long revealed that this is far too simplistic an explanation of what affects people's actions (4).

Today much EE research focuses on investigating the conditions and learning processes that enable citizens, young and old, to (i) develop their own capacity to think critically, ethically, and creatively in appraising environmental situations; (ii) make informed decisions about those situations; and (iii) develop the capacity and commitment to act individually and collectively in ways that sustain and enhance the environment (3).

This new focus implies less emphasis on establishing linkages between educational interventions and behavioral outcomes. More attention is now being given to an understanding of the learning processes and the capacities of individuals and communities needed to help resolve complex socioecological issues. This focus also calls for a better understanding of people's cognitive and emotional responses to environmental issues. These responses are influenced by their worldviews and belief systems, which in turn are linked to identity (5). For example, recent research has rendered problematic a focus solely on better comprehension of the science of climate change owing to "identity-protective cognition theory," which indicates that many people's positions on climate change are largely shaped by their political and religious affiliations and identities (6).

EE research currently offers insights into how to engage the public with environmental issues through participatory action, whereas SE research has traditionally provided insights into learners' understanding of the natural systems and processes that are at the heart of these issues (7). The necessity of connecting the two has been recognized by the North American Association for Environmental Education (NAAEE) in a joint initiative with Underwriters Laboratories to link environmental education with science, technology, engineering, and mathematics (E-STEM) (8). Recently, a number of prominent SE researchers have stressed the importance of science educators engaging with sustainability issues by complementing disciplinary understanding with more integrative thinking and linking scientific knowledge with other forms of knowledge such as indigenous knowledge and local (place-based) understanding (9). So, although SE may have evolved separately from EE, recent research and developments in both EE and SE converge toward generating an interdisciplinary and contextual approach to integrating research in science, education, and the environment (3).

Citizen Science

At the same time, the rise of citizen science (CS) also enables for people to engage with science on relevant environmental issues in collaboration with scientists working in local contexts (10, 11). In a recent review, Dickinson *et al.* reported that the primary impacts of CS are seen in biological studies of global climate change (12). The authors conclude that CS and the resulting ecological data can be viewed as a public good that is generated through increasingly collaborative tools and resources. They consider public participation in science a critical component of what they call Earth stewardship.

Citizen science most often refers to community-based local monitoring of changes in the environment using simple data acquisition devices and communication tools. More recently, CS has taken advantage of the Internet, social media, and mobile applications in crowd-sourcing scientific data (13)—resulting in what we refer to as ICT-supported CS. This trend connects well with recent EE research that identifies the use of social media as well as technology-enhanced citizen data acquisition as a way to enhance the interaction between research in science, education, and the environment (3).

Linking EE, SE, and CS

Creating synergy between EE and SE through ICT-supported CS provides opportunities for new forms of education that can lead to the engagement of seemingly unrelated actors and organizations in making new knowledge and in taking the actions necessary to address socioecological challenges. An example of such synergy emerges from so-called "whole-school approaches" to sustainability and the creation of eco-schools, where different forms of learning (e.g., inquiry-based, disciplinary, and social learning) blend with the use of ICT, citizen science, and community engagement (*14*). Such approaches may involve redesigning school grounds using knowledge from SE—to give such spaces a more central place in teaching about health, food, and ecology—as well as using EE to strengthen community involvement and develop a sense of place.

For instance, by creating "edible gardens" (see the photo), schools can, with the involvement of a wide range of societal actors (e.g., a local garden center, a restaurant, a community organization, and the local government), simultaneously improve the quality and relevance of their education and transform their relationship with the local community (*15*). Soil preparation, seed selection, planting, maintaining, harvesting, and preparing a meal require basic scientific knowledge that connects with the SE curriculum while also creating other benefits, such as community engagement, learner empowerment, improved personal health, and a better connection with food and place (*16*).

Recent EE research indicates that the use of CS in these blended or hybrid learning configurations helps learners contribute to the quality of their local environment (*3*). This process can be enhanced by projects such as YardMap, an ICT-supported CS project, funded by the U.S. National Science Foundation, which enables members of the public not only to increase their appreciation of their local "yard" through the use of mapping software but also to take action for improving the habitats of birds.

Place and Identity

The examples of SE and EE converging, supported by CS, emphasize the importance of place and place-based identity in determining our relations with the planet. The focus on identity is timely: The complexity and uncertainty brought on by globalization and the rapid pace of technological and social change result in substantial cultural shifts, including a search for meaning and affiliation in locally defined identities (*5*). The reasons for the recently established disconnect between people and place that results from a preoccupation with and dependency on ICT (*17*) are underresearched, but there is some evidence that such technologies can actually reconnect people and places (*18*). Numerous examples exist of citizens monitoring changes in the environment (e.g., bird migration patterns and quality of water, soil, and air) using geographic information systems, cell phones, and specially designed monitoring applications (*11*). As such, ICT devices actually get people to go outdoors, even those who normally are not inclined to do so. Participation in scientific studies through ICT-supported CS offers the potential to deepen the experience of the physical place of which people are part and to develop their understanding of how science works.

Society has to learn how to address sustainability challenges. Creating synergy between EE and SE mediated by ICT-supported CS provides an opportunity for such learning. We advocate support for collaborative research efforts among scientists, educators, and the public, linking science and society with place and identity, through more effective processes of public engagement and learning that can result in meaningful socioecological outcomes. The data gathered and shared using ICT can provide useful input to environmental scientists while simultaneously empowering citizens to engage in ongoing debates about local and global sustainability issues and what needs to be done to address them.

References and Notes

1 M. C. Nisbet, *Env. Sci. and Pol. for Sust. Dev.* **51**, 12 (2009).

2 A. Gough, *Int. J. Sci. Educ.* **24**, 1201 (2002).

3 R. B. Stevenson, M. Brady, J. Dillon, A. E. J. Wals, Eds., *International Handbook of Research on Environmental Education* (Routledge, New York, 2013).

4 A. Kollmuss, J. Agyeman, *Environ. Educ. Res.* **8**, 239 (2002).

5 R. B. Stevenson, C. Stirling, in *Engaging Environmental Education: Learning, Culture and Agency*, R. B. Stevenson, J. Dillon, Eds. (Sense Publishers, Rotterdam, 2010), pp. 219–238.

6 D. M. Kahan, H. Jenkins-Smith, D. Braman, *J. Risk Res.* **14**, 147 (2011).

7 R. W. Bybee, *Science* **329**, 996 (2010).

8 NAAEE, E-STEM (2013); www.naaee.net/sites/default/files/E-STEM/ESTEM_NAAEE2013TearSheet.pdf

9 R. Tytler, *Res. Sci. Educ.* **42**, 155 (2012).

10 J. L. Shirk *et al.*, *Ecol. Society* **17**, 29 (2012).

11 R. Bonney, *et al.*, *Science* **343**, 1436 (2014).

12 J. L. Dickinson *et al.*, *Front. Ecol. Environ* **10**, 291 (2012).

13 J. Silvertown, *Trends Ecol. Evol.* **24**, 467 (2009).

14 L. G. Hargreaves, *Educ. Rev.* **6**, 69 (2008); www.developmenteducationreview.com/issue6-perspectives2.

15 J. R. Ruiz-Gallardo, A. Verde, A. Valdes, *J. Environ. Educ.* **44**, 252 (2013).

16 A. C. Bell, J. E. Dyment, *Environ. Educ. Res.* **14**, 77 (2008).

17 P. Zaradic, *Sci. Am.* **18**, 24 (2008).

18 J. L. Dickinson, R. L. Crain, H. K. Reeve, J. P. Schuldt, *Trends Ecol. Evol.* **28**, 561 (2013).

Index